Environmental and Industrial Applications of Organic Chemistry

Environmental and Industrial Applications of Organic Chemistry

Dr. Hirdayesh Kumar Vatsa

RANDOM PUBLICATIONS
NEW DELHI (INDIA)

Environmental and Industrial Applications of Organic Chemistry

ISBN 978-93-5111-620-2

Published in 2015 in India by

RANDOM PUBLICATIONS

4376-A/4B, Gali Murari Lal, Ansari Road
New Delhi-110 002
Phone : +9111-43580356, 011-23289044, 011-43142548
e-mail: sales@randompublications.com,
info@randompublications.com, randomexports@gmail.com

Reprinted 2024

Type Setting by : Friends Media, Delhi-110089
Digitally Printed at: Replika Press Pvt. Ltd.

Preface

The selection of solvents, the chemicals used to dissolve substances into a solution, are a key target within Green Chemistry programs. Solvents are a major source of industrial waste in industrial chemical manufacture but careful selection can also increase reaction rates and lower reaction temperatures. Environmental chemistry is the scientific study of the chemical and biochemical phenomena that occur in natural places. It should not be confused with green chemistry, which seeks to reduce potential pollution at its source.

It can be defined as the study of the sources, reactions, transport, effects, and fates of chemical species in the air, soil, and water environments; and the effect of human activity and biological activity on these. Environmental chemistry is an interdisciplinary science that includes atmospheric, aquatic and soil chemistry, as well as heavily relying on analytical chemistry and being related to environmental and other areas of science.

Organic chemistry defines life. Just as there are millions of different types of living organisms on this planet, there are millions of different organic molecules, each with different chemical and physical properties. The applications of organic chemistry are myriad, and include all sorts of plastics, dyes, flavorings, scents, detergents, explosives, fuels and many, many other products. Although originally defined as the chemistry of biological molecules, organic chemistry has since been redefined to refer specifically to carbon compounds — even those with non-biological origin.

The basic aspects of inorganic chemistry are presented significantly in this book. Many applications and practical problems are described. The order of the techniques included is conventional and would be liked by students.

I would like to thank my team for standing beside me throughout my career and writing this book. My special thanks go to "Random Publications" who have published the book.

– Dr. Hirdayesh Kumar Vatsa

Contents

1

Organic Chemistry

Let us start with the question "What is Organic chemistry?".

The simple answer is: It is the chemistry of carbon containing compounds, which are otherwise known as organic compounds.

So it is pretty easy to recognize that we should start our journey of *organic chemistry* by exploring the chemical nature of carbon.

WHAT IS CARBON

So the next question is: What is carbon?

- Carbon is an element with atomic number (Z) = 6.
- Its ground state electronic configuration can be represented as: $1s^2 2s^2 2p^2$ (or) $1s^2 2s^2 2p_x^1 2p_y^1 2p_z^0$

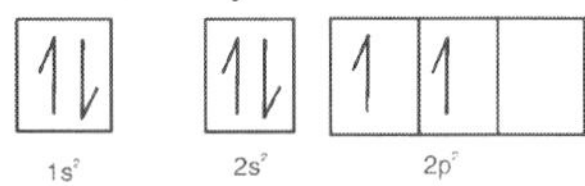

- It is the first element in Group-14 of Long form of Periodic table.
- It is a non-metal.
- On Pauling's scale, its electronegativity value is around 2.5.
- It usually forms covalent bonds.
- Its valency is 4 since there are four electrons in the outer shell *i.e., it can form four covalent bonds with other atoms*.

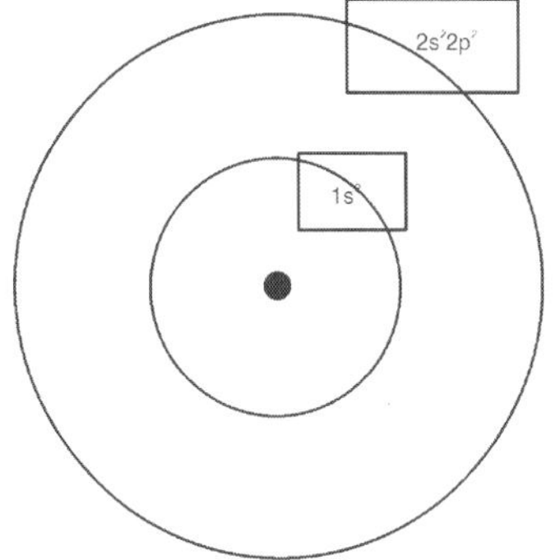

Fig. Representation of Carbon Atom.

ORGANIC MOLECULES

To understand life as we know it, we must first understand a little bit of organic chemistry. Organic molecules contain both carbon and hydrogen. Though many organic chemicals also contain other elements, it is the carbon-hydrogen bond that defines them as organic.

Organic chemistry defines life. Just as there are millions of different types of living organisms on this planet, there are millions of different organic molecules, each with different chemical and physical properties.

There are organic chemicals that make up your hair, your skin, your fingernails, and so on. The diversity of organic chemicals is due to the versatility of the carbon atom. Why is carbon such a special element? Let's look at its chemistry in a little more detail.

Carbon (C) appears in the second row of the periodic table and has four bonding electrons in its valence shell. Similar to other non-metals, carbon needs eight electrons to satisfy its valence shell. Carbon therefore forms four bonds with other atoms (each bond consisting of one of carbon's electrons and one of the bonding atom's electrons).

Every valence electron participates in bonding, thus a carbon atom's bonds will be distributed evenly over the atom's surface. These bonds form a tetrahedron (a pyramid with a spike at the top), as illustrated below:

Fig. Carbon Forms 4 Bonds.

Organic chemicals get their diversity from the many different ways carbon can bond to other atoms. The simplest organic chemicals, called hydrocarbons, contain only carbon and hydrogen atoms;The simplest hydrocarbon (called methane) contains a single carbon atom bonded to four hydrogen atoms:

Fig. Methane - A Carbon Atom Bonded to 4 Hydrogen Atoms.

But carbon can bond to other carbon atoms in addition to hydrogen, as illustrated in the molecule ethane below:

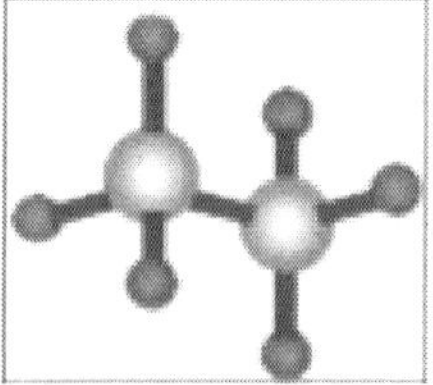

Fig. Ethane - A Carbon-Carbon Bond.

In fact, the uniqueness of carbon comes from the fact that it can bond to itself in many different ways. Carbon atoms can form long chains:

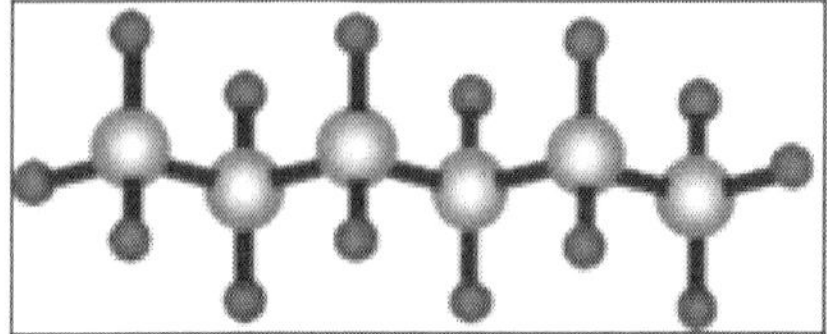

Fig. Hexane - a 6-Carbon Chain.

Branched Chains:

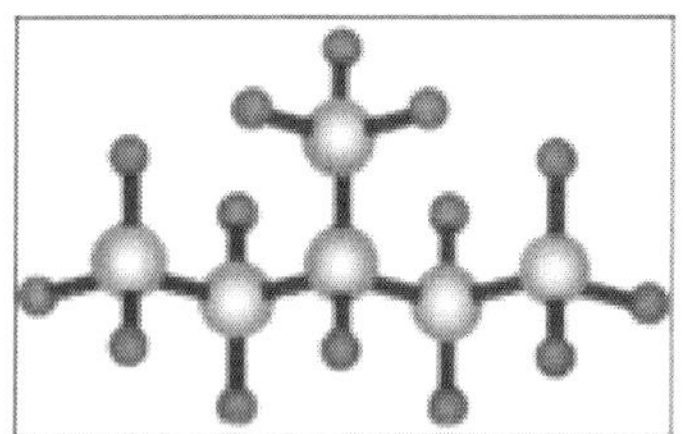

Fig. Isohexane - a Branched-Carbon Chain.

Rings:

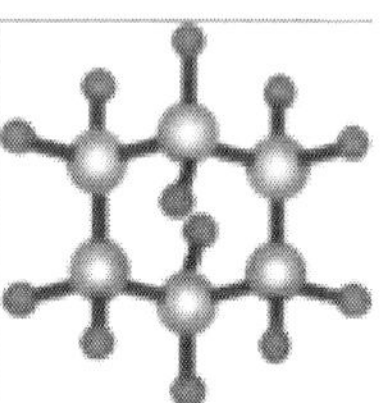

Fig. Cyclohexane - a Ringed Hydrocarbon.

There appears to be almost no limit to the number of different structures that carbon can form. To add to the complexity of organic chemistry, neighboring carbon atoms can form double and triple bonds in addition to single carbon-carbon bonds:

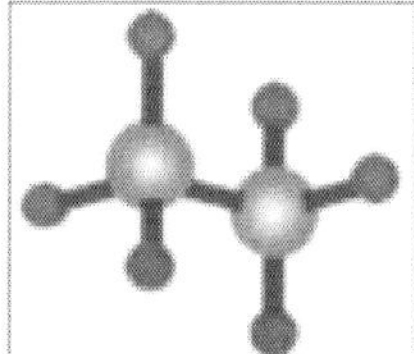

Fig. Single Bonding .

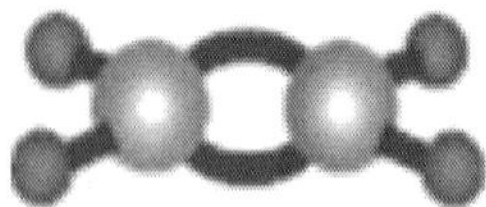

Fig. Double Bonding.

Fig. Triple Bonding

Keep in mind that each carbon atom forms four bonds. As the number of bonds between any two carbon atoms increases, the number of hydrogen atoms in the molecule decreases.

SIMPLE HYDROCARBONS

The simplest hydrocarbons are those that contain only carbon and hydrogen. These simple hydrocarbons come in three varieties depending on the type of carbon-carbon bonds that occur in the molecule. Alkanes are the first class of simple hydrocarbons and contain only carbon-carbon single bonds.

The alkanes are named by combining a prefix that describes the number of carbon atoms in the molecule with the root ending "ane". The names and prefixes for the first ten alkanes are given in the following table.

CarbonAtoms	Prefix	Alkane Name	Chemical Formula	Structural Formula
1	**Meth**	**Methane**	CH_4	CH_4
2	**Eth**	**Ethane**	C_2H_6	CH_3CH
3	**Prop**	**Propane**	C_3H_8	$CH_3CH_2CH_3$
4	**But**	**Butane**	C_4H_{10}	$CH_3CH_2CH_2CH_3$
5	**Pent**	**Pentane**	C_5H_{12}	$CH_3CH_2CH_2CH_2CH_3$
6	**Hex**	**Hexane**	C_6H_{14}	...
7	**Hept**	**Heptane**	C_7H_{16}	
8	**Oct**	**Octane**	C_8H_{18}	
9	**Non**	**Nonane**	C_9H_{20}	
10	**Dec**	**Decane**	$C_{10}H_{22}$	

The chemical formula for any alkane is given by the expression C_nH_{2n+2}. The structural formula, shown for the first five alkanes in the table, shows each carbon atom and the elements that are attached to it. This structural formula is important when we begin to discuss more complex hydrocarbons. The simple alkanes share many properties in common. All enter into combustion reactions with oxygen to produce carbon dioxide and water vapour. In other words, many alkanes are flammable. This makes them good fuels. For example, methane is the principle component of natural gas, and butane is common lighter fluid.

$$CH_4 + 2O_2 \rightarrow CO_2 + 2H_2O$$

The combustion of methane

The second class of simple hydrocarbons, the alkenes, consists of molecules that contain at least one double-bonded carbon pair. Alkenes follow the same naming convention used for alkanes. A prefix (to describe the number of carbon atoms) is combined with the ending "ene" to denote an alkene. Ethene, for example is the two- carbon molecule that contains one double bond. The chemical formula for the simple alkenes follows the expression C_nH_{2n}. Because one of the carbon pairs is double bonded, simple alkenes have two fewer hydrogen atoms than alkanes.

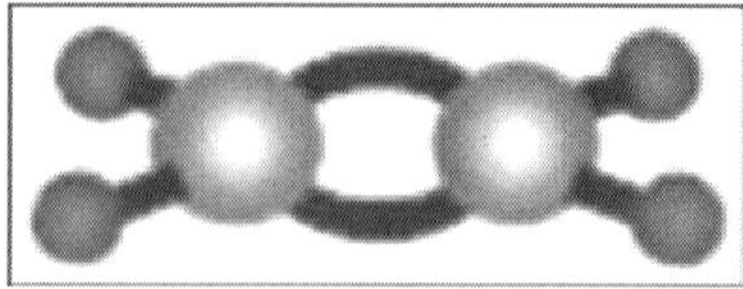

Fig. Ethene

Alkynes are the third class of simple hydrocarbons and are molecules that contain at least one triple-bonded carbon pair. Like the alkanes and alkenes, alkynes are named by combining a prefix with the ending "yne" to denote the triple bond. The chemical formula for the simple alkynes follows the expression C_nH_{2n-2}.

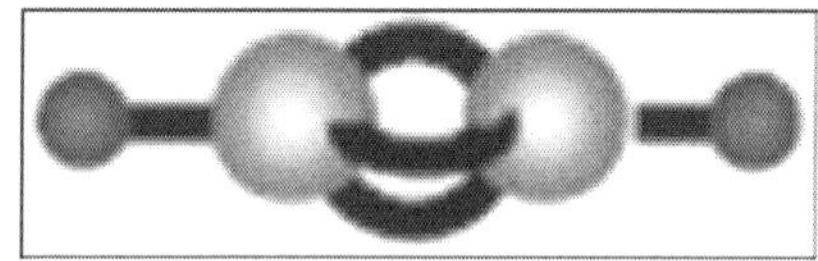

Fig. Ethyne

ISOMERS

Because carbon can bond in so many different ways, a single molecule can have different bonding configurations. Consider the two molecules illustrated in figure.

Isomers of C_6H_{14} are:

- $CH_3CH_2CH_2CH_2CH_2CH_3$

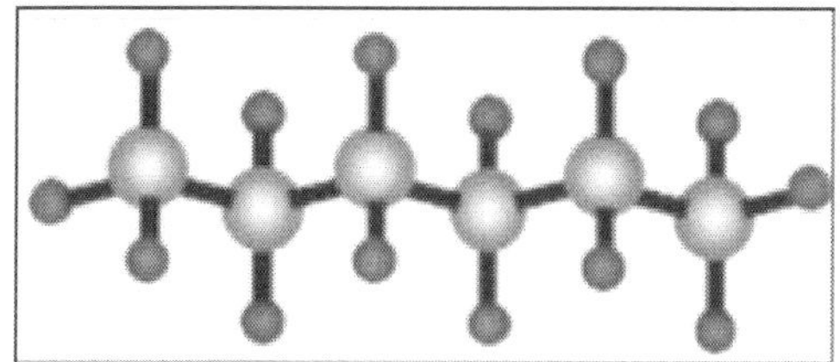

Fig. Molecules Have Identical Chemical Formulas but Different Structural Formulas.

$$\begin{array}{c} \quad CH_3 \\ \quad | \\ CH_3CH_2CHCH_2CH_3 \end{array}$$

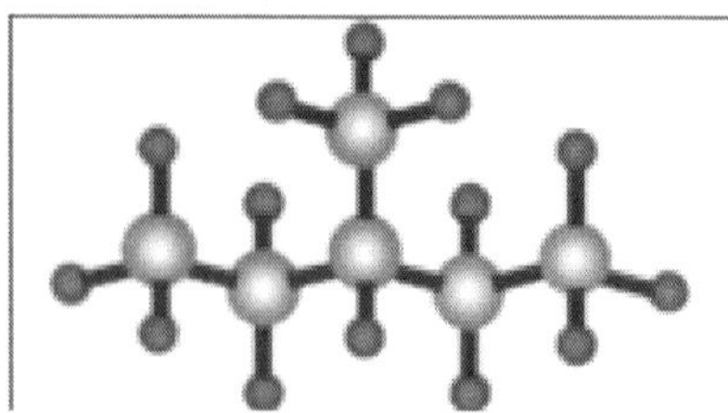

Fig. Molecules Have Identical Chemical Formulas but Different Structural Formulas.

Both molecules have identical chemical formulas; however, their structural formulas (and thus some chemical properties) are different. These two molecules are called isomers. Isomers are molecules that have the same chemical formula but different structural formulas.

FUNCTIONAL GROUPS

In addition to carbon and hydrogen, hydrocarbons can also contain other elements. In fact, many common groups of atoms can occur within organic molecules, these groups of atoms are called functional groups. One good example is the hydroxyl functional group.

The hydroxyl group consists of a single oxygen atom bound to a single hydrogen atom (-OH). The group of hydrocarbons that contain a hydroxyl functional group is called alcohols. The alcohols are named in a similar fashion to the simple hydrocarbons, a prefix is attached to a root ending (in this case "anol") that designates the alcohol.

The existence of the functional group completely changes the chemical properties of the molecule. Ethane, the two-carbon alkane, is a gas at room temperature; ethanol, the two-carbon alcohol, is a liquid. Ethanol, common drinking alcohol, is the active ingredient in "alcoholic" beverages such as beer and wine.

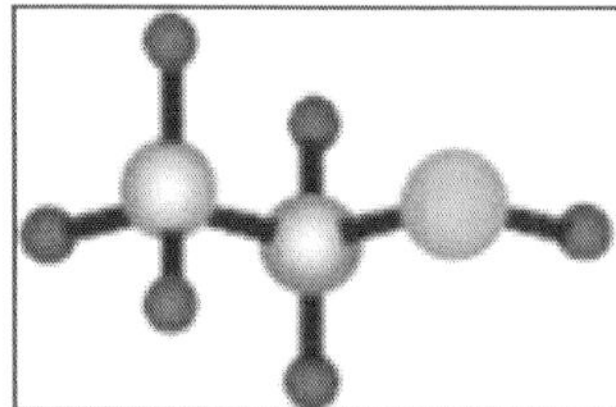

Fig. Ethanol

TOXICITY

Living matter including food, drugs are collections organic compounds, so the potential beneficial and harmful aspects of organic compounds span the entire range required for life to some of the most dangerous materials known.

Indicative of the power of organic chemistry, the biomolecule called botulism toxin ("Botox") is lethal at the level of less than a microgram.

SPECIFIC GROUPS OF ATOMS

Functional groups are specific groups of atoms within molecules, that are responsible for the characteristic chemical reactions of those molecules. The same functional group will undergo the same or similar chemical reaction(s) regardless of the size of the molecule it is a part of. The term 'functional' group is linked to the concept of a homologous series. A homologous series is a group of molecules with the same general formula and the same functional group. They have similar physical and chemical properties (albeit with trends e.g. increasing boiling point with increasing carbon chain length). The terms higher/ lower refer to a larger/smaller carbon chain. Many important organic chemistry molecules contain oxygen or nitrogen. It's a good idea to memorize the names and structures of these functional groups.

—X, X= Cl, Br, I, F	Alkyl Halide Functional Group
—OH	Alcohol Functional Group
—O—	Ether Functional Group
—C—H ‖ C	Aldehyde Functional Group
—C— ‖ O	Keytone Function Group
—C—OH ‖ O	Carboxylic Acid Function Group
—C—O— ‖ O	Ester Function Group
| —N—	Amine Function Group
| —C—N— ‖ O	Amide Function Group

IUPAC NOMENCLATURE

The IUPAC nomenclature of organic chemistry is a systematic way of naming organic chemical compounds as recommended by the International Union of Pure and Applied Chemistry (IUPAC). Ideally, every organic compound should have a name from which an unambiguous structural formula can be drawn. For ordinary communication, to spare a tedious description, the official IUPAC naming recommendations are not always followed in practice except when it is necessary to give a concise definition to a compound, or when the IUPAC name is simpler (viz. ethanol against ethyl alcohol). Otherwise the common or trivial name may be used, often derived from the source of the compound.

Nomenclature of Saturated Hydrocarbons

- Select the longest continuous chain of carbon atoms in the molecule. The compound is named as a derivative of this alkane
- Number the carbon atoms in the parent chain starting from the end which gives lowest possible sum for the numbers of the carbon atoms carrying the substituents
- That set of locants is preferred, which when compared term by term with other set of locants, each in order of increasing magnitude, has the lowest term at the first point of difference. For example the set of locants (2,7,8) is preferred over the set of locants (3,4,9) since 2 comes before 3 even though the sum of locants in the former case is 17 while in the latter case, it is 16

```
1     2      3       4       5       6       7     8      9      10
Ch3 - Ch  - Ch2  - Ch2  - Ch2  - Ch2  - Ch  - Ch  - Ch2  - Ch3
      |                                   |     |
      Ch3                                 Ch3   Ch3
```

- The correct name is: 2,7,8 - Trimethyldecane and not 3,4,9 - Trimethyldecane

Nomenclature of compounds containing functional group or multiple bonds:

- Select the longest continuous chain containing the carbon atoms having the functional group or those involved in the multiple bonds
- The numbering of atoms in the parent chain is done in such a way that the carbon atom bearing the functional group or those carrying the multiple bond gets the lowest possible number
- While writing the name of alkene (double bond) or alkyne (triple bond), the primary suffix 'ane' of the corresponding alkane is replaced by 'ene' and 'yne' respectively. However, if the multiple bond occurs twice or thrice in the parent chain, the prefix di- or tri- is attached to the primary suffix ene or yne
- In naming the organic compounds containing one functional group a suffix known as secondary suffix is added to the primary suffix (giving number of carbon atoms in the chain) to indicate the nature of the functional group. A few important secondary suffixes are:

Functional group	Secondary suffix	Functional group	Secondary suffix
Alcohols (-OH)	-ol	Aldehydes (-CHO)	-al
Ketones (>C=O)	-one	Carboxylic acids (-COOH)	-oic acid
Amines ($-NH_{2)}$	-amine	Acid amides ($-CONH_{2)}$	-amide
Acid chlorides (-COCL)	-oyl chloride	Esters (-COOR)	-oate
Nitrites (-C≡N)	-nitrite	Thioalcohols (-SH)	-thiol

Nomenclature of compounds having polyfunctional groups:

When an organic compound contains two or more functional groups, one group is called the principal functional group while the others are called the secondary functional groups and are treated as substituents: The order of preference for principal group is: Carboxylic acid > acid anhydrides > esters > acid halides > amides > nitrites > aldehydes > ketone > alcohols > amines > double bond > triple bond.

When the functional groups act as substituents, they ar named as:

Functional group	Prefix	Functional group	Prefix
- COOH	Carboxy	-CHO	Formyl
-COOR	Alkoxy cabonyl or Carbalkoxy	>CO	Oxo or Keto
-COCL	Chloroformyl	-OH	Hydroxy
$-CONH_2$	Carbamoyl	-SH	Mecaplo
-CN	Cyano	$-NH_2$	Amino
-OR	Akoxy	=NH	Imino
-X	Halo	$-NO_2$	Nitro

Nomenclature of Simple Aromatic Compounds:

- *Nuclear Substituted*: In these the functional group is directly attached to the benzene ring. Most of these compounds are better known by their common and historical names. In the IUPAC system, they are named as derivatives of benzene.
- *Side Chain Substituted*: In these the functional group is present in the side chain of the benzene ring. Both in the common and IUPAC systems, these are usually named as phenyl derivatives of the corresponding aliphatic compounds.

ATOMIC ORBITALS

Orbits and orbitals sound similar, but they have quite different meanings. It is essential that you understand the difference between them.

ORBITS FOR ELECTRONS

To plot a path for something you need to know exactly where the object is and be able to work out exactly where it's going to be an instant later. You can't do this for electrons. The Heisenberg Uncertainty Principle says that you can't know with certainty both where an electron is and where it's going next. That makes it impossible to plot an orbit for an electron around a nucleus.

HYDROGEN'S ELECTRON - THE 1S ORBITAL

Suppose you had a single hydrogen atom and at a particular instant plotted the position of the one electron. Soon afterwardss, you do the same thing, and find that it is in a new position. You have no idea how it got from the first place to the second. You keep on doing this over and over again, and gradually build up a sort of 3D map of the places that the electron is likely to be found. In the hydrogen case, the electron can be found anywhere within a spherical space surrounding the nucleus. The diagram shows a *cross-section* through this spherical space. 95per cent of the time (or any other per centage you choose), the electron will be found within a fairly easily defined region of space quite close to the nucleus. Such a region of space is called an orbital. You can think of an orbital as being the region of space in which the electron lives. What is the electron doing in the orbital? We don't know, we can't know, and so we just ignore the problem! All you can say is that if an electron is in a particular orbital it will have a particular definable energy.

Each Orbital has a name.

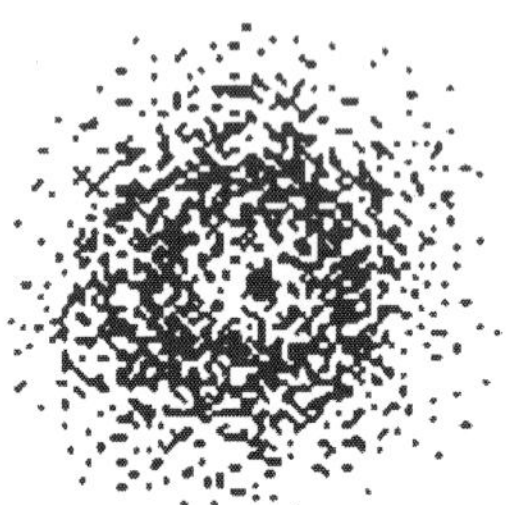

Fig. 1s orbital

The orbital occupied by the hydrogen electron is called a 1s orbital. The "1" represents the fact that the orbital is in the energy level closest to the nucleus. The "s" tells you about the shape of the orbital. s orbitals are spherically symmetric around the nucleus - in each case, like a hollow ball made of rather chunky material with the nucleus at its centre.

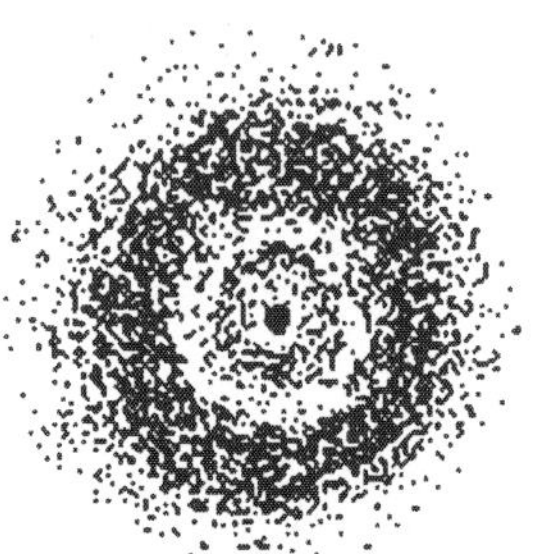

Fig. 2d orbital

The orbital above is a 2s orbital. This is similar to a 1s orbital except that the region where there is the greatest chance of finding the electron is further from the nucleus - this is an orbital at the second energy level.

If you look carefully, you will notice that there is another region of slightly higher electron density (where the dots are thicker) nearer the nucleus. ("Electron density" is another way of talking about how likely you are to find an electron at a particular place.) 2s (and 3s, 4s, etc) electrons spend some of their time closer to the nucleus than you might expect. The effect of this is to slightly reduce the energy of electrons in s orbitals. The nearer the nucleus the electrons get, the lower their energy. 3s, 4s (etc) orbitals get progressively further from the nucleus.

ORBITALS

Not all electrons inhabit s orbitals (in fact, very few electrons live in s orbitals). At the first energy level, the only orbital available to electrons is the 1s orbital, but at the second level, as well as a 2s orbital, there are also orbitals called 2p orbitals.

Fig. p Orbital

A p orbital is rather like 2 identical balloons tied together at the nucleus. The diagram on the right is a cross-section through that 3-dimensional region of space. Once again, the orbital shows where there is a 95per cent chance of finding a particular electron.

Unlike an s orbital, a p orbital points in a particular direction. At any one energy level it is possible to have three absolutely equivalent p orbitals pointing mutually at right angles to each other.

These are arbitrarily given the symbols p_x, p_y and p_z. This is simply for convenience - what you might think of as the x, y or z direction changes constantly as the atom tumbles in space. The p orbitals at the second energy level are called $2p_x$, $2p_y$ and $2p_z$. There are similar orbitals at subsequent levels - $3p_x$, $3p_y$, $3p_z$, $4p_x$, $4p_y$, $4p_z$ and so on.

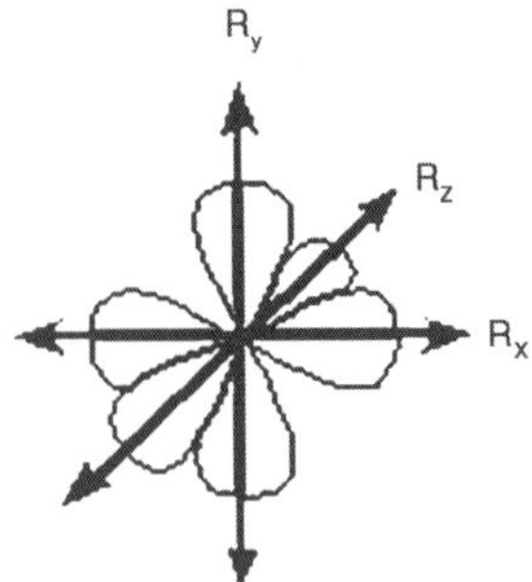

All levels except for the first level have p orbitals. At the higher levels the lobes get more elongated, with the most likely place to find the electron more distant from the nucleus.

Fitting electrons into orbitals

Because for the moment we are only interested in the electronic structures of hydrogen and carbon, we don't need to concern ourselves with what happens beyond the second energy level.

Remember

At the first level there is only one orbital - the 1s orbital.

At the second level there are four orbitals - the 2s, $2p_x$, $2p_y$ and $2p_z$ orbitals. Each orbital can hold either 1 or 2 electrons, but no more.

"Electrons-in-boxes"

Orbitals can be represented as boxes with the electrons in them shown as arrows. Often an up-arrow and a down-arrow are used to show that the electrons are in some way different.

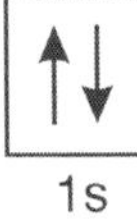

A 1s orbital holding 2 electrons would be drawn as shown on the right, but it can be written even more quickly as $1s^2$. This is read as "one s two" - not as "one s squared".

You mustn't confuse the two numbers in this notation:

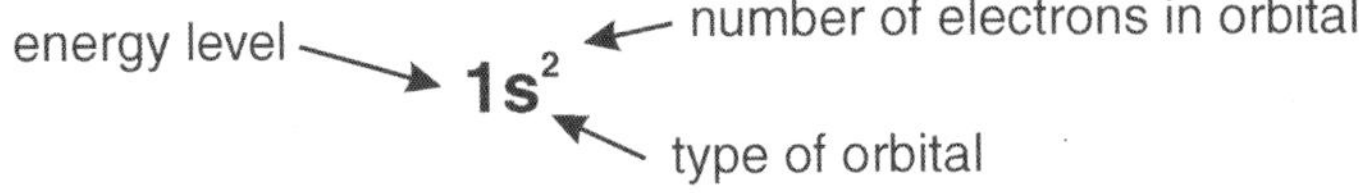

Order of Filling Orbitals

Electrons fill low energy orbitals (closer to the nucleus) before they fill higher energy ones. Where there is a choice between orbitals of equal energy, they fill the orbitals singly as far as possible.

The diagram (not to scale) summarises the energies of the various orbitals in the first and second levels.

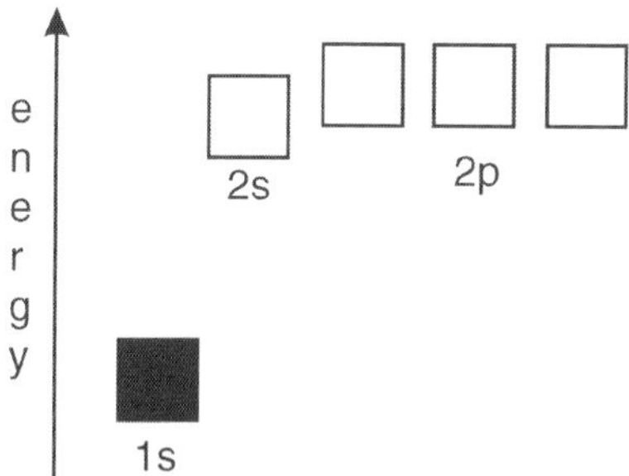

Notice that the 2s orbital has a slightly lower energy than the 2p orbitals. That means that the 2s orbital will fill with electrons before the 2p orbitals. All the 2p orbitals have exactly the same energy.

Electronic Structure of Hydrogen

Hydrogen only has one electron and that will go into the orbital with the lowest energy - the 1s orbital. Hydrogen has an electronic structure of $1s^1$. We have already described this orbital earlier.

Electronic Structure of Carbon

Carbon has six electrons. Two of them will be found in the 1s orbital close to the nucleus. The next two will go into the 2s orbital. The remaining ones will be in two separate 2p orbitals. This is because the p orbitals all have the same energy and the electrons prefer to be on their own if that's the case.

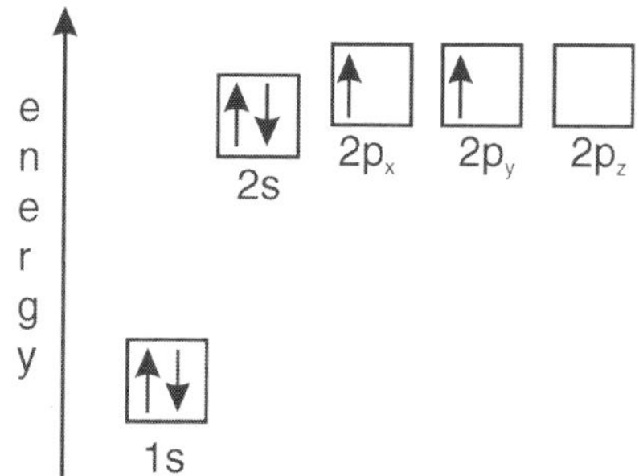

The electronic structure of carbon is normally written $1s^2 2s^2 2p_x^1 2p_y^1$.

OPEN-CHAIN AND RING COMPOUNDS

In organic chemistry, there are a few basic structural shapes that you will encounter. They are *chains* and *rings*.

There are also two types of chains, a *straight chain*, and a *branched chain*. In a straight chain, one carbon atom holds no more than two other carbon atoms. As its name implies, the straight chain is a straight link of carbon, sometimes oxygen or nitrogen, atoms, in structural formula that is. Because of twisting and contouring, they chain may have several *conformations.* Branched chains have at least one carbon holding more than two other carbon atoms. It will, as its name implies, have branches of other chains coming off another chain. Branching is one of the reasons why there are so many isomers for each compound.

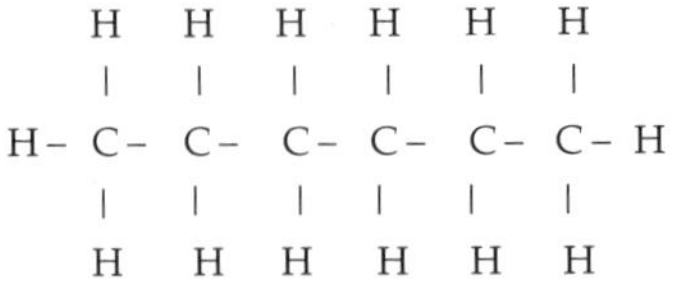

```
     H   H   H   H   H   H
     |   |   |   |   |   |
 H - C - C - C - C - C - C - H
     |   |   |   |   |   |
     H   H   H   H   H   H
```

Straight Chain

```
                  H
                  |
    H      H H  H-C-H  H H H
    |      | |    |    | | |
H---C---C-C---C---C-C-C-H
    |      | |    |    | | |
  H-C-H  H H      H    H H H
    |
  H-C-H
    |
    H
```

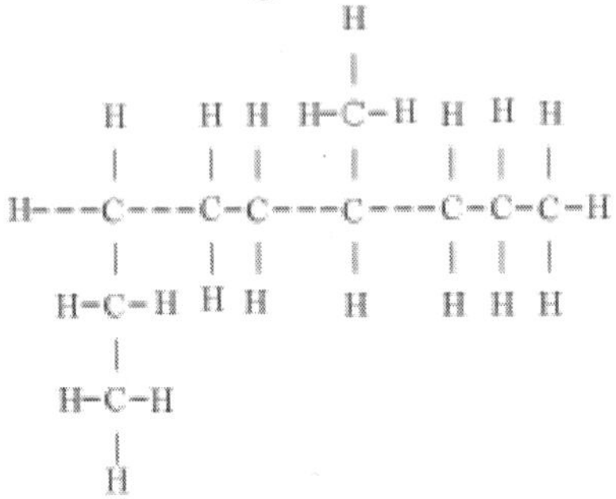

Branched Chain

Rings (or cyclic compounds) are composed of rings of carbon and sometimes oxygen or nitrogen. *For example, cyclohexane has a ring of six carbon atoms*:

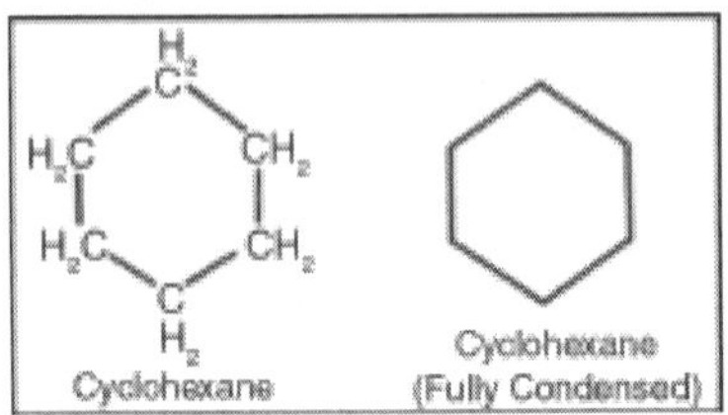

In the fully condensed version of the structural formula of cyclohexane, or any cyclic compound for that matter, the conventions for drawing them are:

- A C occurs at every corner unless O or N (or another atom) is explicitly written at the corner.
- A line connecting two corners is a covalent bond between adjacent ring atoms.
- Remaining bonds are understood to hold H atoms (C normally has 4 total bonds).
- Double bonds are always explicitly shown.

To illustrate the rules, here are some types of cyclic compounds:

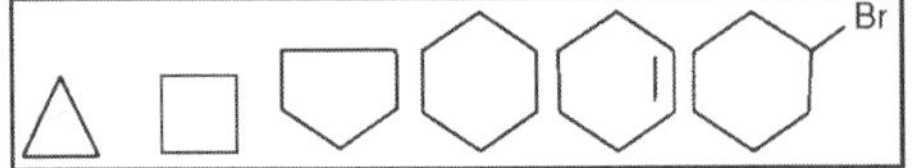

Fig. Cyclopropane,Cyclobutane, Cyclopentane, Cyclohexane,Cyclohexene, Bromocyclohexane.

There is no theoretical upper limit to the size of the ring.

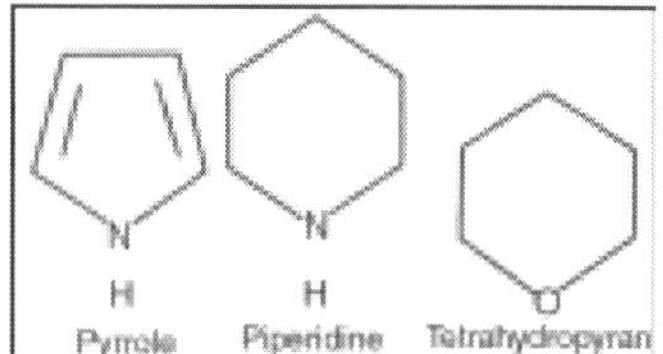

Heterocyclic Compounds are those Which have an atom other than Carbon in the Ring.

Examples are: The atom that is not carbon is called the *heteroatom*. The heteroatom's bonds must be shown or clearly understood. For example, nitrogen normally forms three bonds, so three bonds must be indicated for it. Oxygen normally forms two, so two bonds must be shown. Rings may have more than one heteroatom.

2

Foundational Concepts of Organic Chemistry

Jöns Jacob Berzelius, a physician by trade, first coined the term "organic chemistry" in 1807 for the study of compounds derived from biological sources. Up through the early 19th century, naturalists and scientists observed critical differences between compounds that were derived from living things and those that were not.

Chemists of the period noted that there seemed to be an essential yet inexplicable difference between the properties of the two different types of compounds. The vital force theory (sometimes called "vitalism") was therefore proposed (and widely accepted) as a way to explain these differences. Vitalism proposed that there was something called a "vital force" which existed within organic material but did not exist in any inorganic materials.

SYNTHESIS OF UREA

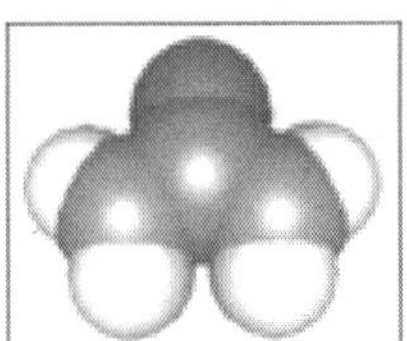

Fig. Urea

Friedrich Wöhler is widely regarded as a pioneer in organic chemistry as a result of his synthesizing of the biological compound urea (a component of urine in many animals) utilizing what is now called "the Wöhler synthesis."

Wöhler mixed silver or lead cyanate with ammonium nitrate; this was supposed to yield ammonium cyanate as a result of an exchange reaction, according to Berzelius's dualism theory. Wöhler, however, discovered that the end product of this reaction is *not* ammonium cyanate (NH_4OCN), an inorganic salt, but urea ($(NH_2)_2CO$), a biological compound. (Furthermore, heating ammonium cyanate turns it into urea.) Faced with this result, Berzelius had to concede that $(NH_2)_2CO$ and NH_4OCN were *isomers*. Until this discovery in the year 1828, it was widely believed by chemists that organic substances could only be formed under the influence of the "vital force" in the bodies of animals and plants. Wöhler's synthesis dramatically proved that view to be false. Urea synthesis was a critical discovery for biochemists because it showed that a compound known to be produced in nature only by biological organisms could be produced in a laboratory under controlled conditions from inanimate matter. This "in vitro" synthesis of organic matter disproved the common theory (vitalism) about the vis vitalis, a transcendent "life force" needed for producing organic compounds.

ORGANIC VS INORGANIC CHEMISTRY

Although originally defined as the chemistry of biological molecules, organic chemistry has since been redefined to refer specifically to carbon compounds — even those with non-biological origin. Some carbon molecules are not considered organic, with carbon dioxide being the most well known and most common inorganic carbon compound, but such molecules are the exception and not the rule.

Organic chemistry focuses on carbon and following movement of the electrons in carbon chains and rings, and also how electrons are shared with other carbon atoms and heteroatoms. Organic chemistry is primarily concerned with the properties of covalent bonds and non-metallic elements, though ions and metals do play critical roles in some reactions. The applications of organic chemistry are myriad, and include all sorts of plastics, dyes, flavorings, scents, detergents, explosives, fuels and many, many other products. Read the ingredient list for almost any kind of food that you eat — or even your shampoo bottle — and you will see the handiwork of organic chemists listed there.

MAJOR ADVANCES IN THE FIELD OF ORGANIC CHEMISTRY

Of course a chemistry text should at least mention Antoine Laurent Lavoisier. The French chemist is often called the "Father of Modern Chemistry" and his place is first in any pantheon of great chemistry figures. Your general chemistry textbook should contain information on the specific work and discoveries of Lavoisier — they will not be repeated here because his discoveries did not relate directly to organic chemistry in particular.

Berzelius and Wöhler are discussed above, and their work was foundational to the specific field of organic chemistry. After those two, three more scientists are famed for independently proposing the elements of structural theory. Those chemists were August Kekulé, Archibald Couper, and Alexander Butlerov.

Kekulé was a German, an architect by training, and he was perhaps the first to propose that isomerism was due to carbon's proclivity towards forming four bonds. Its ability to bond with up to four other atoms made it ideal for forming long chains of atoms in a single molecule, and also made it possible for the same number of atoms to be connected in an enormous variety of ways. Couper, a Scot, and Butlerov, a Russian, came to many of the same conclusions at the same time or just a short time after.

Through the nineteenth century and into the twentieth, experimental results brought to light much new knowledge about atoms, molecules, and molecular bonding. In 1916 it was Gilbert Lewis of U.C. Berkeley who described covalent bonding largely as we know it today (electron-sharing). Nobel laureate Linus Pauling further developed Lewis' concepts by proposing resonance while he was at the California Institute of Technology. At about the same time, Sir Robert Robinson of Oxford University focused primarily on the electrons of atoms as the engines of molecular change. Sir Christopher Ingold of University College, London, organized what was known of organic chemical reactions by arranging them in schemes we now know as mechanisms, in order to better understand the sequence of changes in a synthesis or reaction.

The field of organic chemistry is probably the most active and important field of chemistry at the moment, due to its extreme applicability to both biochemistry (especially in the pharmaceutical industry) and petrochemistry (especially in the energy industry). Organic chemistry has a relatively recent history, but it will have an enormously important future, affecting the lives of everyone around the world for many, many years to come.

ATOMIC STRUCTURE

Atoms are made up of a nucleus and electrons that orbit the nucleus. The nucleus consists of protons and neutrons. An atom in its natural, uncharged state has the same number of electrons as protons.

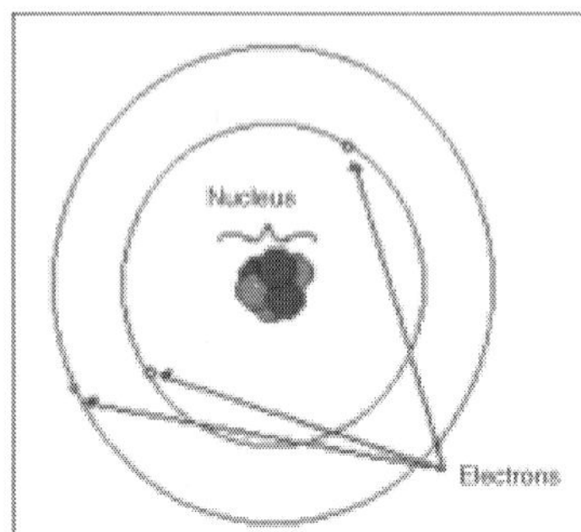

Fig. A Simple Model of a Lithium Atom.

THE NUCLEUS

The nucleus is made up of protons, which are positively charged, and neutrons, which have no charge. Neutrons and protons have about the same mass, and together account for most of the mass of the atom.

ELECTRONS

The electrons are negatively charged particles. The mass of an electron is about 2000 times smaller than that of a proton or neutron at 0.00055 amu. Electrons circle so fast that it cannot be determined where electrons are at any point in time. The image depicts the old Bohr model of the atom, in which the electrons inhabit discrete "orbitals" around the nucleus much like planets orbit the sun. This model is outdated. Current models of the atomic structure hold that electrons occupy fuzzy clouds around the nucleus of specific shapes, some spherical, some dumbbell shaped, some with even more complex shapes.

SHELLS AND ORBITALS

Electron Shells

Electrons orbit atoms in clouds of distinct shapes and sizes. The electron clouds are layered one inside the other into units called shells, with the electrons occupying the simplest orbitals in the innermost shell having the lowest energy state and the electrons in the most complex orbitals in the outermost shell having the highest energy state. The higher the energy state, the more energy the electron has, just like a rock at the top of a hill has more potential energy than a rock at the bottom of a valley.

The main reason why electrons exist in higher energy orbitals is because only two electrons can exist in any orbital. So electrons fill up orbitals, always taking the lowest energy orbitals available. An electron can also be pushed to a higher energy orbital, for example by a photon. Typically this is not a stable state and after a while the electron descends to lower energy states by emitting a photon spontaneously.

These concepts will be important in understanding later concepts like optical activity of chiral compounds as well as many interesting phenomena outside the realm of organic chemistry (for example, how lasers work).

Electron Orbitals

Each different shell is subdivided into one or more orbitals, which also have different energy levels, although the energy difference between orbitals is less than the energy difference between shells.

Longer wavelengths have less energy; the s orbital has the longest wavelength allowed for an electron orbiting a nucleus and this orbital is observed to have the lowest energy. Each orbital has a characteristic shape which shows where electrons most often exist. The orbitals are named using letters of the alphabet. In order of increasing energy the orbitals are: s, p, d, and f orbitals.

As one progresses up through the shells (represented by the principal quantum number n) more types of orbitals become possible. The shells are designated by numbers. So the 2s orbital refers to the s orbital in the second shell.

S Orbital

The s orbital is the orbital lowest in energy and is spherical in shape. Electrons in this orbital are in their fundamental frequency. This orbital can hold a maximum of two electrons.

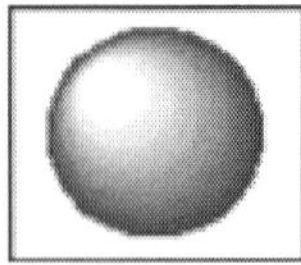

P Orbital

The next lowest-energy orbital is the p orbital. Its shape is often described as like that of a dumbbell. There are three p-orbitals each oriented along one of the 3-dimensional coordinates x, y or z. Each of these three "p" orbitals can hold a maximum of two electrons.

These three different p orbitals can be referred to as the p_x, p_y, and p_z. The s and p orbitals are important for understanding most of organic chemistry as these are the orbitals that are occupied in the type of atoms that are most common in organic compounds.

D and F Orbitals

There are also D and F orbitals. D orbitals are present in transition metals. Sulphur and phosphorus have empty D orbitals. Compounds involving atoms with D orbitals do come into play, but are rarely part of an organic molecule. F are present in the elements of the lanthanide and actinide series. Lanthanides and actinides are mostly irrelevant to organic chemistry.

FILLING ELECTRON SHELLS

ELECTRON CONFIGURATION

In atomic physics and quantum chemistry, electron configuration is the arrangement of electrons of an atom, a molecule, or other physical structure. It concerns the way electrons can be distributed in the orbitals of the given system (atomic or molecular for instance).

Like other elementary particles, the electron is subject to the laws of quantum mechanics, and exhibits both particle-like and wave-like nature. Formally, the quantum state of a particular electron is defined by its wave function, a complex-valued function of space and time. According to the Copenhagen interpretation of quantum mechanics, the position of a particular electron is not well defined until an act of measurement causes it to be detected.

The probability that the act of measurement will detect the electron at a particular point in space is proportional to the square of the absolute value of the wavefunction at that point.

An energy is associated with each electron configuration and, upon certain conditions, electrons are able to move from one orbital to another by emission or absorption of a quantum of energy, in the form of a photon. Knowledge of the electron configuration of different atoms is useful in understanding the structure of the periodic table of elements. The concept is also useful for describing the chemical bonds that hold atoms together. In bulk materials this same idea helps explain the peculiar properties of lasers and semiconductors.

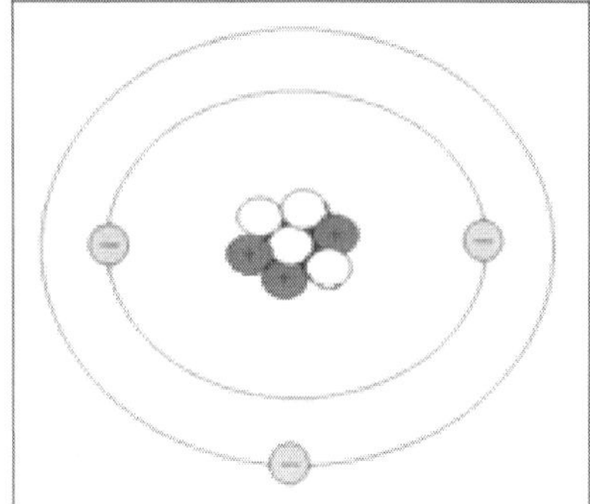

Fig. Simple Electron Shell Diagram of Lithium

SHELLS AND SUBSHELLS

Electron Shell

An electron shell may be thought of as an orbit followed by electrons around an atom's nucleus. The closest shell to the nucleus is called the "1 shell" (also called "K shell"), followed by the "2 shell" (or "L shell"), then the "3 shell" (or "M shell"), and so on further and further from the nucleus. The shell letters K,L,M,... are alphabetical.

Each shell can contain only a fixed number of electrons: The 1 shell can hold up to two electrons, the 2 shell can hold up to eight electrons, and in general, the n shell can hold up to $2n^2$ electrons. Since electrons are electrically attracted to the nucleus, an atom's electrons will generally occupy outer shells only if the more inner shells have already been completely filled by other electrons. However, this is not a strict requirement: Atoms may have two or even three outer shells that are only partly filled with electrons.

The electrons in the partially-filled outermost shell (or shells) determine the chemical properties of the atom; it is called *valence shell*. Each shell consists of one or more *subshells*, and each subshell consists of one or more atomic orbitals. The shell terminology comes from Arnold Sommerfeld's modification of the Bohr model. Sommerfeld retained Bohr's planetary model, but added mildly elliptical orbits (characterized by additional quantum numbers *l* and *m*) to explain the fine spectroscopic structure of some elements. The multiple electrons with the same principal quantum number (n) had close orbits that formed a "shell" of finite thickness instead of the infinitely thin circular orbit of Bohr's model.

The existence of electron shells was first observed experimentally in Charles Barkla's and Henry Moseley's X-ray absorption studies. Barkla labeled them with the letters K, L, M, N, O, P, and Q. The origin of this terminology was alphabetic.

A *J* series was also suspected, though later experiments indicated that the *K* absorption lines are produced by the innermost electrons. These letters were later found to correspond to the *n*-values 1, 2, 3, etc. They are used in the spectroscopic Siegbahn notation.

The physical chemist Gilbert Lewis was responsible for much of the early development of the theory of the participation of valence shell electrons in chemical bonding. Linus Pauling later generalized and extended the theory while applying insights from quantum mechanics.

SHELLS

The electron shells are labeled K, L, M, N, O, P, and Q; or 1, 2, 3, 4, 5, 6, and 7; going from innermost shell outwards. Electrons in outer shells have higher average energy and travel farther from the nucleus than those in inner shells. This makes them more important in determining how the atom reacts chemically and behaves as a conductor, because the pull of the atom's nucleus upon them is weaker and more easily broken. In this way, a given element's reactivity is highly dependent upon its electronic configuration.

Subshells

Each shell is composed of one or more subshells, which are themselves composed of atomic orbitals. For example, the first (K) shell has one subshell, called "1s"; the second (L) shell has two subshells, called "2s" and "2p"; the third shell has "3s", "3p", and "3d"; and so on. The various possible subshells are shown in the following table:

Subshell Label	e	Max Electrons	Shell Containing it	Historical Name
S	0	2	Every shell	Sharp
P	1	6	2nd shell and higher	Principal
D	2	10	3rd shell and higher	Diffuse
F	3	14	4th shell and higher	Fundamental
G	4	18	5th shell and higher after f)	(Next in Alphabet

- The first column is the "subshell label", a lowercase-letter label for the type of subshell. For example, the "4s subshell" is a subshell of the fourth (N) shell, with the type ("s") described in the first row.
- The second column is the azimuthal quantum number of the subshell. The precise definition involves quantum mechanics, but it is a number that characterizes the subshell.
- The third column is the maximum number of electrons that can be put into a subshell of that type. For example, the top row says that each s-type subshell ("1s", "2s", etc.) can have at most two electrons in it. In each case the figure is 4 greater than the one above it.
- The fourth column says which shells have a subshell of that type. For example, looking at the top two rows, every shell has an s subshell, while only the second shell and higher have a p subshell (i.e., there is no "1p" subshell).
- The final column gives the historical origin of the labels s, p, d, and f. They come from early studies of atomic spectral lines. The other labels, namely g, h and i, are an alphabetic continuation following the last historically originated label of f.

Although it is commonly stated that all the electrons in a shell have the same energy, this is an approximation. However, the electrons in a *subshell* do have exactly the same level of energy, with later subshells having more energy per electron than earlier ones. This effect is great enough that the energy ranges associated with shells can overlap (see Valence shells and Aufbau Principle).

Number of Electrons in Each Shell

An atom's electron shells are filled according to the following theoretical constraints:

- Each s subshell holds at most 2 electrons
- Each p subshell holds at most 6 electrons
- Each d subshell holds at most 10 electrons
- Each f subshell holds at most 14 electrons
- Each g subshell holds at most 18 electrons

Therefore, the K shell, which contains only an s subshell, can hold up to 2 electrons; the L shell, which contains an s and a p, can hold up to 2+6=8 electrons; and so forth. The general formula is that the nth shell can in principle hold up to $2n^2$ electrons.

Although that formula gives the maximum in principle, in fact that maximum is only *achieved* (by known elements) for the first four shells (K,L,M,N). No known element has more than 32 electrons in any one shell. This is because the subshells are filled according to the Aufbau principle. The first elements to have more than 32 electrons in one shell would belong to the g-block of period 8 of the periodic table. These elements would have some electrons in their 5g subshell and thus have more than 32 electrons in the O shell (fifth principal shell).

	Subshell	**Shell**	
Shell Name	**Subshell Name**	**Max Electrons**	**Max Electrons**
K	1s	2	2
	2s	2	
L	2p	6	2+6=8
	3s	2	
M	3p	6	2+6+10=18
	3d	10	
	4s	2	
N	4p	6	2+6+10+14=32
	4d	10	
	4f	14	

Valence Shells

The valence shell is the outermost shell of an atom. It is usually (and misleadingly) said that the electrons in this shell make up its valence electrons, that is, the electrons that determine how the atom behaves in chemical reactions. Just as atoms with complete valence shells (noble gases) are the most chemically non-reactive, those with only one electron in their valence shells (alkalis) or just missing one electron from having a complete shell (halogens) are the most reactive.

However, this is a simplification of the truth. The electrons that determine how an atom reacts chemically are those that travel farthest from the nucleus, that is, those with the most energy. As stated in Subshells, electrons in the inner subshells have less energy than those in outer subshells. This effect is great enough that the 3d electrons have more energy than 4s electrons, and are therefore more important in chemical reactions, making them valence electrons although they are not in the so-called valence shell.

NOTATION

Physicists and chemists use a standard notation to indicate the electron configurations of atoms and molecules. For atoms, the notation consists of a sequence of atomic orbital labels (e.g. for phosphorus the sequence 1s, 2s, 2p, 3s, 3p) with the number of electrons assigned to each orbital (or set of orbitals sharing the same label) placed as a superscript. For example, hydrogen has one electron in the s-orbital of the first shell, so its configuration is written $1s^1$. Lithium has two electrons in the 1s-subshell and one in the (higher-energy) 2s-subshell, so its configuration is written $1s^2\ 2s^1$ (pronounced "one-s-two, two-s-one"). Phosphorus (atomic number 15), is as follows: $1s^2\ 2s^2\ 2p^6\ 3s^2\ 3p^3$.

For atoms with many electrons, this notation can become lengthy and so an abbreviated notation is used, since all but the last few subshells are identical to those of one or another of the noble gases. Phosphorus, for instance, differs from neon ($1s^2\ 2s^2\ 2p^6$) only by the presence of a third shell. Thus, the electron configuration of neon is pulled out, and phosphorus is written as follows: [Ne] $3s^2\ 3p^3$. This convention is useful as it is the electrons in the outermost shell which most determine the chemistry of the element.

The order of writing the orbitals is not completely fixed: some sources group all orbitals with the same value of *n* together, while other sources (as here) follow the order given by Madelung's rule. Hence the electron configuration of iron can be written as [Ar] $3d^6\ 4s^2$ (keeping the 3d-electrons with the 3s- and 3p-electrons which are implied by the configuration of argon) or as [Ar] $4s^2\ 3d^6$ (following the Aufbau principle, see below).

The superscript 1 for a singly occupied orbital is not compulsory. It is quite common to see the letters of the orbital labels (s, p, d, f) written in an italic or slanting typeface, although the International Union of Pure and Applied Chemistry (IUPAC) recommends a normal typeface (as used here). The choice of letters originates from a now-obsolete system of categorizing spectral lines as "sharp", "principal", "diffuse" and "fundamental" (or "fine"), based on their observed fine structure: their modern usage indicates orbitals with an azimuthal quantum number, *l*, of 0, 1, 2 or 3 respectively. After "f", the sequence continues alphabetically "g", "h", "i"... (l = 4, 5, 6...), skipping "j", although orbitals of these types are rarely required.

The electron configurations of molecules are written in a similar way, except that molecular orbital labels are used instead of atomic orbital labels.

Energy — Ground State and Excited States

The energy associated to an electron is that of its orbital. The energy of a configuration is often approximated as the sum of the energy of each electron, neglecting the electron-electron interactions. The configuration that corresponds to the lowest electronic energy is called the ground state. Any other configuration is an excited state.

As an example, the ground state configuration of the sodium atom is $1s^2 2s^2 2p^6 3s$, as deduced from the Aufbau principle (see below). The first excited state is obtained by promoting a 3s electron to the 3p orbital, to obtain the $1s^2 2s^2 2p^6 3p$ configuration, abbreviated as the 3p level. Atoms can move from one configuration to another by absorbing or emitting energy. In a sodium-vapor lamp for example, sodium atoms are excited to the 3p level by an electrical discharge, and return to the ground state by emitting yellow light of wavelength 589 nm.

Usually the excitation of valence electrons (such as 3s for sodium) involves energies corresponding to photons of visible or ultraviolet light. The excitation of core electrons is possible, but requires much higher energies generally corresponding to x-ray photons. This would be the case for example to excite a 2p electron to the 3s level and form the excited $1s^2 2s^2 2p^5 3s^2$ configuration. The remainder of this article deals only with the ground-state configuration, often referred to as "the" configuration of an atom or molecule. Niels Bohr was the first to propose (1923) that the periodicity in the properties of the elements might be explained by the electronic structure of the atom. His proposals were based on the then current Bohr model of the atom, in which the electron shells were orbits at a fixed distance from the nucleus. Bohr's original configurations would seem strange to a present-day chemist: sulfur was given as 2.4.4.6 instead of $1s^2\ 2s^2\ 2p^6\ 3s^2\ 3p^4$ (2.8.6).

The following year, E. C. Stoner incorporated Sommerfeld's third quantum number into the description of electron shells, and correctly predicted the shell structure of sulfur to be 2.8.6. However neither Bohr's system nor Stoner's could correctly describe the changes in atomic spectra in a magnetic field (the Zeeman effect).

Bohr was well aware of this shortcoming (and others), and had written to his friend Wolfgang Pauli to ask for his help in saving quantum theory (the system now known as "old quantum theory"). Pauli realized that the Zeeman effect must be due only to the outermost electrons of the atom, and was able to reproduce Stoner's shell structure, but with the correct structure of subshells, by his inclusion of a fourth quantum number and his exclusion principle (1925):

It should be forbidden for more than one electron with the same value of the main quantum number n *to have the same value for the other three quantum numbers* k *[*l*],* j *[*m_l*] and* m *[*m_s*].*

The Schrödinger equation, published in 1926, gave three of the four quantum numbers as a direct consequence of its solution for the hydrogen atom: this solution yields the atomic orbitals which are shown today in textbooks of chemistry (and above). The examination of atomic spectra allowed the electron configurations of atoms to be determined experimentally, and led to an empirical rule for the order in which atomic orbitals are filled with electrons.

Aufbau Principle and Madelung Rule

The Aufbau principle (from the German *Aufbau*, "building up, construction") was an important part of Bohr's original concept of electron configuration. It may be stated as: A maximum of two electrons are put into orbitals in the order of increasing orbital energy: the lowest-energy orbitals are filled before electrons are placed in higher-energy orbitals.

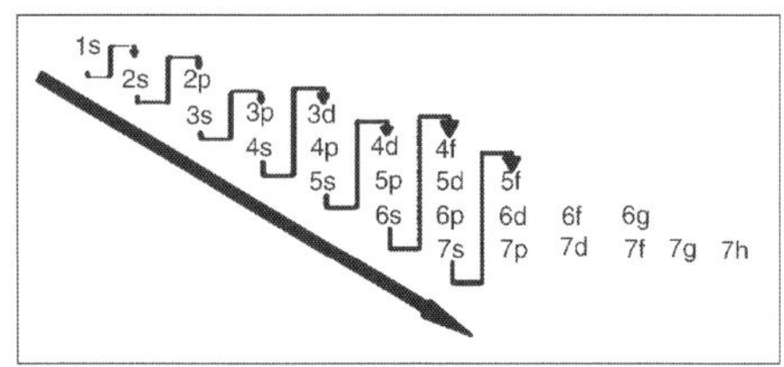

Fig. The Approximate order of Filling of Atomic Orbitals, Following the Arrows from 1s to 7p. After 7p the Order is Wrong because Orbitals with n > 7 are not Shown. Also, the 5g and 6h are Missing.

The principle works very well (for the ground states of the atoms) for the first 18 elements, then decreasingly well for the following 100 elements. The modern form of the Aufbau principle describes an order of orbital energies given by Madelung's rule (or Klechkowski's rule). This rule was first stated by Charles Janet in 1929, rediscovered by Erwin Madelung in 1936, and later given a theoretical justification by V.M. Klechkowski

- Orbitals are filled in the order of increasing $n + l$;
- Where two orbitals have the same value of $n + l$, they are filled in order of increasing n.

This gives the following order for filling the orbitals:

1s, 2s, 2p, 3s, 3p, 4s, 3d, 4p, 5s, 4d, 5p, 6s, 4f, 5d, 6p, 7s, 5f, 6d, 7p, (8s, 5g, 6f, 7d, 8p, and 9s)

In this list the orbitals in parentheses are not occupied in the ground state of the heaviest atom now known (Uuo, Z = 118). The Aufbau principle can be applied, in a modified form, to the protons and neutrons in the atomic nucleus, as in the shell model of nuclear physics and nuclear chemistry.

The Periodic table

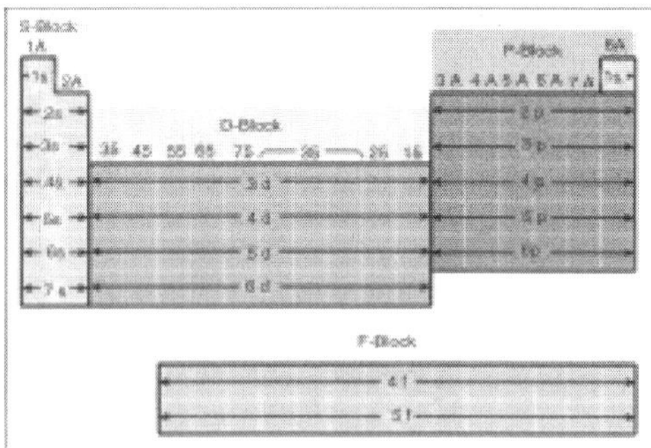

Fig. Electron Configuration Table

The form of the periodic table is closely related to the electron configuration of the atoms of the elements. For example, all the elements of group 2 have an electron configuration of [E] ns^2 (where [E] is an inert gas configuration), and have notable similarities in their chemical properties. In general, the periodicity of the periodic table in terms of periodic table blocks is clearly due to the number of electrons (2, 6, 10, 14...) needed to fill s, p, d, and f subshells.

The outermost electron shell is often referred to as the "valence shell" and (to a first approximation) determines the chemical properties. It should be remembered that the similarities in the chemical properties were remarked more than a century before the idea of electron configuration, It is not clear how far Madelung's rule *explains* (rather than simply describes) the periodic table, although some properties (such as the common +2 oxidation state in the first row of the transition metals) would obviously be different with a different order of orbital filling.

Shortcomings of the Aufbau Principle

The Aufbau principle rests on a fundamental postulate that the order of orbital energies is fixed, both for a given element and between different elements: neither of these is true (although they are approximately true enough for the principle to be useful). It considers atomic orbitals as "boxes" of fixed energy into which can be placed two electrons and no more.

However the energy of an electron "in" an atomic orbital depends on the energies of all the other electrons of the atom (or ion, or molecule, etc.). There are no "one-electron solutions" for systems of more than one electron, only a set of many-electron solutions which cannot be calculated exactly (although there are mathematical approximations available, such as the Hartree–Fock method).

The fact that the Aufbau principle is based on an approximation can be seen from the fact that there is an almost-fixed filling order at all, that, within a given shell, the s-orbital is always filled before the p-orbitals. In a hydrogen-like atom, which only has one electron, the s-orbital and the p-orbitals of the same shell have exactly the same energy, to a very good approximation in the absence of external electromagnetic fields. (However, in a real hydrogen atom, the energy levels are slightly split by the magnetic field of the nucleus, and by the quantum electrodynamic effects of the Lamb shift).

Ionization of the Transition Metals

The naive application of the Aufbau principle leads to a well-known paradox (or apparent paradox) in the basic chemistry of the transition metals. Potassium and calcium appear in the periodic table before the transition metals, and have electron configurations [Ar] $4s^1$ and [Ar] $4s^2$ respectively, i.e. the 4s-orbital is filled before the 3d-orbital. This is in line with Madelung's rule, as the 4s-orbital has $n + l = 4$ ($n = 4, l = 0$) while the 3d-orbital has $n + l = 5$ ($n = 3, l = 2$).

However, chromium and copper have electron configurations [Ar] $3d^5\ 4s^1$ and [Ar] $3d^{10}\ 4s^1$ respectively, i.e. one electron has passed from the 4s-orbital to a 3d-orbital to generate a half-filled or filled subshell. In this case, the usual explanation is that "half-filled or completely filled subshells are particularly stable arrangements of electrons".

The apparent paradox arises when electrons are *removed* from the transition metal atoms to form ions. The first electrons to be ionized come not from the 3d-orbital, as one would expect if it were "higher in energy", but from the 4s-orbital. The same is true when chemical compounds are formed.

Chromium hexacarbonyl can be described as a chromium atom (not ion, it is in the oxidation state 0) surrounded by six carbon monoxide ligands: it is diamagnetic, and the electron configuration of the central chromium atom is described as $3d^6$, i.e. the electron which was in the 4s-orbital in the free atom has passed into a 3d-orbital on forming the compound. This interchange of electrons between 4s and 3d is universal among the first series of the transition metals.The phenomenon is only paradoxical if it is assumed that the energies of atomic orbitals are fixed and unaffected by the presence of electrons in other orbitals.

If that were the case, the 3d-orbital would have the same energy as the 3p-orbital, as it does in hydrogen, yet it clearly doesn't. There is no special reason why the Fe^{2+} ion should have the same electron configuration as the chromium atom, given that iron has two more protons in its nucleus than chromium and that the chemistry of the two species is very different. When care is taken to compare "like with like", the paradox disappears.

Other Exceptions to Madelung's Rule

There are several more exceptions to Madelung's rule among the heavier elements, and it is more and more difficult to resort to simple explanations such as the stability of half-filled subshells. It is possible to predict most of the exceptions by Hartree–Fock calculations, which are an approximate method for taking account of the effect of the other electrons on orbital energies. For the heavier elements, it is also necessary to take account of the effects of Special Relativity on the energies of the atomic orbitals, as the inner-shell electrons are moving at speeds approaching the speed of light. In general, these relativistic effects tend to decrease the energy of the s-orbitals in relation to the other atomic orbitals.

ELECTRON CONFIGURATION IN MOLECULES

In molecules, the situation becomes more complex, as each molecule has a different orbital structure. The molecular orbitals are labelled according to their symmetry, rather than the atomic orbital labels used for atoms and monoatomic ions: hence, the electron configuration of the dioxygen molecule, O_2, is $1\sigma_g^2\ 1\sigma_u^2\ 2\acute{\sigma}_g^2\ 2\sigma_u^2\ 1\pi_u^4\ 3\sigma_g^2\ 1\pi_g^2$. The term $1\pi_g^2$ represents the two electrons in the two degenerate π*-orbitals (antibonding). From Hund's rules, these electrons have parallel spins in the ground state, and so dioxygen has a net magnetic moment (it is paramagnetic). The explanation of the paramagnetism of dioxygen was a major success for molecular orbital theory.

Electron Configuration in Solids

In a solid, the electron states become very numerous. They cease to be discrete, and effectively blend into continuous ranges of possible states (an electron band). The notion of electron configuration ceases to be relevant, and yields to band theory.

APPLICATIONS

The most widespread application of electron configurations is in the rationalization of chemical properties, in both inorganic and organic chemistry.

In effect, electron configurations, along with some simplified form of molecular orbital theory, have become the modern equivalent of the valence concept, describing the number and type of chemical bonds that an atom can be expected to form.

This approach is taken further in computational chemistry, which typically attempts to make quantitative estimates of chemical properties. For many years, most such calculations relied upon the "linear combination of atomic orbitals" (LCAO) approximation, using an ever larger and more complex basis set of atomic orbitals as the starting point. The last step in such a calculation is the assignment of electrons among the molecular orbitals according to the Aufbau principle. Not all methods in calculational chemistry rely on electron configuration: density functional theory (DFT) is an important example of a method which discards the model.

A fundamental application of electron configurations is in the interpretation of atomic spectra. In this case, it is necessary to convert the electron configuration into one or more term symbols, which describe the different energy levels available to an atom. Term symbols can be calculated for any electron configuration, not just the ground-state configuration listed in tables, although not all the energy levels are observed in practice. It is through the analysis of atomic spectra that the ground-state electron configurations of the elements were experimentally determined.

ELECTRONS INTO ITS ORBITALS

When an atom or ion receives electrons into its orbitals, the orbitals and shells fill up in a particular manner.

There are three principles that govern this process:

1. The Pauli exclusion principle,
2. The Aufbau (build-up) principle, and
3. Hund's rule.

Pauli Exclusion Principle

No two electrons in an atom can have all four quantum numbers the same. What this translates to in terms of our pictures of orbitals is that each orbital can only hold two electrons, one "spin up" and one "spin down".

Hund's Rule

This states that filled and half-filled shells tend to have additional stability. In some instances, then, for example, the 4s orbitals will be filled before the 3d orbitals.

This rule is applicable only for those elements that have d electrons, and so is less important in organic chemistry (though it is important in organometallic chemistry).

OCTET RULE

The octet rule states that atoms tend to prefer to have eight electrons in their valence shell, so will tend to *combine* in such a way that each atom can have eight electrons in its valence shell, similar to the electronic configuration of a noble gas. In simple terms, molecules are more stable when the outer shells of their constituent atoms are empty, full, or have eight electrons in the outer shell.

The main exception to the rule is helium, which is at lowest energy when it has two electrons in its valence shell.

Other notable exceptions are aluminum and boron, which can function well with six valence electrons; and some atoms beyond group three on the periodic table that can have over eight electrons, such as sulfur. Additionally, some noble gasses can form compounds when expanding their valence shell.

HYBRIDIZATION

Hybridization refers to the combining of the orbitals of two or more covalently bonded atoms. Depending on how many free electrons a given atom has and how many bonds it is forming, the electrons in the s and the p orbitals will combine in certain manners to form the bonds.

It is easy to determine the hybridization of an atom given a Lewis structure. First, you count the number of pairs of free electrons and the number of *sigma* bonds (single bonds). Do not count double bonds, since they do not affect the hybridization of the atom. Once the total of these two is determined, the hybridization pattern is as follows:

Sigma Bonds + Electron Pairs Hybridization

2 sp

3 sp^2

4 sp^3

The pattern here is the same as that for the electron orbitals, which serves as a memory guide.

ELECTRONEGATIVITY

Whenever two atoms form a bond, the nucleus of each atom attracts the other's electrons. Electronegativity is a measure of the strength of this attraction.

Periodic Trends

Several traits of atoms are said to have "periodic trends", meaning that different atoms in a period have identifiable relationships to one another based on their position. Is that confusing? Think of the periodic table as a group picture, maybe of a very large basketball team. Each period is a row of players in the picture, and the "photographer" has decided to arrange the "players" by their characteristics. Of course, no conscious effort was made to arrange the periodic table by any characteristic other than number of protons, but some properties are consistent in its layout regardless.

Atomic size is one characteristic that shows a periodic trend. In case of atomic radius the "photographer" (Mendeleev and others since) decided to arrange "players" (atoms) by size with the very shortest and smallest players at the top right. As you go left to right along a row (a period) the atoms get sequentially smaller and smaller. Fluorine is smaller than carbon, and carbon is smaller than magnesium. This is due to the number of protons in the nucleus increasing, while the increasing number of electrons are unable to shield one another from the attractive force of the positive charge from the nucleus.

REMEMBER: *largest* > Li > Be > B > C > N > O > F > Ne > *smallest*

Another characteristic with a periodic trend is ionization energy. This is the amount of energy necessary to remove one electron from an atom. Since all the atoms favour an electron configuration of a noble gas, the atoms at the extreme left of the table will give up their first electron most readily. (In almost all cases, a metal will readily give up its first electron.) The halogens, which need only one more electron to fill their outer shells, require a great deal of energy to give up an electron because they would be much more stable if they gained one electron instead. Ionization energy is the opposite of atomic radius, therefore, because it increases from left to right across a period.

REMEMBER: *least energy to ionize* < Li < Be < B < C < N < O < F < Ne < *most energy to ionize*

Electronegativity is perhaps the most important periodic trend, and it is not related to ionization energy directly — but its trend is the same, increasing from left to right. Also, the elements in a group (like the halogen group) gain stability as they grow in atomic number, so the smallest member of an electronegative group is often the most electronegative. In general, it can be said that among periods (rows) or groups (columns) of the periodic table, the closer an element is to fluorine, the more electronegative it will be. For Group VIIA (the aforementioned halogens) of the periodic table, you memorize the following relationships:

REMEMBER: *most electronegative* > F > Cl > Br > I > *least electronegative*

And REMEMBER: *least electronegative* < Li < Be < B < C < N < O < F < *most electronegative*

(Notice that the noble gas Neon is not on the electronegativity chart. In its non-ionized form, a noble gas is usually treated as if it has no electronegativity at all.)

Electronegativities of Atoms Common in Organic Chemistry

- C – 2.55
- H – 2.20

- N – 3.04
- O – 3.44
- P – 2.19
- S – 2.58
- Cl – 3.16
- Br – 2.96
- F – 3.98

Higher numbers represent a stronger attraction of electrons.

When atoms of similar electronegativity bond, a non-polar covalent bond is the result.

- Common non-polar bonds
  - C–C
  - H–C

When atoms of slightly different electronegativities bond, a polar covalent bond results:

- Common polar bonds
  - δ^+ C–O δ^-
  - δ^+ C–N δ^-
  - δ^- O–H δ^+
  - δ^- N–H δ^+
  - δ^- and δ^+ represent partial charges

When atoms of very different electronegativities bond, an ionic bond results.

ELECTRONEGATIVITY

Electronegativity is a measure of the ability of an atom or molecule to attract electrons in the context of a chemical bond. The type of bond formed is largely determined by the difference in electronegativity between the atoms involved. Atoms with similar electronegativities will share an electron with each other and form a covalent bond. However, if the difference is too great, the electron will be permanently transferred to one atom and an ionic bond will form. Furthermore, in a covalent bond if one atom pulls slightly harder than the other, a polar covalent bond will form.

The reverse of electronegativity, the ability of an atom to lose electrons, is known as electropositivity. Two scales of electronegativity are in common use: the Pauling scale (proposed in 1932) and the Mulliken scale (proposed in 1934). Another proposal is the Allred-Rochow scale.

Pauling Scale

The Pauling scale was devised in 1932 by Linus Pauling. On this scale, the most electronegative chemical element (fluorine) is given an electronegativity value of 3.98 (textbooks often state this value to be 4.0); the least electronegative element (francium) has a value of 0.7, and the remaining elements have values in between. On the Pauling scale, hydrogen is arbitrarily assigned a value of 2.1 or 2.2. 'δEN' is the difference in electronegativity between two atoms or elements. Bonds between atoms with a large electronegativity difference (greater than or equal to 1.7) are usually considered to be ionic, while values between 1.7 and 0.4 are considered polar covalent. Values below 0.4 are considered non-polar covalent bonds, and electronegativity differences of 0 indicate a completely non-polar covalent bond.

Mulliken Scale

The Mulliken scale was proposed by Robert S. Mulliken in 1934. On the Mulliken scale, numbers are obtained by averaging ionization potential and electron affinity. Consequently, the Mulliken electronegativities are expressed directly in energy units, usually electron volts.

Electronegativity Trends

Each element has a characteristic electronegativity ranging from 0 to 4 on the Pauling scale. The most strongly electronegative element, fluorine, has an electronegativity of 3.98 while weakly electronegative elements, such as lithium, have values close to 1.

The least electronegative element is francium at 0.7. In general, the degree of electronegativity decreases down each group and increases across the periods, as shown below.

Across a period, non-metals tend to gain electrons and metals tend to lose them due to the atom striving to achieve a stable octet. Down a group, the nuclear charge has less effect on the outermost shells. Therefore, the most electronegative atoms can be found in the upper, right hand side of the periodic table, and the least electronegative elements can be found at the bottom left. Consequently, in general, atomic radius decreases across the periodic table, but ionization energy increases.

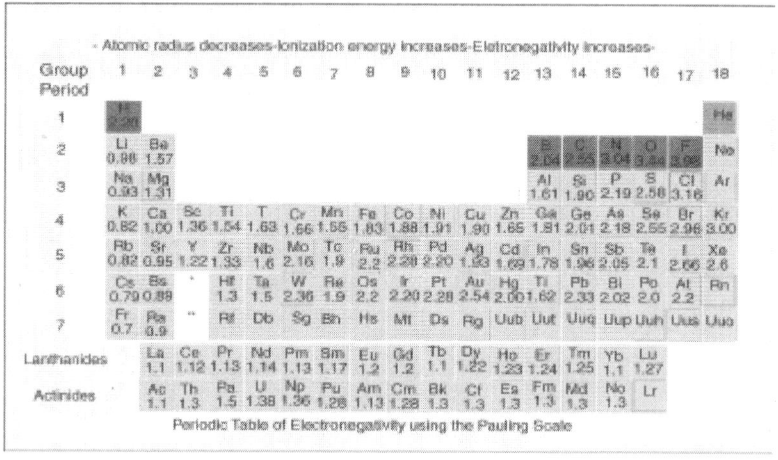

- Atomic radius decreases-Ionization energy increases-Eletronegativity increases-

Group / Period	1	2	3	4	5	6	7	8	9	10	11	12	13	14	15	16	17	18
1	H 2.20																	He
2	Li 0.98	Be 1.57											B 2.04	C 2.55	N 3.04	O 3.44	F 3.98	Ne
3	Na 0.93	Mg 1.31											Al 1.61	Si 1.90	P 2.19	S 2.58	Cl 3.16	Ar
4	K 0.82	Ca 1.00	Sc 1.36	Ti 1.54	V 1.63	Cr 1.66	Mn 1.55	Fe 1.83	Co 1.88	Ni 1.91	Cu 1.90	Zn 1.65	Ga 1.81	Ge 2.01	As 2.18	Se 2.55	Br 2.96	Kr 3.00
5	Rb 0.82	Sr 0.95	Y 1.22	Zr 1.33	Nb 1.6	Mo 2.16	Tc 1.9	Ru 2.2	Rh 2.28	Pd 2.20	Ag 1.93	Cd 1.69	In 1.78	Sn 1.96	Sb 2.05	Te 2.1	I 2.66	Xe 2.6
6	Cs 0.79	Ba 0.89	*	Hf 1.3	Ta 1.5	W 2.36	Re 1.9	Os 2.2	Ir 2.20	Pt 2.28	Au 2.54	Hg 2.00	Tl 1.62	Pb 2.33	Bi 2.02	Po 2.0	At 2.2	Rn
7	Fr 0.7	Ra 0.9	**	Rf	Db	Sg	Bh	Hs	Mt	Ds	Rg	Uub	Uut	Uuq	Uup	Uuh	Uus	Uuo

Lanthanides	La 1.1	Ce 1.12	Pr 1.13	Nd 1.14	Pm 1.13	Sm 1.17	Eu 1.2	Gd 1.2	Tb 1.1	Dy 1.22	Ho 1.23	Er 1.24	Tm 1.25	Yb 1.1	Lu 1.27
Actinides	Ac 1.1	Th 1.3	Pa 1.5	U 1.38	Np 1.36	Pu 1.28	Am 1.13	Cm 1.28	Bk 1.3	Cf 1.3	Es 1.3	Fm 1.3	Md 1.3	No 1.3	Lr

Periodic Table of Electronegativity using the Pauling Scale

BONDING

Ionic Bonding

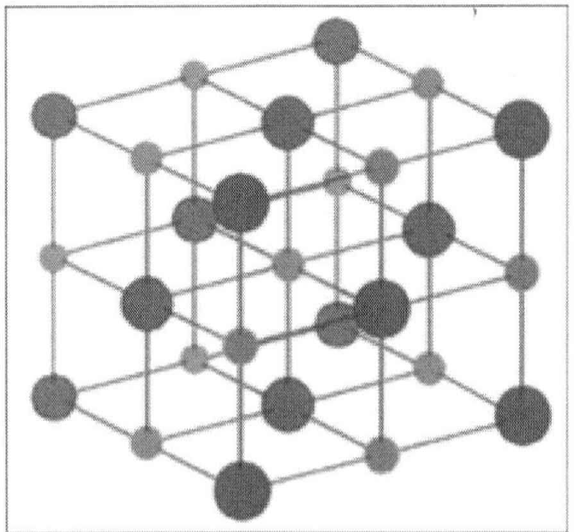

Fig. The Sodium Chloride Crystal Structure.

Each atom has six nearest behaviours, with octahedral geometry. This arrangement is known as *cubic close packed* (ccp).

Light blue = Na^+

Dark green = Cl^-

Ionic bonding is when positively and negatively charged ions stick to each other through electrostatic force. These bonds are *slightly weaker than covalent bonds* and stronger than Van der Waals bonding or hydrogen bonding.

In ionic bonds the electronegativity of the negative ion is so much stronger than the electronegativity of the positive ion that the two ions do not share electrons. Rather, the more electronegative ion assumes full ownership of the electron(s). Sodium chloride forms crystals with cubic symmetry. In these, the larger chloride ions are arranged in a cubic close-packing, while the smaller sodium ions fill the octahedral gaps between them. Each ion is surrounded by six of the other kind. This same basic structure is found in many other minerals, and is known as the halite structure.

Perhaps the most common example of an ionically bonded substance is NaCl, or table salt. In this, the sodium (Na) atom gives up an electron to the much more electronegative chlorine (Cl) atom, and the two atoms become ions, Na^+ and Cl^-. The electrostatic bonding force between the two oppositely charged ions extends outside the local area attracting other ions to form giant crystal structures. For this reason most ionically bonded materials are solid at room temperature.

Covalent Bonding

Covalent bonding is close to the heart of organic chemistry. This is where two atoms share electrons in a bond. The goal of each atom is to *fill its octet* as well as have a *formal charge of zero*. To do this, atomic nuclei share electrons in the space between them. This sharing also allows the atoms to reach a lower energy state, which stabilizes the molecule. Most reactions in chemistry are due to molecules achieving a lower energy state. Covalent bonds are most frequently seen between atoms with similar electronegativity. In molecules that only have one type of atom, e.g. H_2 or O_2, the electronegativity of the atoms is necessarily identical, so they cannot form ionic bonds. They always form covalent bonds.

Carbon is especially good at covalent bonding because its electronegativity is intermediate relative to other atoms. That means it can give as well as take electrons as needs warrant.

Covalently bonded compounds have strong internal bonds but weak attractive forces between molecules. Because of these weak attractive forces, the melting and boiling points of these compounds are much lower than compounds with ionic bonds. Therefore, such compounds are much more likely to be liquids or gases at room temperature than ionically bonded compounds.

In molecules formed from two atoms of the same element, there is no difference in the electronegativity of the bonded atoms, so the electrons in the covalent bond are shared equally, resulting in a completely non-polar covalent bond.

In covalent bonds where the bonded atoms are different elements, there is a difference in electronegativities between the two atoms. The atom that is more electronegative will attract the bonding electrons more towards itself than the less electronegative atom. The difference in charge on the two atoms because of the electrons causes the covalent bond to be polar.

Greater differences in electronegativity result in more polar bonds. Depending on the difference in electronegativities, the polarity of a bond can range from non-polar covalent to ionic with varying degrees of polar covalent in between. An overall imbalance in charge from one side of a molecule to the other side is called a dipole moment. Such molecules are said to be polar. For a completely symmetrical covalently bonded molecule, the overall dipole moment of the molecule is zero. Molecules with larger dipole moments are more polar. The most common polar molecule is water.

Bond Polarity and Dipole Moment

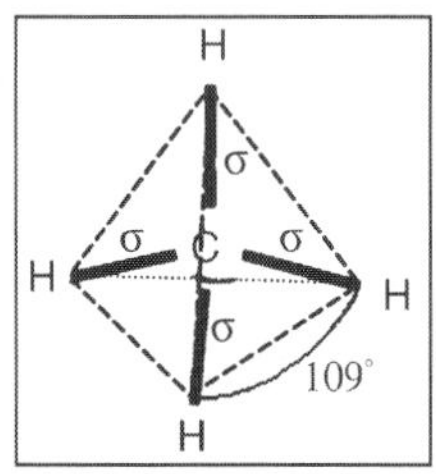

Fig. Methane

The ideas of bond polarity and dipole moment play important roles in organic chemistry. If you look at the image of methane on the right, the single most important aspect of it in terms of bond polarity is that it is a symmetric molecule. It has 4 hydrogens, all bonded at 109.5° from the other, and all with precisely the same bond angle. Each carbon-hydrogen bond is slightly polar (hydrogen has an electronegativity of 2.1, carbon 2.5), but because of this symmetry, the polarities cancel each other out and overall, methane is a non-polar molecule.

The distinction is between Bond Polarity and Molecular polarity. The total polarity of a molecule is measured as Dipole Moment. The actual calculation of dipole moment isn't really necessary so much as an understanding of what it means.

Frequently, a guesstimate of dipole moment is pretty easy once you understand the concept and until you get into the more advanced organic chemistry, exact values are of little value.

Basically, the molecular polarity is, essentially, the summation of the vectors of all of the bond polarities in a molecule.

Van der Waals Bonding

Van der Waals bonding is the collective name for three types of interactions:

1. Permanent Dipole interactions: these are the electrostatic attractive forces between two dipoles, these are responsible for fluromethane's (CH_3F) high boiling point (about –15 °C) compared to Nitrogen (about –180 °qC).
2. Permanent dipole/ induced dipole: these are the interactions between a permanent dipole and another molecule, causing the latter molecule's electron cloud to be distorted and thus have an induced dipole itself. These are much weaker than the permanent dipole/ dipole interactions. These forces occur in permanent dipole-molecules, and in mixtures of permanent dipole and dipole free molecules.
3. Instantaneous dipole/ induced dipole: At any specific moment the electron cloud is not necesarily symetrical, this instantaneous dipole then induces a dipole in another molecule and they are attracted, this is the weakest of all molecular interactions.

A Dipole is caused by an atom or molecule fragment having a higher electronegativity (this is a measure of its effective nuclear charge, and thus the attraction of the nucleus by electrons) than one to which it is attached. This means that it pulls electrons closer to it, and has a higher share of the electrons in the bond. Dipoles can cancel out by symmetry, eg: Carbon dioxide (O=C=O) is linear so there is no dipole, but the charge distribution is asymmetric causing a quadripole moment (this acts similarly to a dipole, but is much weaker).

Organometallic Compounds and Bonding

Organometallic chemistry combines aspects of inorganic chemistry and organic chemistry, because organometallic compounds are chemical compounds containing bonds between carbon and a metal or metalloid element. Organometallic bonds are different from other bonds in that they are not either truly covalent or truly ionic, but each type of metal has individual bond character.

Cuprate (copper) compounds, for example, behave quite differently than Grignard reagents (magnesium), and so beginning organic chemists should concentrate on how to use the most basic compounds mechanistically, while leaving the explanation of exactly what occurs at the molecular level until later and more in-depth studies in the subject.

THE CHEMICAL BOND

ATOMIC THEORY

The atomic theory of electrons began in the early 1900s and gained acceptance around 1926 after Heisenberg and Schroedinger found mathematical solutions to the electronic energy levels found in atoms, the field is now called quantum mechanics.

Electrons exist in energy levels that surround the nucleus of the atom. The energy of these levels increases as they get farther from the nucleus. The energy levels are called shells, and within these shells are other energy levels, called subshells or orbitals., that contain up to two electrons. The calculations from atomic theory give the following results for electron energy and orbitals. The results for the first two energy levels (shells 1 and 2) are the most important for bonding in organic chemistry.

	Orbitals				
Shell	**s**	**p**	**d**	**f**	**Total Electrons Possible**
1	1				2
2	2	3			8
3	3	3	5		18
4	1	3	5	7	32

- Energy level 1 contains up to two electrons in a spherical orbital called a 1s orbital.
- Energy level 2 contains up to eight electrons; two in an 2s-orbital and two in each of three orbitals designated as 2p-orbitals.

The p-orbitals have a barbell type shape and are aligned along the x, y, and z axes. They are thus called the px, py, and pz orbitals.

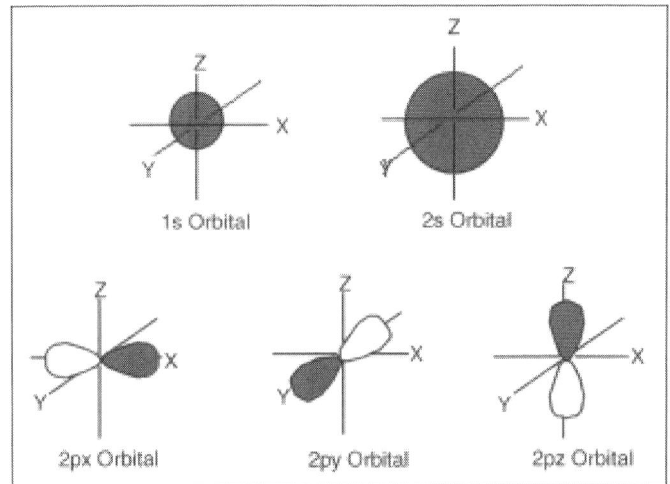

- Energy level 3 contains up to eighteen electrons, two electrons in a 3s orbital, six electrons in the three 3p orbitals, and ten electrons in the five 3d orbitals.
- Energy level 4 contains up to thirty-two electrons, two electrons in a 4s-orbital, six electrons in the three 4p-orbitals, ten electrons in the five 4d-orbitals, and fourteen electrons in the seven 4f-orbitals.

Electrons fill the lower energy levels first until all of the electrons are used (Aufbau Principle). An element contains the number of electrons equal to its atomic number. For the first and second row elements the electron configurations are relatively simple.

Element (Atomic Number)	Electron Configuration
H (1)	$1s^1$ (1st shell, s orbital, one electron)
He (2)	$1s^2$
Li (3)	$1s^2$, $2s^1$
Be (4)	$1s^2$, $2s^2$
B (5)	$1s^2$, $2s^2$, $2p^1$
C (6)	$1s^2$, $2s^2$, $2p^2$
N (7)	$1s^2$, $2s^2$, $2p^3$
O (8)	$1s^2$, $2s^2$, $2p^4$
F (9)	$1s^2$, $2s^2$, $2p^5$
Ne (10)	$1s^2$, $2s^2$, $2p^6$ (inert, completely filled)

ELECTRONEGATIVITY

Electronegativity is the ability of an atom to attract electrons to itself, and generally increases as one moves from left to the right across the periodic table.

least electronegative $Li < Be < B < C < N < O < F$
most electronegative

Electronegativity also increases as we go from the bottom to the top of a column in the periodic table.

least electronegative $I < Br < Cl < F$
most electronegative

Elements that easily lose electrons and attain a positive charge are called electropositive elements. Alkali metals are electropositive elements.

BONDING

Atoms can become bonded with each other, and their electronic structure governs the type of bond formed. The main two types of bonds that are formed are called ionic and covalent.

Ionic Bond

Ionic bonding is important between atoms of vastly different electronegativity. The bond results from one atom giving up an electron while another atom accepts the electron. Both atoms attain a stable nobel gas configuration.

In the compound lithium fluoride, the $2s^1$ electron of lithium is transferred to the $2p^5$ orbital of fluorine. The lithium atom gives up an electron to form the positively charged lithium cation with $1s^2$, $2s^0$ configuration, and the fluorine atom receives an electron to form a fluoride anion with $1s^2$, $2s^2$, $2p^6$ configuration. Thus the outer energy levels of both ions are completely filled. The ions are held together by the electrostatic attraction of the positive and negative ions.

Li	+	F	$\longrightarrow$	Li^+	+	F^-
$1s^2$		$1s^2$		$1s^2$		$1s^2$
$2s^1$		$2s^2$ $2p^5$		$2s^0$		$2s^2$ $2p^6$

Covalent Bond

A covalent bond is formed by a sharing of two electrons by two atoms. A hydrogen atom possessing the $1s^1$ electron joins with another hydrogen atom with its $1s^1$ configuration. The two atoms form a covalent bond with two electrons by sharing their electrons.

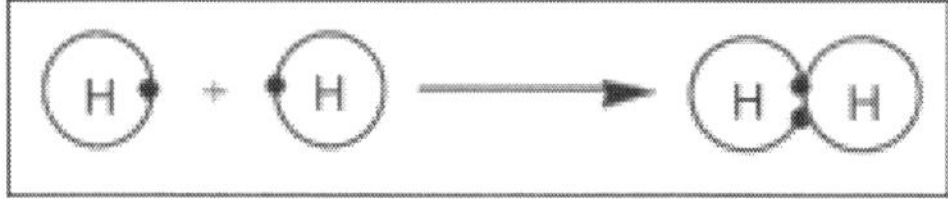

In hydrogen fluoride, HF, the hydrogen 1s electron is shared with a $2p^5$ electron in fluorine ($1s^2$, $2s^2$, $2p^5$), and the molecule is now held together by a covalent bond. In this case, the fluorine atom is much more electronegative than the hydrogen atom and the electrons in the bond tend to stay closer to the fluorine atom. This is called a polar covalent bond, and the atoms possess a small partial charge denoted by the Greek δ symbol.

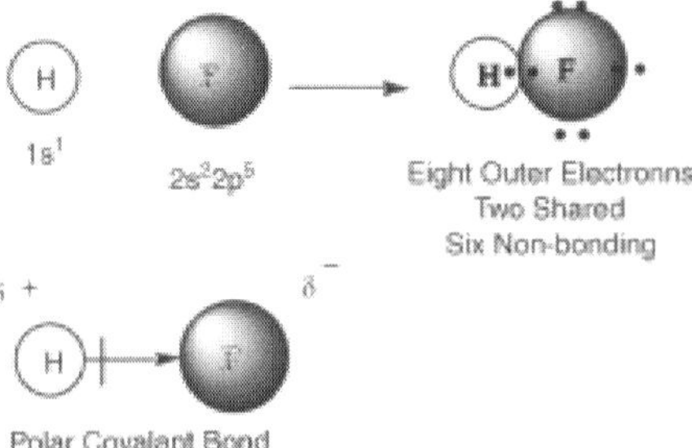

BONDING IN CARBON COMPOUNDS

The property of carbon that makes it unique is its ability to form bonds with itself and therefore allows a large number of organic chemicals with many diverse properties. Carbon has the property of forming single, double and triple bonds with itself and with other atoms. This multiple bond ability allows carbon compounds to have a variety of shapes. *In all carbon compounds, carbon forms four bonds. The types of bonds used by the carbon atom are known as sigma (σ) and pi (π) bonds. Different combinations of these bonds lead to carbon single bonds, double bonds and triple bonds.*

The Carbon-Hydrogen Single Bond-The Sigma (σ) Bond

By far most of the bonds in carbon compounds are covalent bonds found commonly in the carbon-hydrogen single bond.

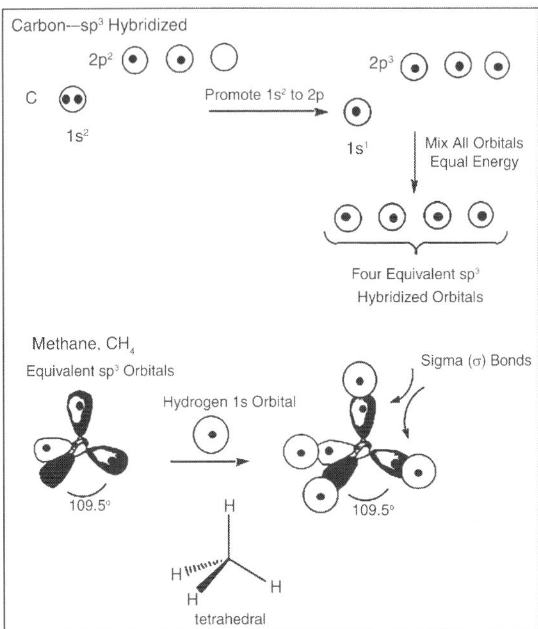

In carbon ($1s^2$, $2s^2$, $2p^2$) one of the electrons of the $2s^2$ orbital is promoted to the third $2p^0$ orbital. The s and three p orbitals hybridize to form four new orbitals of equal energy called sp3 hybrid orbitals. The electrons in the four sp3 hybridized orbitals bond by overlap with the 1s1 hydrogen orbital. The single covalent bond is called a sigma (σ) bond. The sp3 bonds arrange themselves as far from each other as possible, the shape of a molecule of methane, CH_4, is tetrahedral with 109.5° bond angles.

The unique property of carbon that differentiates it from the other elements and allows the formation of so many different organic compounds is the ability of carbon to bond with itself through covalent bonding. Thus, addition of another carbon atom to methane results in ethane which has covalent sigma bonds to the hydrogen atoms and a covalent sigma bond between the carbon atoms. Addition of more carbon atoms leads to many more compounds.

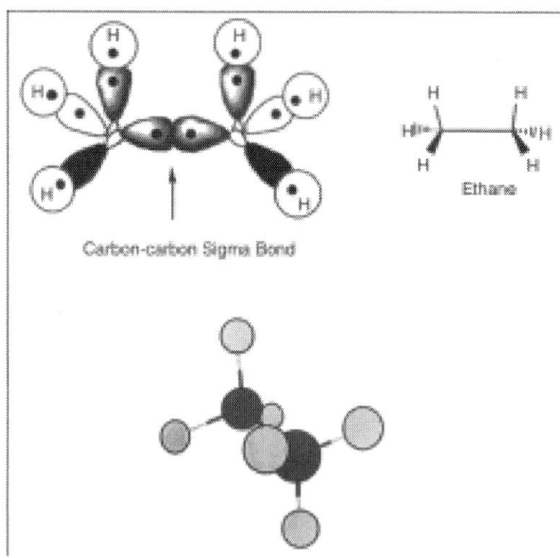

The Carbon-Carbon Double Bond-The Pi (π) Bond

Carbon forms a wide variety of compounds that contain carbon bonded to another carbon with a double bond between the two atoms. These compounds are classified as alkenes (older naming calls them olefins). The orbital model below explains the carbon-carbon double bond. The carbon electron configuration shows one s electron being promoted to a p orbital. But now only three orbitals are mixed, a s orbital and two p orbital, that are called sp^2 hybrid orbitals and are used to form single bonds (sigma bonds). The p orbital contains one electron.

The combination of two of the sp^2 hybridized carbon atom leads to two carbon atoms being joined by overlap of sp^2 orbitals to form a C-C single bond, and the side-to-side overlap of the p orbitals to form another bond known as a pi (π) bond. In the molecule of ethene shown below there are a total of 5 sigma bonds and one pi bond. As a result of the bonding in an ethene, the molecule is planar with bond angles of 120° and a C=C bond length that is longer than the C-H bond length.

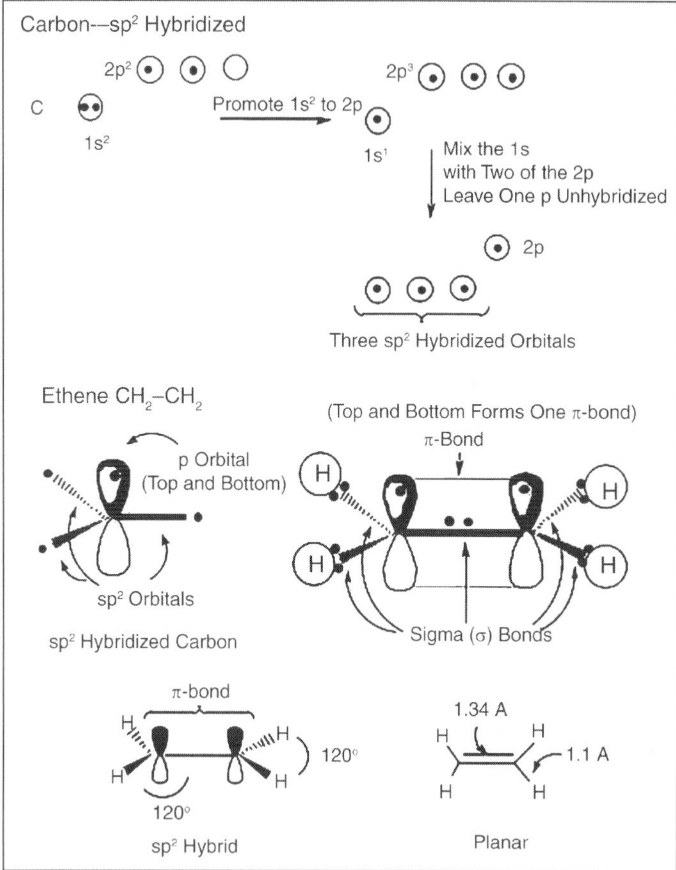

When carbon forms a bond with an electronegative atom such as oxygen, nitrogen, sulfur or a halogen, the bond is a polar covalent bond with the electrons of the bond residing closer to the electronegative atom.

The Carbon-Carbon Triple Bond

Another type of bond that carbon forms with itself is the triple bond found in a class of compounds called alkynes. After promotion of the 2s electron to a 2p orbital, one s orbital mixes with one p orbital to give two hybrid sp orbitals. The two remaining p orbitals are used to make p bonds.

Thus the carbon is bound by a sigma bond to hydrogen from one of the sp hybrid orbital, to the other carbon atom by a sigma bond from one of the sp hybrid orbitals, and the two carbon atoms are bound by two pi bonds from side-to-side overlap of the two p orbitals. The sp hybrid orbitals position themselves 180° apart and thus a molecule of ethyne is linear with the hydrogen atoms 180° apart.

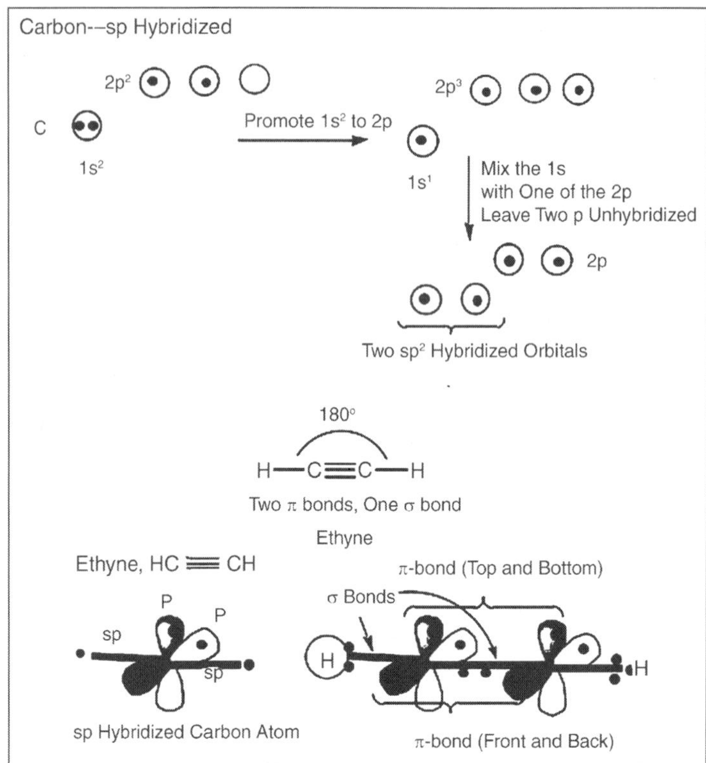

Polar Covalent Bonds in Carbon

Carbon forms single, double and triple bonds with elements other than carbon. The atoms involved in the bonding are usually oxygen, nitrogen, sulfur and the halogens. These elements are more electronegative than carbon and thus attract the electrons to themselves. The bonds are therefore polar covalent bonds. Bonds that contain a separation of charge possess a dipole moment, a property that contributes to the overall polarity of the molecule.

H_3C—OH	H_3C—C1	H_2C=O
Alcohol	Alkyl Chloride	Aldehyde

Hydrogen Bonds and Bond Polarity

The bonds O-H, N-H and F-H are highly polar covalent bonds because the electronegative draws electrons away from the hydrogen atom. In every case the hydrogen atom has a partial positive charge.

$O^{\delta-}$ -$H^{\delta+}$ $N^{\delta-}$ -$H^{\delta+}$ $F^{\delta-}$ -$H^{\delta+}$

A result of molecules having these highly polarized bonds with the hydrogen atom partly positive in nature, the hydrogen atom is attracted to the basic site in other molecules, such as the non-bonding electrons on oxygen and nitrogen (non-bonding electrons are electrons belong to an element that complete the octet but do not participate in bonding). This attraction is called hydrogen bonding and is useful for explaining high boiling points and high melting points of fairly low mass molecules. Thus hydrogen bonding in water explains why the compound with only three atoms boils relatively high when compared with other molecules of similar mass. Extra energy is required to break the hydrogen bonds during the boiling process.

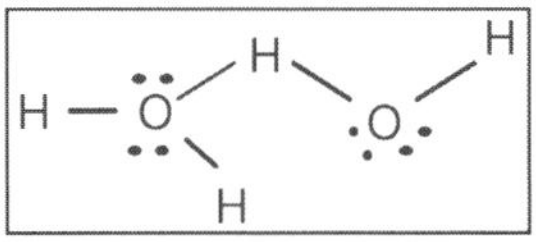

Fig. Hydrogen Bonding in Water

In organic compounds hydrogen bonding is very important for describing the boiling and solubility characteristics of alcohols and acids, and the concept will be given in more detail in chapters dealing with those types of molecules.

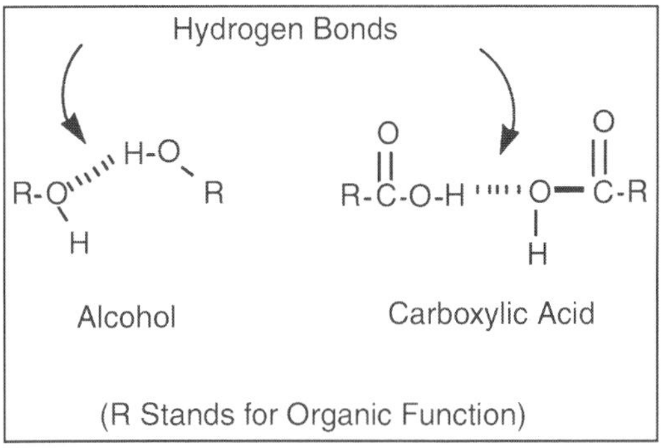

Hydrogen Bonds

Alcohol

Carboxylic Acid

(R Stands for Organic Function)

3

Chemicals in the Environment

CHEMICALS AND THEIR CHEMISTRY

Everything around us - animal, vegetable and mineral - is made up of chemicals. Some are simple substances such as the water molecule, which is made up of hydrogen and oxygen atoms, and others are very complicated compounds made up of many different chemical elements.

As well as naturally occurring chemical compounds, scientists have created millions more in the past 100 years that do not exist in nature. Chemicals come as powders, pellets, dusts, liquids, vapours, gases, etc.

Chemists tend to divide the world into inorganic and organic compounds, largely depending on whether they contain the element carbon.

Carbon-containing compounds were originally described as organic because many of them came from nature. Since then, chemists have synthesised many organic compounds, many of which are important in industrial products and processes.

The organic solvents and organophosphate pesticides involved in the case studies are just a few examples of these organic compounds. Inorganic materials, comprising metals, minerals and salts, among other types, are much less likely to contain carbon.

WHAT'S IN A NAME?

Individual chemical compounds are described by their formulae, such as H_2O for water and CH_4 for methane. A chemical formula describes the number of atoms of each element making up the compound, whereas the structure describes the arrangement of these atoms. The molecule is not flat but a three-dimensional tetrahedron, with the carbon atom in the centre and a hydrogen at each of the four corners.

Chemical structure is not just of academic interest. It affects how chemicals react with each other, and with biological systems, including humans and the environment. It is common for chemicals with the same formula to have different structures (called isomers), and therefore have different properties and toxicity.

In fact, the seemingly small difference of one molecule being the mirror image of the other (called stereoisomers) can cause major differences in their biological effects.

Although certain rules apply to naming chemicals, most chemicals are known by more than one name, or synonym, and some also have trade names. This means that finding information on specific chemicals can appear complicated but useful information can be gleaned by non-chemists and this becomes easier with familiarity and practice.

When searching for information on a chemical, it is always useful to know its CAS registry number. This number is assigned by the US Chemical Abstracts Service, and each chemical has its own unique number. This system is accepted globally. Many books and databases of chemicals are indexed by CAS registry number.

PROPERTIES

As well as their formula and structure, chemists describe a great many properties of a chemical. Some of these are also important to the toxicologist or industrial hygienist because they tell us how a chemical behaves in various situations. According to its boiling point and melting point, a chemical will be a solid, liquid or gas at room temperature and pressure. This is known as the chemical's physical state, and this changes with temperature and pressure. Water, for example, changes from a solid to a liquid to a gas between 0 and 100°C. A chemical's physical state, coupled with other properties such as its volatility, will influence the likelihood of its entry into the human body by various routes of exposure.

Several other properties influence the effect of chemicals:

- Vapour density: The weight of a volume of a gaseous chemical compared with the weight of the same volume of air. If the vapour density is greater than 1, the gas is heavier than air and will tend to collect at floor level where it can be a fire or explosion hazard. In confined spaces, gases with vapour density less than 1 may replace air in the space, so that there is a danger of suffocation.
- Vapour pressure: Describes how fast a solid or liquid evaporates, and increases with increasing temperature. A liquid which evaporates easily into the air is more likely to be inhaled.
- Flash point: Describes the temperature at which a substance gives off enough vapour to form a mixture with air which can be ignited by a spark or flame.
- *Autoignition temperature*: The lowest temperature at which a substance burns without a spark or flame.
- *Explosive or flammability V limits*: An upper and lower concentration of a gas or vapour in air, between which it may explode if ignited by a spark or flame.

TOXICITY

How poisonous - or toxic - a chemical is depends on the nature of the chemical itself, including its structure and properties. But its potential to cause harm to workers also depends on the dose and duration of exposure, as well as other factors.

The word 'toxic' comes from the Greek toxicon or arrow-poison. The word toxic is the root of many terms covered in the following sections, such as toxicity (how toxic a substance is) and toxicology (the study of poisons).

"Toxicology is the study of the harmful effects of chemicals on biological systems.

It is a hybrid science built on advances in biochemistry, physiology, pathology, physical chemistry, pharmacology, and public health." "It is probably safe to say that the mechanism of action of no chemical is understood in every detail. Toxicologists know a great deal about a few chemicals, a little about many, and next to nothing about most".

The following sections describe how chemicals enter the body (routes of exposure), and the effects that they can have on various parts of the body. We will look at the time-scale over which these effects operate, so-called acute and chronic effects, and how certain combinations of chemicals are much more toxic together than would be predicted by adding up the damage they do on their own.

Finally, we will look at the many ways in which scientists study and measure toxicity, so that toxicity information in books, journals and safety data sheets can be interpreted more easily.

Routes of Exposure

Chemicals enter the body by three main routes: through the lungs (inhalation); through the skin (dermal absorption); and by being swallowed (ingestion). They can also enter through the eyes. Chemicals can have local effects, such as damaging the skin, but may also be absorbed into the body and affect other, distant parts of the body (systemic effects).

The Lungs

The job of the lungs is to exchange gases, allowing oxygen to be absorbed into the blood stream from the air we breathe, and get rid of waste carbon dioxide from the body. In the same way, the lungs will readily absorb other chemicals found in workplace air.

As well as gases, solids can reach the lungs when present as small particles suspended in air, such as dusts and fumes. One of our case studies involves fumes from solder flux, and dusts from asbestos, silica, wood, cotton and flour have killed thousands of workers. Some dusts may also cause a particular kind of explosion.

The Skin

Although the skin acts as a protective barrier against many micro-organisms and chemicals, some chemicals can penetrate the skin and enter the blood stream. Whether or not a chemical is absorbed through the skin depends on its structure: chemicals need to be able to dissolve in both water and fat (lipids) to get through the skin. Those that are insoluble, or dissolve only in fats or water, and chemicals made up of very large molecules, tend not to penetrate the skin. Chemicals are more easily absorbed where the skin is thin, such as on the forearms, than through the thick skin covering the palms of the hands and soles of the feet.

Chemicals are also more easily absorbed if skin is moist or damaged. Some chemicals, e.g. organic solvents, cause 'defatting' of the skin, making it a less effective barrier against further chemical exposure. Methyl ethyl ketone (MEK), the organic solvent used by Tony Bradshaw, is rapidly absorbed through the skin. In a study using human volunteers, MEK was applied to the skin of their forearms, and could be detected in the air they breathed out just three minutes later. During that time, the MEK had made its way through the skin, into the blood, and then to the lungs. Because MEK is soluble in water, it is absorbed through the skin even faster if the skin is sweaty. However, the amount of MEK in exhaled air accounted for only 10per cent of the amount applied to their skin. The other 90per cent was excreted in the urine, both as MEK and as its metabolites.

Ingestion

As well as the lungs and skin, chemicals can also enter the body if swallowed. In the workplace, this can occur if areas used to eat, drink or smoke are contaminated with chemicals, or if workers do not wash their hands or remove their gloves before eating or smoking.

Transportation

Once absorbed into the blood, the chemical is carried around the body in the blood stream, and where it ends up is influenced by its structure and properties. However, some barriers exist in the body which can keep out some (but not other) chemicals, such as the blood-brain barrier which helps protect the brain, and the placenta which helps protect a foetus.

Storage

Inside the body, some chemicals are stored in certain tissues, such as fat or bone, and while they remain bound up there, they may do little damage. However, under certain conditions such as rapid weight loss, large amounts of the chemical are released into the blood. How long such chemicals remain in the body varies, but some, like the pesticide DDT, remain for years. One of the reasons why DDT stopped being used in the developed world was because of this persistence in the environment, and even though it has not been used for years in the developed world, most of us have DDT in our bodies.

Metabolism

If chemicals are not stored, the body deals with them by metabolising (changing their structure) and excreting them. This occurs mainly in the liver, but also the skin, lungs, gut and kidneys, by similar processes used by our bodies to metabolise the chemicals which make up our food.

The products of metabolism are known as metabolites, and these can be more or less toxic than the original chemical. In fact, many of the adverse effects of chemical exposure are due to the effects of metabolites.

The pathways involved in metabolising chemicals vary greatly between species, and also between individuals, which explains why some people are harmed by very low levels of chemicals that others seem able to tolerate.

Excretion and Biological Monitoring

Chemicals and their metabolites are excreted from the body, mainly via urine produced by the kidneys. Small amounts are also excreted by the lungs, and in sweat, semen, milk, saliva and bile. The amount of a chemical a worker has been exposed to can sometimes be estimated by measuring how much of the chemical, or certain metabolites, is found in urine. This is known as biological monitoring.

Toxic Effects

We need to look more at what happens when chemicals enter the body and at the range of effects they might have. Toxicologists often quote a 16th Century Swiss doctor, Paracelsus, to illustrate that it is incorrect to divide chemicals into those which are toxic, and those which are not. Paracelsus said, "All substances are poisons; there is none which is not a poison. The right dose differentiates a poison and a remedy." Certain substances, including some naturally occurring plant and animal poisons, are lethal at tiny doses. Others, such as some of the things we eat, drink, or work with, can be lethal but only in massive quantities.

Acute and Chronic Exposure

As well as the dose of a chemical, its toxicity also depends on how long exposure lasts, the duration of exposure. This duration is one way that the toxicity of a chemical can be categorised. Single exposures are referred to as acute exposure, and repeated exposure over a longer time as chronic exposure. Duration of exposure should be reported with toxicity data in safety data sheets, etc. The other way toxicity is categorised is according to the chemical's target, the organ or body system which it damages.

The terms 'acute' and 'chronic' are also used to describe how long it takes for the effect of a chemical to occur, and it is important to be aware of these two uses. An acute effect is one which happens immediately on exposure, whereas a chronic effect does not. It may take years to appear.

The time between exposure and the onset of disease is referred to as the latency period. Because there is such a long latency between certain chemical exposures and diseases such as cancer, it is often difficult to link them to occupational exposures. This is one reason why workers should insist that the chemicals they work with are recorded on their medical records.

These two uses of the terms 'acute' and 'chronic' are not connected in the sense that an acute exposure only leads to an acute effect. In fact, an acute exposure can lead to either an acute or a chronic effect.

When talking about toxicity, the easiest way to divide chemicals is by the organ or system they damage. These target organs or systems are often referred to in safety data sheets and toxicology books. The commonest are: the lungs, the skin, the gut, the liver, the kidneys, the nervous system, the blood, the cardiovascular system, the immune system, and the reproductive system. There are even chemicals which can affect hearing.

Chemicals causing liver damage are sometimes called 'hepatotoxins', those which damage the kidneys 'renal toxins', and those harming the nervous system 'neurotoxins'. Chemicals that cause cancer, although they may affect either one or several organs, are lumped together and described as 'carcinogens'. Those that cause birth defects are called 'teratogens'.

"A toxin is generally understood to be a substance that is harmful to biological systems, but within this simple concept lies a great deal of variability. A substance that is harmful at a high dose may be innocuous or even essential at a lower dose.

A toxin may damage a specific body system, or it may exert a general effect on an organism. A substance that is toxic to one species may not be toxic to another because of different metabolic pathways or protective mechanisms. And the biologic damage may be temporary, permanent over the organism's lifetime, or expressed over subsequent generations"

Respiratory System

The respiratory system includes the nose and the tubes leading to the lungs (the trachea and bronchi), as well as the lungs themselves. Because their job is to exchange gases, the lungs are vulnerable to chemicals both in the form of small particles, and as gases, vapours and mists.

The lungs have a surface area of about 500 square feet (Rodricks) and come into contact with 20 kg of air each day (Stacey). Chemicals can cause a variety of damage to the lungs, including irritation, structural changes such as fibrosis, and cancer.

Irritation

Many chemicals irritate the lungs, such as ammonia, chlorine, hydrogen chloride and sulphur dioxide gases. The respiratory system responds to these irritants by tightening the bronchi, which results in a feeling of tight-chestedness.

This is sometimes referred to as 'dyspnoea' in safety data sheets. If the irritation is very severe, cells can be damaged and fluid released (oedema). Such damage can make the lungs more vulnerable to infection.

Emphysema and Fibrosis

Chronic exposure to some chemicals, such as certain forms of some metals, can cause structural changes in the lungs like emphysema. Dusts can also produce a particular type of lung damage called fibrosis. Some dusts, like crystalline silica, cause cells in the lungs to produce fibrous materials which can build up, making the lung rigid and unable to work properly.

Asbestos also causes a kind of fibrosis, known as asbestosis, as well as mesothelioma and lung cancer. As well as asbestos, other chemicals that cause lung cancer include: radon, arsenic, some forms of chromium, nickel, cadmium, bischloromethyl ether (BCME), beryllium, soot and environmental tobacco smoke.

Asphyxia

Some gases, like methane and nitrogen, although they have little effect on the body itself, can displace oxygen in air and cause suffocation or asphyxia. Other chemicals like carbon monoxide work like this but on a cellular level, where they stop cells taking up or using oxygen in the blood.

Asthma

Other chemicals, like colophony, cause allergic reactions in the respiratory system, including occupational asthma. Occupational asthma is the most commonly reported occupational disease according to SWORD, the UK respiratory disease monitoring scheme. In 1995, a National Asthma Campaign telephone survey of over 300 small and medium-sized businesses found 59per cent did not know what occupational asthma was, 62per cent of firms who understood the condition had done nothing to control the hazards, and 81per cent had taken no action to control exposure to occupational asthmagens.

In a study of 100 patients of Birmingham Heartlands Hospital's occupational lung disease unit, 91per cent said they had never been informed about the risks of getting asthma at work, and 73per cent had never seen a safety data sheet. Most worked in the car industry, hospitals, foundries or with wood.

Many worked with chemicals well known to cause asthma, such as isocyanates, colophony and wood dust. Although almost half worked for firms which had an occupational health service, less than one-third had pre-employment screening. The results of this study paint a very different picture of COSHH compliance compared with the HSE's 1991-92 evaluation survey which found 80per cent of employers provided workers with adequate information and training.

According to the Labour Force Survey, 70,000 people say that their work either caused their asthma, or made it worse. Over 1,000 new cases of occupational asthma are reported to the UK's Surveillance of Work-related and Occupational Respiratory Diseases (SWORD) scheme each year. SWORD was set up in 1989, funded by the HSE, and pools data from chest clinics and occupational physicians.

This is likely to be a significant underestimate and TUC figures suggest that up to 400,000 people suffer from asthma because of their work. Occupational asthma causes over a million days' sick leave each year, four times as many in 1994 as were lost in strikes.

Skin

The skin is the largest organ of the body, covering about 9 m^2, the size of a tennis court. Skin effects are much more obvious than damage to other organs, and in the USA, for example, skin diseases account for a third of all reported occupational diseases (Levy and Wegman). Occupational skin diseases can be classified into several types, or reaction patterns: contact dermatitis (irritant contact dermatitis and allergic contact dermatitis), urticaria, follicle abnormalities, infections, pigment disorders, and neo-plasms.

Dermatitis

Contact dermatitis is the most common occupational skin reaction. In the UK, it accounts for 80per cent of the cases reported to EPI-DERM, the skin disease monitoring scheme. Dermatitis is an inflammation reaction, where the area which comes into contact with the chemical becomes red, swollen or blistered, and feels itchy or burning.

It can be caused by irritants or result from an allergic reaction. The HSE says: "At worst, contact dermatitis can be as disabling at work as the loss of a limb." The HSE estimates 132,000 working days are lost every year because of occupational dermatitis, and many people are forced to change jobs as a result.

Irritant contact dermatitis can be caused by many substances. Some, like soaps, detergents and many solvents, are mild irritants to which the skin needs to be exposed in large amounts or for long periods to cause dermatitis. Others are very strong irritants, such as hydrogen fluoride and sulphuric acid (both strong acids), or sodium hydroxide (a strong alkali).

These chemicals cause what are sometimes called 'chemical burns', destroying the skin and causing ulcers and scarring. Irritant contact dermatitis is much more successfully treated if detected early.

Allergic contact dermatitis results from a specific allergy, where an individual is sensitised by an initial exposure, and then reacts when re-exposed to even small amounts. Chemicals which often cause allergic contact dermatitis are: latex (in surgical gloves), formaldehyde, rubber additives, nickel or chromium compounds, bactericides and fungicides, adhesives and sealants, plants and wood, and hairdressing chemicals.Because people vary in their susceptibility to substances that can cause allergic contact dermatitis, it may occur in only a few people in a workplace. Just because only one or two people are affected does not mean it is not work-related.

Chloracne

Some chemicals like creosote, oils and greases can make existing acne worse, especially affecting skin beneath clothing which is saturated with the oils. Other chemicals cause a specific type of acne called 'chloracne'. Chloracne is small straw-coloured cysts and inflamed follicles, often behind the ears and at the outer corner of the eyes. It is caused by chlorinated hydrocarbons, found in herbicide manufacturing and cable splicing, and by polychlorinated biphenyls. As well as its effect on the skin, chloracne should be treated as evidence that these chemicals have been absorbed into the body through the skin.

Pigment Changes

Although changes in skin pigmentation can be due to any skin injury, particular kinds of pigment loss can be caused by chemical exposure. One such is the monobenzyl ether of hydroquinone, an antioxidant used in rubber manufacture.

Skin Cancer

The final, and most potentially serious, kinds of skin disease are neo-plasms (growths). These may be benign or cancerous. Discovery of skin cancer of the scrotums of chimney sweeps in London in 1775 by Sir Percival Potts was the first recorded case of occupational cancer.

Nervous System

The nervous system is usually described in two parts, the central nervous system or CNS (brain and spinal cord), and the peripheral nervous system (the other nerves which control the muscles and senses). Chemicals may affect the CNS or the peripheral nervous system or both.

The Peripheral Nervous System

Chemicals that damage the peripheral nervous system do so in one or both of two ways. Chemicals can damage the outer covering of the nerve (the myelin sheath), which can re-grow quite quickly, or the nerve (axon) itself. Axons can regenerate, but only slowly.

The solvents n-hexane and methyl butyl ketone (or more precisely, their metabolite hexane-2,5-dione) are examples of chemicals which damages axons. n-Hexane has been used in paints, glues, varnishes, plastics and rubber and its effects on the peripheral nervous system have been reported in shoe makers and cabinet makers (Stacey).

Damage to the peripheral nervous system causes symptoms like pins and needles (parasthesia), numbness and weakness in the hands and feet. The onset of these symptoms can be delayed for several months. Although many chemicals which affect the peripheral nervous system cause very similar effects, some produce more specific symptoms.

Arsenic, for example, causes painful limbs and sensitive feet. Trichloroethylene, an organic solvent, affects nerves in the face, causing facial numbness and weakness, while the insecticide chlordecone (also known as Kepone) causes abnormal eye movements. Recovery from peripheral nervous system damage depends on the chemical involved, and the intensity of the exposure (Levy and Wegman).

The Liver

As well as storing vitamins and iron, regulating blood sugar levels and its role in digestion, the liver metabolises foreign chemicals. Metabolism sometimes results in a chemical being changed into a more toxic metabolite, and this is one way in which chemicals can damage the liver.

As with other organs, liver toxins can be grouped together according to the kind of liver disease they cause, including acute hepatitis (inflammation of the liver) or chronic diseases like cirrhosis and cancer. These diseases can also be caused by viruses such as hepatitis B, an important risk for health care workers, as well as non-occupational factors.

Chemicals that cause acute hepatitis include carbon tetrachloride, chloroform, dinitrophenol, dinitrobenzene, dioxin, polychlorobi-phenyls, the pesticide DDT, chlordecone, chlorobenzenes, the anaesthetic halothane, the dye feedstock methylenedianiline and the explosive TNT. Symptoms of acute hepatitis include headache, nausea, vomiting, dizziness and drowsiness.

As well as alcohol, cirrhosis is also caused by arsenic and vinyl chloride. Vinyl chloride is well known for causing liver angiosarcoma, a rare kind of cancer. After clusters of cases were reported in the 1970s, studies were done in vinyl chloride plants which found high levels of this liver cancer. Vinyl chloride is metabolised in the liver to an epoxide that causes the cancer. As well as a Maximum Exposure Limit, a yearly exposure limit of 3 ppm applies to exposed workers.

High levels of chronic liver disease have been reported in refrigeration engineers, chemists, dry cleaners, rubber manufacturers, and workers exposed to carbon tetrachloride and plutonium.

The Kidney and Urinary Tract

Proven or suspected kidney toxins (nephrotoxins) include arsenic, beryllium, lead, cadmium, mercury and uranium plus their compounds, solvents, and pesticides. Kidney failure from exposure to lead was common earlier this century, and kidney disease is the best known effect of chronic exposure to cadmium. The kidneys, via urine, are the major route by which toxic chemicals are excreted from the body. Because of this, and the way the kidneys do their job, they are vulnerable to the toxic effects of chemicals. The damage caused is complicated, and in many cases still not well understood, but can result from acute and chronic exposure.

According to the US National Bladder Cancer Study, up to a quarter of bladder cancers are caused by work. Aromatic amines are one group of chemicals known to cause bladder cancer. Occupationally-induced bladder cancer was first reported in 1895 in German dye-manufacturing workers.

Because they use dyes, jobs where workers are at risk of developing bladder cancer include textile, fur and leather dyeing as well as aromatic amine manufacturing. Aromatic amines are also used in rubber and plastics manufacture. Even though bladder cancer usually takes around 20 years to develop after exposure to aromatic amines, this can range from 4-40 years, and can result from exposures as short as 19 weeks (Levy and Wegman).

MbOCA is one aromatic amine used as a stabiliser in plastics manufacturing, often by small manufacturing plants. After an HSE national enforcement project in 1996-97, the HSE described conditions in many of these workplaces as "rough and ready." MbOCA is easily absorbed through the skin, and in many of these small factories even the eating areas and maintenance workers' tools were contaminated.

Blood

Chemicals which affect the blood can do so either by being directly toxic to blood cells, or by preventing it from delivering oxygen to the rest of the body. Haemoglobin is the molecule which carries oxygen to the body's tissues, and by combining with it to form carboxyhaemoglobin, carbon monoxide prevents oxygen reaching the tissues. Carbon monoxide is produced when organic substances such as gas and petrol are incompletely burnt, and as it is present in car exhausts, fire fighters, garage workers and traffic police can be exposed to high concentrations. Carbon monoxide is also produced by poorly maintained gas fires, and can also be produced in the body by metabolism of some chemicals like methylene chloride. Blood carboxyhaemoglobin levels are around 0.5per cent in non-smokers, and 5per cent in smokers. Headaches occur at levels of 10-20per cent, and a level of 60-70per cent is fatal within hours.

Other substances combine with haemoglobin to form methaemoglobin, which is also unable to deliver oxygen to the tissues. Workplace chemicals which have this effect include aniline dyes, nitrous gases, potassium chlorate, nitrobenzenes, phenylenediamine, and toluenediamine. Methaemoglobinaemia also results from environmental exposures to foods high in nitrates or nitrites, or well water contaminated with nitrates (Levy and Wegman). As well as binding with red blood cells to interfere with the carriage of oxygen, other chemicals can break up red blood cells. This is known as haemolysis, and is caused by chemicals like naphthalene, copper, organic compounds of metals such as tributyltin, and arsine. Arsine gas is an inorganic arsenic compound, and although it is used in the electronics industry, it is most dangerous because it can be formed accidentally when arsenic-contaminated metals or coal come into contact with acids. There have been cases of sewer workers being gassed by arsine after using acids to clear drains previously contaminated with substances containing arsenic. As well as its effects on red blood cells, arsine also causes kidney failure.

Lead also affects the blood by interfering with red blood cell manufacture. These cells are produced in lower quantities and do not live as long as normal, causing anaemia. Workers exposed to lead, such as welders, painters, jewellers, and in smelters, foundries and potteries, are covered by special legislation in the UK and should have biological monitoring to assess their exposure. Benzene is another well known example of a chemical which causes blood toxicity. It is metabolised in the body to a chemical which damages bone marrow, the site where blood cells are produced. Leukaemia, a form of cancer, was first linked with exposure to benzene in the 1920s. It also causes aplastic anaemia which, like leukaemia, is often fatal.

The Heart and Blood Vessels

Much less work has been done on occupational, as opposed to hereditary or life-style, factors associated with heart disease. However, because heart disease is the largest cause of death among both men and women in the UK, even a small reduction in risk due to occupational exposure could involve large numbers of people, and be an important public health measure. There is good evidence that occupational exposure to certain materials, such as the solvent carbon disulphide, is linked with heart disease.

Explosives manufacturing workers exposed to nitroglycerin and ethylene glycol dinitrate can suffer from angina when away from work, because these chemicals (like those used to treat angina) cause the heart's blood vessels to expand. Acute exposure to some solvents has also been associated with sudden death, probably due to changes in the heart's rhythm.

TOXICITY TESTING

Understanding how chemicals are tested for toxicity is very important for trade union safety reps and members as users. The results of these tests often appear in safety data sheets, they are used to decide how chemicals are used, controlled, labelled, and, most importantly, in setting occupational exposure limits. Newly introduced chemicals are now required to have a certain set of toxicity data, but these do not exist for many chemicals already in use before legislation was introduced. It is important to remember that absence of evidence of risk is not the same thing as evidence of absence of risk.

In 1992, the United Nations Conference on Environment and Development called on member countries and the Organisation for Economic Co-operation and Development to complete toxicity testing for 2,550 industrial chemicals produced in high volumes (at least 1,000 tonnes per year in any member country). In 1998 the OECD announced it had completed 109 tests.

How much (or how little) is known about the toxicity of even the most widely used chemicals sparked a major debate in the US press during 1997, between the Environmental Defence Fund (EDF) and the Chemical Manufacturers' Association (CMA). The EDF said that toxicity data was inadequate for 71per cent of the 3,000 chemicals made and used in the USA in the highest amounts.

The CMA disagreed, saying the figure was only 53per cent, but only because they included industrial data that had not been made publicly available. The US Environmental Protection Agency said, "There is a problem with public availability of basic screening information on chemicals." To try and speed up this voluntary approach to testing, the EDF mounted a 'naming and shaming' campaign. They asked the top 100 chemical companies if they would find and disclose basic toxicity data on the high-volume chemicals they produced by January 2000. The names of those which would not commit themselves were then listed in US newspaper advertisements.

According to UK academic Professor Andrew Watterson of the Centre for Occupational and Environmental Health, "The demands of carrying out complete health assessments on the tens of thousands of chemicals, metals and other substances used in the world today are beyond the resources and abilities of the global scientific community." Instead, he argues for a precautionary approach such as toxics use reduction.

Testing for the Effects of Acute Exposure

Toxicity studies can be divided into laboratory studies using live animals (*in vivo* studies) or groups of cells (*in vitro* studies), and studies of human populations (epidemiological studies). Animal studies are used to test chemicals for their acute and chronic toxicity by various routes of exposure.

The standard way of measuring a chemical's acute toxicity is to feed it at a range of single doses to groups of laboratory animals, such as rats. The dose that kills 50per cent of the group, the LD50 (lethal dose-50), is then recorded as well as the effects noticed in the animals. One use of LD50 values is in deciding how to label chemicals under the Chemicals (Hazard Information and Packaging for Supply) (CHIP) Regulations.

Table. Examples of Toxicity Measures

Category	LD50 oral rat mg/kg
very toxic	Less than 25
Toxic	From 25 to 200
Harmful	From 200 to 2000

Acute toxicity of methyl ethyl ketone (MEK)

LD50 oral rat	2,740 mg/kg
LD50 dermal (skin) rabbit	13 g/kg
LC50 (2 hour) inhalation mouse	40 g/m3
LDLO intraperitoneal (abdominal) guinea pig	2000 mg/kg

Table lists various measures of the acute toxicity of MEK, and illustrates some of the differences that are used in these tests. Table lists four routes of exposure, oral, skin, inhalation and intraperitoneal injection, and four of the most commonly used laboratory species.

Inhalation toxicity is described as a lethal concentration-50 (LC50) rather than LD50, and concentrations are expressed as milligrams or grams of substance per m^3 of air, or in parts per million (ppm). The LC50 is only meaningful if the duration of exposure is known. The other doses are expressed as milligrams (mg) or grams (g) of substance per kilogram (kg) of body weight.

A chemical's irritancy is tested by applying it to the eyes and skin of animals. One such test is known as the Draize test, which involves applying chemicals to rabbits' skin and eyes, rabbits being chosen because of their large and exposed eyeballs. The Draize test has frequently been criticised by animal rights campaigners, particularly when used to assess the irritancy of cosmetic products.

Testing for the Effects of Chronic Exposure

The effects of chronic exposure are also tested in animals. In these tests, the animals are fed, inhale, or have the chemical painted on their skins throughout their lives. The type and amount of disease they develop are then compared with the effects in a control group. The controls should be the same as the exposed group in every respect except the chemical exposure.

They should, for example, be the same strain of the same species, and be fed the same diet. The differences in rates of various diseases are then tested by various statistical means to see if they are significant. Having an adequate control group is just one element of good experimental design, as is the number of animals used. If a study is badly designed, its results will not be reliable. For example, when testing carcinogens on animals the US National Cancer Institute says that the chemical should be tested at two doses in both sexes of two species of rodent, and each group should contain at least 50 animals. One dose is often the highest dose that will not kill, or acutely poison, the animal.

Although this means that positive results will not be missed just because a high enough dose was not used, it means that predicting the effects of very low levels of the chemical is difficult. Trying to extrapolate (predict) what will happen at low doses, or what will happen in humans, are two of the major problems of animal testing.

Dose-response

When the results of these tests are plotted on a graph, with dose along the bottom (horizontal axis) and response up the side (vertical axis), a dose-response curve is obtained. (It is still called a curve even if it is a straight line!)

For example, the dose could be milligrams of chemical per kilogram of body weight per day (mg/kg/day) and the response could be the percentage of animals that got a certain type of cancer, or suffered a certain degree of liver damage. The question of whether there is a threshold dose, below which there is no toxic effect (no observed effect level, or NOEL) is controversial, especially for carcinogens. Because many carcinogens act by damaging DNA, and once this damage has been done it is permanent and by definition results in an increased risk of cancer, many argue that there is no such thing as a safe dose of these carcinogens.

Also, because cancer is such a serious disease, and so many gaps exist in our knowledge of how carcinogens act, it makes sense to take a precautionary approach. In the Carcinogens Approved Code of Practice the HSE says, "the risk of cancer from exposure to a substance cannot in most cases be presumed to be zero except by eliminating exposure."

NOELs are also used in setting occupational exposure limits. "We assume that every molecule of benzene to which an individual is exposed has some finite risk, albeit small, of producing a mutation that may result in acute myelogenous leukaemia".

Usefullness of Animal Studies

Some researchers in Canada and the USA have done interesting work comparing the attitudes of toxicologists and the public towards chemical risks. The public is more likely than toxicologists to think chemicals pose greater risks, and also finds it difficult to understand the concept of dose-response relationships. The public is much more likely than toxicologists to think the results of animal carcinogenicity studies can be applied to humans.

The study also found much disagreement between toxicologists about how to interpret various results. No wonder the public is confused, when the US study says, "Among the most important findings in this study was the high percentage of toxicologists who doubted the validity of the animal and bacterial studies that form the backbone of their science." Fewer toxicologists in industry than in university or government jobs agreed that animal carcinogens could reasonably be expected to cause cancer in humans.

"Good toxicology recognises its limits, its gaps in knowledge and the debates about the validity of its understanding of the mechanisms of toxicity in the context of moving towards practical public health precautionary policies on toxics use in our workplaces,".

Genetic Toxicology Tests

Tests for gene mutations in bacteria	*Salmonella*
	Escherichia coli
Tests for gene and chromosome mutations in mammal cells	Chinese hamster ovary (CHO) cell assay
	V79 Chinese hamster lung fibroblast assay
	L5178Y/tk$^{+/-}$ mouse lymphoma cell assay
Tests for chromosome mutations	Chromosome aberration in CHO cells *in vitro*
	Chromosome aberration in human lymphocytes *in vitro*
	Micronucleus test in erythrocytes from mouse bone marrow *in vivo*
	dominant lethal test in rodents
Tests to measure repair of DNA damage	sister chromatid exchange *in vitro* or *in vivo*
	Unscheduled DNA synthesis *in vitro*

Testing for Carcinogens

As well as long-term animal tests, a range of other tests are used to predict whether a chemical is likely to cause cancer. These are done with cell cultures, for example using mouse lymphoma cells, and micro-organisms such as the bacteria *Salmonella* (the Ames test) and *Escherichia coli*, and are known as *in vitro* (literally, in glass) tests. They are quicker, cheaper, and more ethically acceptable than animal tests.

They test a chemical's ability to damage genetic material (genotoxicity), although as mentioned before, most but not all genotoxins cause cancer, and (more importantly) not all carcinogens act by damaging genes.

These would not, therefore, be detected if only *in vitro* tests were used. While animal and *in vitro* testing might provide regulators with a lot of data, because the mechanisms leading from exposure to effect are often unknown, there is still much educated guesswork involved in risk assessment.

HAZARDOUS CHEMICALS

Chemicals, for which there is statistically evidence based on at least one study conducted in accordance with established scientific principles that acute or chronic health effects may occur in exposed employees, is classifies as a health hazard. Health hazards include chemicals that are carcinogens, toxin or highly toxic agents, reproductive toxins, irritants, corrosives, sensitizers, hepatotoxins, nephrotoxins, neurotoxins, agents which act on the hematopietic systems; and agents that damage the lungs, skin, eye or mucous membranes are hazardous chemicals.

Many chemicals can cause harm by virtue of their toxicity. In this case, the toxicity of a chemical refers to its ability to damage an organ system, such as the liver and the kidney or to disrupt a biochemical process such as the blood clotting mechanism or to disturb an enzyme system at some sites in the body removed from the site of contact. In actual practice, hazards do not group themselves in neat categories, but usually occur in combination and/or sequence. The following are some characteristics of flammability, a common chemical hazard.

MERCURY

Exposure to mercury occurs from breathing contaminated air, ingesting contaminated water and food, and having dental and medical treatments. Mercury, at high levels, may damage the brain, kidneys, and developing fetus.

when Mercury Enters the Environment

- Inorganic mercury (metallic mercury and inorganic mercury compounds) enters the air from mining ore deposits, burning coal and waste, and from manufacturing plants.
- It enters the water or soil from natural deposits, disposal of wastes, and volcanic activity.
- Methylmercury may be formed in water and soil by small organisms called bacteria.
- Methylmercury builds up in the tissues of fish. Larger and older fish tend to have the highest levels of mercury.

CHROMIUM

Exposure to chromium occurs from ingesting contaminated food or drinking water or breathing contaminated workplace air. Chromium (VI) at high levels can damage the nose and can cause cancer

When Chromium Enters the Environment

- Chromium enters the air, water, and soil mostly in the chromium (III) and chromium(VI) forms.
- In air, chromium compounds are present mostly as fine dust particles which eventually settle over land and water.
- Chromium can strongly attach to soil and only a small amount can dissolve in water and move deeper in the soil to underground water.
- Fish do not accumulate much chromium in their bodies from water.

CADMIUM

Exposure to cadmium happens mostly in the workplace where cadmium products are made. The general population is exposed from breathing cigarette smoke or eating cadmium contaminated foods. Cadmium damages the lungs, can cause kidney disease, and may irritate the digestive tract.

When Cadmium Enters the Environment

- Cadmium enters air from mining, industry, and burning coal and household wastes.
- Cadmium particles in air can travel long distances before falling to the ground or water.
- It enters water and soil from waste disposal and spills or leaks at hazardous waste sites.
- it binds strongly to soil particles.
- Some cadmium dissolves in water.
- It doesn't break down in the environment, but can change forms.
- Fish, plants, and animals take up cadmium from the environment.
- Cadmium stays in the body a very long time and can build up from many years of exposure to low levels.

ARSENIC

Exposure to higher than average levels of arsenic occur mostly in the workplace, near hazardous waste sites, or in areas with high natural levels. At high levels, inorganic arsenic can cause death. Exposure to lower levels for a long time can cause a discoloration of the skin and the appearance of small corns or warts.

When Arsenic Enters the Environment

- Arsenic occurs naturally in soil and minerals and it therefore may enter the air, water, and land from wind-blown dust and may get into water from run-off and leaching.
- Arsenic cannot be destroyed in the environment. It can only change its form.
- Rain and snow remove arsenic dust particles from the air.
- Many common arsenic compounds can dissolve in water. Most of the arsenic in water will ultimately end up in soil or sediment.
- Fish and shellfish can accumulate arsenic; most of this arsenic is in an organic form called arsenobetaine that is much less harmful.

LEAD

Lead is a heavy, low melting, bluish-gray metal that occurs naturally in the Earth's crust. However, it is rarely found naturally as a metal. It is usually found combined with two or more other elements to form lead compounds

When lead enters the environment?

Lead is a naturally occurring element that people have used almost since the beginning of civilization. Human activities have spread lead widely throughout the environment-the air, water, soil, plants, animals, and man-made constructions.

Because lead is spread so widely throughout the environment, it can now be found in everyone's bodies; most people have lead levels that are orders of magnitude greater than that of ancient times and within an order of magnitude of levels that have resulted in adverse health effects.

CARCINOGENIC CHEMICALS

According to a recent study, published in a scientific journal, burning of incense in temples and homes releases Polycyclic Aromatic Hydrocarbons (PAHs), which are carcinogenic chemicals. If homes and temples are poorly ventilated, as generally is the case, the level of these toxic chemicals at times can be even more than that at a busy traffic intersection.

The study conducted by Ta Chang Ling of National Cheng Kung University of Taiwan points out that incense is often mixed with carcinogenic chemicals that cause cellular changes and which in turn can lead to cancer.

Ling says that the risk of cancer in poorly ventilated houses and temples is 40 times higher. Besides even very low levels of certain other chemicals in incense, fragrance and scented products can lead to a wide range of deleterious effects on the human health. We use a variety of scented products like soap, deodorants, hair sprays, lotions, cosmetics, after-shave lotions and fragrances everyday. The effect of chemicals in these and other scented products becomes more pronounced when we spend much of our time indoors.

Perfumes and scented products are composed of toxic chemicals like acetone, toluene, ethyl acetate, formaldehyde and ethanol that are found on any list of toxic and hazardous chemicals. All these and other chemicals used in the manufacture of personal care products have been shown in animal studies to cause serious health problems and even death.

Many of these chemicals are also respiratory irritants and can trigger asthma, allergies, migraines, eye problems, nausea and fatigue. These chemicals can also affect people who do not have any pre-existing health problems. On many occasions it so happens that we may not have used any of these products personally but we may suffer discomfort due to our co-workers, colleagues or family members having used such a product.

Much work has been done on passive smoking but no attention is being paid to the fact that what is fragrance and perfume to one man could be injurious to the other person. All fragrance chemicals are volatile and quickly become airborne once released from their receptacles. Besides being volatile these chemicals are also affected by heat, light and moisture. Once in the air these chemicals, due to the presence of external agents, break down into several other compounds that could be more irritating than the original compounds. At one time it was thought that the chemicals used in cosmetics and perfumes affect only the skin. However, now it is known that the effects are not limited to skin alone. From the skin these chemicals are absorbed into the blood stream and from there they move on to other organs. Several of these chemicals are known to have got into the human food chains.

Man has used perfumes and fragrances since long, but such problems never cropped up as these products were made from natural ingredients like flowers and other plant parts. But today the use of these vanity products has increased tremendously and most of these products are manufactured using chemicals that are harmful to human health and environment.

Around 4000 chemicals are used in our day-today cosmetics and other personal products and 95 percent of these trace their origin to petroleum. More often than not manufacturers of these products also do not reveal the actual components of their products under the guise of 'trade secret'. This makes it difficult to study the adverse impacts of these chemicals on man.

All of us are exposed to a variety of deleterious and harmful chemicals in our environment, so why increase the chances of falling prey to them by using these perfumes, fragrances and so called "personal products".

Most of these products do not in any way either improve the quality of air or contribute to our hygiene. They rather add more pollutants to the air. And mind you that higher price of a particular product is no guarantee of its being user friendly.

We can protect ourselves, others in our vicinity and above all our very fragile environment, which is already under stress, by avoiding these products. We can make a beginning by observing these simple and easy to follow rules.

- Stop using all scented products unless you are sure that they contain only biological products.
- Do not have your clothes dry-cleaned.
- Do not use air-fresheners at home or in your car. Try ventilating them in natural air.
- Do not burn plastics, cardboard, magazines and newspapers.
- Keep your homes, offices clean and well ventilated; it will solve most of the problems.
- Grow more plants in and around your homes.

COMMON HAZARDOUS AND TOXIC AGENTS

Many people don't realise it but there are a lot of common household items that are considered to be hazardous materials. These include medications, paint, motor oil, antifreeze, auto batteries, lawn care products, pest control products, drain cleaners, pool care products such as chlorine and acids, and household cleaners. Some household cleaners may be harmful separately or when combined such as ammonia and bleach.

The vast majority of chemical cleaners pose environmental, health and safety concerns to the air, water, animals, plants and humans. Pollution caused by chemical run-off has contaminated 40per cent of the nation's waterways remain too polluted for fishing and swimming.

HOUSEHOLD HAZARDOUS MATERIALS

A wide range of health effects are linked to materials used, depending on the substance, the dose, the duration of exposure, and the susceptibility of the person exposed. Many solvents affect the central nervous system and are skin and eye irritants. Most are flammable; many are linked to long-term adverse health effects such as liver damage. Several are known or suspected carcinogens such as benzene and toluene.

Dusts/fibres are eye and respiratory irritants, and may aggravate asthma and provoke allergies. Specific hazards: silica in clay dust causes lung disease over years of exposure; talc (white clays) may be contaminated with asbestos, a known carcinogen; some hardwood dusts lead to nasal and sinus cancers in woodworkers. Heavy metals are hazardous both as dusts and as fumes. Lead affects the nerves, digestive system, muscles and joints. Arsenic, cadmium and chromium are known carcinogens. Mercury, copper, cobalt, silver, manganese, selenium and zinc are all acutely toxic.

Acids are corrosive to skin and eyes. Acid vapors are irritating to the lungs and inhalation of small amounts may damage lung tissue. Concentrated acids can react with many other materials. Gases generated from kilns, welding or sculpting with plastics are acutely toxic; some may lead to long-term lung damage with repeated exposure.

BASIC SAFETY RULES

Know the hazards of the materials you're working with. Read the labels, request material safety data sheets (MSDS) on new products, know what precautions, safety gear and clean up procedures are advised.

Use the safest materials and procedures possible. Stay current on the new developments in your art or craft; safer, less-toxic alternatives are being devised for many activities. Use good ventilation at all times. Local exhaust is the best, such as a hood or spray booth that vents to the outside. Next best is to use exhaust fans that pull the contaminated air away from you and exhaust it outside (an air-conditioning system is not adequate, since it recirculates most of the air). An open window usually does not provide adequate ventilation; toxins may be blown back into your face. Use good hygiene and housekeeping. Separate work and living areas; avoid eating, drinking or smoking in the work area; don't store materials in food containers; and wash and change clothes after working. Wet mop or vacuum for cleanup of dusts.

Special precautions are needed for children's art. In general, children over the age of 12 can understand and consistently follow safety instructions for the more toxic materials; younger children cannot and should use only the safest materials.

Products Hazardous Ingredients Dangers Alternatives

- Aerosol sprays butanol, butane, propanol flammable, irritant, explosive pump-type sprays, potpourri.
- Ammonia-based cleaners ammonia, ethanol irritant, toxic, corrosive (Forms poison gas when mixed with bleach.) vinegar, salt and water for surfaces and baking soda and water for the bathroom.
- Antifreeze ethylene glycol toxic (especially to pets) unknown (use caution, take to a collection centre).
- Batteries sulfuric acid, lead corrosive, toxic unknown (use caution, recycle).
- Brake fluid glycol ethers, heavy metals flammable, toxic unknown (take to a collection centre).
- Disinfectants diethylene glycol, sodium, hypochlorite, phenols corrosive, toxic 1/2 cup borax in 1 gallon of water.
- Drain opener sodium hypochlorite, sodium/potassium hydroxide corrosive, toxic plunger, flush w/boiling water, 1/4 cup baking soda.
- Flea repellant carbamates, organophosphate, pyrethrins toxic eucalyptus leaves where pet sleeps, brewer's yeast in diet.
- Floor/furniture polishes diethylene glycol, petroleum distillates, nitrobenzene flammable, toxic 1 part lemon juice w/2 parts olive or vegetable oil.
- Furniture stripper acetone, methyl ethyl ketone, toluene xylenes flammable, toxic sandpaper.
- Latex paint resins, glycol, ethers, esters flammable lime stone based white wash or casein based paint.
- Oil based paints ethylene, aliphathydro-carbons, petroleum distillates flammable, toxic latex or water based paints.
- Oven cleaner potassium, sodium hydroxide, ammonia, lye corrosive, toxic baking soda and water, salt on spills that are still warm.
- Photographic chemicals silver, acetic acid, ferrocyanide, hydro quinone corrosive, toxic, irritant unknown (use caution, take to collection centre).
- Pool chemicals muriatic acid, sodium hypochlorite algicide corrosive, toxic unknown (use caution, use until gone take to a collection centre).
- Rat and mouse killer lead arsenate, coumarins (warfarin) strychnine toxic remove food and water sources, clear harborage, cover holes and drains rats may enter, use mechanical traps. Get a cat.

- Roach and ant killer organo-phosphates, carbamates toxic roaches: traps, boric acid. Ants: chili pepper/cream of tartar in ants' path.
- Rug and upholstery cleaners naphthalene, oxalic acid, diethylene glycol irritant, toxic, corrosive dry corn starch sprinkled on rug then vacumed up.
- Toilet bowl cleaner muriatic or oxalic paradi-chlorobenzene calcium hypochlorite irritant, toxic, corrosive toilet brush and baking soda; mild detergent.
- Thinners and turpentine n-butyl alcohol, isobutyl keytone, petroleum distillates flammable use water with water based paints.
- Transmission fluid hydrocarbons mineral spirits flammable, toxic unknown (take to a collection centre).
- Used oil hydrocarbons, (e.g. benzene) heavy metals flammable, toxic unknown (recycle).

Note: It is illegal to dispose of oil on/in the ground.

CHLORINE (SODIUM HYPOCHLORITE)

In paper products, such as toilet paper and paper towels:

- Bleaching paper products with chlorine bleach causes the formation of dioxin, an extremely toxic and persistent chemical known to cause cancer and disrupt the endocrine system.
- Chlorine-free toilet paper and paper towels are available at many natural food stores. Additionally, newspaper can be used in place of paper towels for cleaning windows, and rags can be used for other surfaces.

In Cleaning Products

Many household cleaners contain chlorine bleach. Chlorine bleach, or sodium hypochlorite, is a lung and eye irritant. If mixed with ammonia or acid-based cleaners (including vinegar), chlorine bleach releases toxic chloramine gas. Short-term exposure to this gas may cause mild asthmatic symptoms or more serious respiratory problems.

- To be on the safe side, don't mix chlorine bleach with anything — or just avoid chlorine bleach altogether. The EPA recommends using a non-chlorine bleach such as hydrogen peroxide to bleach clothes.

PHOSPHATES

Phosphates are minerals that act as water softeners. Although they are very effective cleaners, phosphates also act as fertilizers. When cleaning products go down the drain, phosphates are discharged into rivers, lakes, estuaries, and oceans. In lakes and rivers especially, phosphates cause a rapid growth of algae, resulting in pollution of the water.

Many states have banned phosphates from household laundry detergents and some other cleaning products. Automatic dishwasher detergents are usually exempt from phosphate restrictions, and most major brands contain phosphates.

Some phosphate-free alternatives are available and hand dishwashing liquids do not contain phosphates.

ALKYLPHENOLS AND THEIR DERIVATIVES

Alkylphenol Ethoxylates are found in some laundry detergents, disinfecting cleaners, all-purpose cleaners, spot removers, hair colours and other hair-care products, and spermicides.

- Alkylphenol Ethoxylates are endocrine disruptors.
- Alkylphenols are produced in the environmental breakdown of alkylphenol ethoxylate surfactants, are slow to bio-degrade and have been shown to disrupt the endocrine systems of fish, birds, and mammals.

VOLATILE ORGANIC COMPOUNDS

Organic chemicals are widely used as ingredients in household products. All of these products can release pollutants while you are using them, and, to some degree, when they are stored.

- EPA's Total Exposure Assessment Methodology (TEAM) studies found levels of about a dozen common organic pollutants to be 2 to 5 times higher inside homes than outside, regardless of whether the homes were located in rural or highly industrial areas.
- Additional TEAM studies indicate that while people are using products containing organic chemicals, they can expose themselves and others to very high pollutant levels, and elevated concentrations can persist in the air long after the activity is completed.
- Many organic compounds are known to cause cancer in animals; some are suspected of causing, or are known to cause, cancer in humans.
- Some of the hazardous volatile organic compounds (VOCs) that frequently pollute indoor air — such as toluene, styrene, xylenes, and trichloroethylene — may be emitted from aerosol products, dry-cleaned clothing, paints, varnishes, glues, art supplies, cleaners, spot removers, floor waxes and polishes and air fresheners.

High levels of toluene can put pregnant woman at risk of having babies with neurological problems, retarded growth, and developmental problems. Xylenes may also cause birth defects. Trichloroethylene is one of the chemicals suspected of causing a cluster of childhood leukemia cases due to drinking water contamination in the town of Woburn, Massachusetts, in the early 1980s. The subsequent lawsuit against the polluting company was the subject of the 1995 book and 1998 film, A Civil Action.

Styrene is a suspected endocrine disruptor, a chemical that can interfere, block or mimic hormones in humans or animals.

- VOCs such as xylene, ketones, and aldehydes are found in many aerosol products and air fresheners. Researchers found that babies less than six months old in homes where air fresheners are used on most days had 30 percent more ear infections than those exposed less than once a week.
- Levels of formaldehyde in air as low as 0.1 ppm (0.1 part formaldehyde per million parts of air) can cause watery eyes, burning sensations in the eyes, nose and throat, stuffy nose, nausea, coughing, chest tightness, wheezing, skin rashes and allergic reactions.
- Babies frequently exposed to aerosols had a 22 percent increase in diarrhea, and pregnant women frequently exposed to these products had 25 percent more headaches and a 19 percent increase in postnatal depression compared to those less frequently exposed.

Definitions:

Corrosive: A chemical, (solid, liquid or gas), that can cause destructive damage to body tissues at the site of contact. It can cause severe burns to the skin and can "eat through" clothing, metal and other materials.

Flammable: Can be ignited at almost any temperature. Spontaneously react with oxides.

Irritant: Causes soreness or inflammation of the skin, eyes, mucous membranes or respiratory system.

Oxidizer: An unstable chemical that can spontaneously react with flammables and releases oxygen.

Toxic: May cause injury or death upon ingestion (eating/drinking), absorption (touching) or inhalation (breathing into lungs).

IMPACT OF PESTICIDE ON HEALTH

A pesticide is any chemical which is used by man to control pests. The pests may be insects, plant diseases, fungi, weeds, nematodes, snails, slugs, etc. Therefore, insecticides, fungicides, herbicides, etc., are all types of pesticides. Some pesticides must only contact (touch) the pest to be deadly. Others must be swallowed to be effective.

The way that each pesticide attacks a pest suggests the best way to apply it; to reach and expose all the pests. For example, a pesticide may be more effective and less costly as a bait, rather than as a surface spray.

BACTERICIDES

A bactericide or bacteriocide is a substance that kills bacteria and, preferably, nothing else. Bactericides are either disinfectants, antiseptics or antibiotics.

FUNGICIDES

Fungicides are chemical compounds used to prevent the spread of fungi or plants in gardens and crops, which can cause serious damage resulting in loss of yield and thus profit. Though oomycetes are not fungi, they use the same mechanisms to infect plants and therefore in phytopathology chemicals used to control oomycetes are also referred to as fungicides.

Fungicides are also used to fight fungal infections.Fungicides can either be contact or systemic. A contact fungicide kills fungi when sprayed on its surface; a systemic fungicide has to be absorbed by the plant.Fungicide residues have been found on food for human consumption, mostly from post-harvest treatments. Some fungicides are dangerous to human health, such as Vinclozolin, which has now been removed from use.Like other pesticides, fungicides can induce pesticide resistance. Equivalently, antifungal drugs can induce drug resistance.

HERBICIDES

A herbicide is used to kill unwanted plants. Selective herbicides kill specific targets while leaving the desired crop relatively unharmed. Some of these act by interfering with the growth of the weed and are often based on plant hormones. Herbicides used to clear waste ground are non-selective and kill all plant material with which they come into contact.

Some plants produce natural herbicides, such as the genus Juglans (walnuts). Herbicides are widely used in agriculture and in landscape turf management. They are applied in total vegetation control (TVC) programmes for maintenance of highways and railroads. Smaller quantities are used in forestry, pasture systems, and management of areas set aside as wildlife habitat.

INSECTICIDES

An insecticide is a pesticide used against insects in all developmental forms. They include ovicides and larvicides used against the eggs and larvae of insects. Insecticides are used in agriculture, medicine, industry and the household. The use of insecticides is believed to be one of the major factors behind the increase in agricultural productivity in the 20th century.

Nearly all insecticides have the potential to significantly alter ecosystems; many are toxic to humans; and others are concentrated in the food chain. It is necessary to balance agricultural needs with environmental and health issues when using insecticides.

MITICIDES

Miticides or acaricides are pesticides that kill mites. Antibiotic miticides, carbamate miticides, formamidine miticides, mite growth regulators, organochlorine, permethrin and organophosphate miticides are all in this category. Diatomaceous earth will also kill mites by cutting through the skin which drys out the mite. Ivermectin can be prescribed by a medical doctors to rid humans of mite and lice infestations and there are agricultural formulations for birds and rodents that are infested.

COMMON MITICIDES

Methoprene is virtually harmless to non-insects, and the US EPA has exempted it from tolerance. It is widely available in supermarkets, ctc. Hydroprene is toxic to fish and perhaps birds. Both are for indoor use only, as they break down in sunlight. Methoprene is applied as a wetting spray, hydroprene as an aerosol space spray. Neither will affect adult insects; they work on future generations by preventing growth or maturation. Permethrin can be applied as a spray or in more targeted forms (e.g. Damminix TickTubes) that attack the ticks and mites on mammalian hosts. Their effects are not limited to mites: lice, cockroaches, fleas, mosquitos, and other insects will be affected. Permethrin, however, is not known to harm mammals or birds, as it has a low mammalian toxicity and is poorly absorbed by skin.

MOLLUSCICIDES

Molluscicides are pesticides used to control molluscs, such as motts, slugs and snails. These substances include metaldehyde, methiocarb and aluminium sulfate. They should be used with caution, as they can be harmful to non-target animals. Most molluscicides are not used in organic gardening, though there are exceptions, such as iron phosphate.

NEMATICIDES

A nematicide is a type of chemical pesticide used to kill parasitic nematodes (roundworms). One common nematicide is obtained from neem cake, the residue obtained after cold-pressing the fruit and kernels of the neem tree. Known by several names in the world, the tree was first cultivated in India since ancient times and is now widely distributed throughout the world.

RODENTICIDES

Rodenticides are a category of pest control chemicals intended to kill rodents.Single feed baits are chemicals sufficiently dangerous that the first dose is sufficient to kill.Rodents are difficult to kill with poisons because their feeding habits reflect their place as scavengers. They will eat a small bit of something and wait, and if they don't get sick, they continue. An effective rodenticide must be tasteless and odorless in lethal concentrations, and have a delayed effect.

VIRUCIDES

It is for the control of viruses. Pesticides can also be classed as synthetic pesticides or biological pesticides, although the distinction can sometimes blur.

Broad-Spectrum Pesticides are those that kill an array of species, while narrow-spectrum, or selective pesticides only kill a small group of species

A Systemic Pesticide moves inside a plant following absorption by the plant. This movement is usually upward (through the xylem) and outward. Increased efficiency may be a result. Systemic insecticides which poison pollen and nectar in the flowers may kill needed pollinators such as bees.

USES AND BENEFITS

Pesticides are used to control organisms which can otherwise result in harm.For example, they are used to kill mosquitoes that can transmit potentially deadly diseases like west nile virus and malaria and bees, wasps or ants that can cause allergic reactions. Insecticides can protect animals, because infestations by parasites such as fleas may cause them illness.Pesticides can prevent sickness in humans that could be caused by moldy food or diseased produce. Herbicides can prevent accidents by clearing roadside trees and brush, which may block visibility. They can also kill invasive weeds in parks and wilderness areas which may cause environmental damage. Uncontrolled pests such as termites and mold can damage structures such as houses

.Pesticides are often very cost-effective for farmers. Pesticides are used in grocery stores and food storage facilities to manage rodents and insects that infest food such as grain. Each use of a pesticide carries some associated risk. Proper pesticide use decreases these associated risks to a level deemed acceptable and increases quality of life and protects property and the environment.

In 2006, the World Health Organization suggested the resumption of the limited use of DDT to fight malaria. They called for the use of DDT to coat the inside walls of houses in areas where mosquitoes are prevalent. Dr. Arata Kochi, WHO's malaria chief, said, "One of the best tools we have against malaria is indoor residual house spraying. Of the dozen insecticides WHO has approved as safe for house spraying, the most effective is DDT."

Scientists estimate that DDT and other chemicals in the organophosphate class of pesticides have saved 7 million human lives since 1945 by preventing the transmission of diseases such as malaria, bubonic plague, sleeping sickness, and typhus.

BANNED PESTICIDES

Pesticides Banned for manufacture, import and use (**25** Nos.)

- Aldrin
- Benzene Hexachloride
- Calcium Cyanide
- Chlordane
- Copper Acetoarsenite
- CIbromochloropropane

- Endrin
- Ethyl Mercury Chloride
- Ethyl Parathion
- Heptachlor
- Menazone
- Nitrofen
- Paraquat Dimethyl Sulphate
- Pentachloro Nitrobenzene
- Pentachlorophenol
- Phenyl Mercury Acetate
- Sodium Methane Arsonate
- Tetradifon
- Toxafen
- Aldicarb
- Chlorobenzilate
- Dieldrine
- Maleic Hydrazide
- Ethylene Dibromide
- TCA (Trichloro acetic acid)

Pesticide / Pesticide formulations banned for use but their manufacture is allowed for export (2 Nos.)

- Nicotin Sulfate
- Captafol 80per cent Powder

Pesticide formulations banned for import, manufacture and use (4 Nos)

- Methomyl 24per cent L
- Methomyl 12.5% L
- Phosphamidon 85% SL
- Carbofuron 50% SP

Pesticide Withdrawn(7 Nos)

- Dalapon
- Ferbam
- Formothion
- Nickel Chloride
- Paradichlorobenzene (PDCB)
- Simazine
- Warfarin

Pesticides Restricted for use in India

- Aluminium Phosphide
- DDT
- Lindane
- Methyl Bromide
- Methyl Parathion
- Sodium Cyanide

- Methoxy Ethyl Merciru Chloride (MEMC)
- Monocrotophos(ban for use on vegetables)

Effect of Pesticides on Human Health

The effect of pesticides on human health is worst.Due to pesticides there can be number of diseases which are lungs cancer Chronic liver damage cirrhosis and chronic hepatitis, endocrine and reproductive disorders, immuno suppression, cytogenic effects, breast cancer, Non hodkins lymphoma, polyneuritis, etc, etc.

CHEMICALS IN EVERYDAY PRODUCTS

HARMFUL CHEMICALS FOUND IN SOAPS AND SHAMPOOS

Sodium Lauryl Sulfate (SLS) and Sodium Laureth Sulfate (SLES)

Both Sodium Lauryl Sulfate (SLS) and its close relative Sodium Laureth Sulfate (SLES) are commonly used in many soaps, shampoos, detergents, toothpastes and other products that we expect to "foam up". Both chemicals are very effective foaming agents, chemically known as surfactants.

SLS and SLES are esters of Sulphuric acid - SLS is also known as "Sulfuric acid monododecyl ester sodium salt", however there are over 150 different names by which it is. In fact, SLES is commonly contaminated with dioxane, a known carcinogen. Although SLES is somewhat less irritating than Sodium Lauryl Sulfate, it cannot be metabolised by the liver and its effects are therefore much longer-lasting.

A report published in the Journal of The American College of Toxicology in 1983 showed that concentrations as low as 0.5per cent could cause irritation and concentrations of 10-30per cent caused skin corrosion and severe irritation. National Institutes of Health "Household Products Directory" of chemical ingredients lists over 80 products that contain sodium lauryl sulfate. Some soaps have concentrations of up to 30per cent, which the ACT report called "highly irritating and dangerous".

Shampoos are among the most frequently reported products to the FDA. Reports include eye irritation, scalp irritation, tangled hair, swelling of the hands, face and arms and split and fuzzy hair. The main cause of these problems is sodium lauryl sulfate. Both SLS and SLES are known to have many effects that can potentially be detrimental to health. Among the possible dangers are the following

- Skin irritation / skin corrosion.
- Hormone Imbalance.
- Eye irritation / eye deformities in children.
- Protein Denaturing.
- Carcenogenicity (potential to cause cancer).

The AJT report staes that "Other studies have indicated that Sodium Lauryl Sulfate enters and maintains residual levels in the heart, the liver, the lungs and the brain from skin contact. This poses question of it being a serious potential health threat to its use in shampoos, cleansers, and tooth pastes."

Skin Irritation

SLS is used routinely in clinical studies. This may suggest a level of comfort, however, the way in which it is used is disturbing. Despite being the number one active ingredient in virtually all soaps, shampoos and cleansers, the sole purpose of using SLS in clinical studies is to cause skin irritation that can then be used to identify the properties of other chemicals!

Amazing isn't it? For years, we have been applying known irritants to our skin on a daily basis. To quote the ACT report "The abbreviated symbol for Sodium Lauryl Sulfate is used around the world in clinical studies as a skin irritant. SLS is the universal standard, by which a measured percentage is evaluated to promote a given level of irritation and reaction. By this SLS standard level of irritation, it is then possible to evaluate the healing or modifying characteristics of any ingredient or formula used on the SLS irritated skin."

Most worryingly, irritation has been shown to occur at concentrations of 0.5per cent, which is $1/60^{th}$ the concentration found in some hand soaps. Caveat emptor!

Hormone Imbalance

In the last 100 years or so, many new health problems have come to light. These include PMS / PMT, the so-called "menopausal symptoms" which never used to exist, and more recently a massive drop in male fertility which threatens our continued existence in many western countries. SLS is most likely a major contributor to all of these problems due to its oestrogen mimicking activity.

Oestrogen is a hormone found quite normally in both men and women. Like all other hormones, it's circulating levels are rigidly controlled by the glands of the body due to the potent effect of its presence on virtually all cells. Not only does SLS irritate the skin, it is also absorbed through the skin (high levels of skin penetration may occur at even low concentration). Once in the body, the SLS molecule attaches to oestrogen receptors, mimicking the effects of the hormone in various body systems.

The result is hormonal chaos. The body can no longer control it's own oestrogen levels (or at least, what it sees as it's own oestrogen levels - it can't tell the difference between endogenous oestrogen and SLS) and therefore loses control of many normal endocrine (hormonal) functions.

In men, whose oestrogen levels are normally extremely low, this massive increase causes breast enlargement, reduction of male hormone levels and a massive drop in both sperm count and sperm motility (ability of the sperm to fertilise an ovum). Gender confusion may also be related to SLS levels, either in the male himself or in his mother during pregancy. In women, the reproductive system, which is totally controlled by oestrogen and progesterone, goes haywire. Rapidly shifting oestrogen levels and their effect on progesterone levels mean that the body is totally confused, leading to menstrual problems, menopausal symptoms and potentially infertility. Because this subject is so important, we have devoted a whole section of this site to womens health.

Eye Irritation / eye Deformities in Children

Have you ever got shampoo in your eyes? Yes, so have I - not pleasant is it? However, the potential effects of SLS on the eye are much more worrying. In animal studies, 10per cent SLS caused acute corneal damage. However, it is not just direct eye contact that is the problem. According to the American College of Toxicology, "tests show permanent eye damage in young animals from skin contact in non-eye areas".

In other words, because SLS is absorbed through the skin, it can cause permanent eye damage without ever directly coming into contact with your eyes. As a result, you would expect that childrens products would be SLS-free. Unfortunately not, most childrens shampoos contain just as much SLS as those for adults. Thankfully, alternatives DO exist, though you would be hard-pressed to find them in your local chemist or supermarket.

Protein Denaturing

Our cells are made from protein: The development of those cells is strictly regulated by the reproductive processes that are continually at work removing damaged and old cells and replacing them with healthy new ones. Virtually every cell in the body is replaced at least every **7** years.

SLS exerts its effects on proteins by forming a chemical bridge between the fat-soluble and water-soluble parts of the protein moecule. This disrupts the hydrophobic forces needed to maintain the protein structure and the molecule collapses, rendering it useless. This effect is usually irriversible.

The result of this is two-fold. Firstly, existing proteins are damaged, leading to an increase in the amount of healing required by the body. Secondly, new proteins can be damaged and cells disrupted while they are under construction. It is exactly this type of activity that can lead to the early stages of skin cancer.

In the skin, this process can be so severe, that skin layers may separate and inflame due to its (SLS's) protein denaturing properties.

Carcenogenicity

Quite apart from it's potential to cause pre-cancerous conditions by denaturing proteins, the oestrogen mimicking effects of SLS also offers massive potential to cause cancer. It is known that many cancers, not least breast and ovarian cancer are directly related to oestrogen levels, in fact some cancer cells actually secrete their own oestrogen, which contributes to the growth of the tumour. Clearly, by disrupting normal oestrogen levels AND by causing similar effects at a cellular level as endogenous oestrogen, SLS exhibits MASSIVE potential to both cause and worsen cancerous states. The incidence of breast cancer has increased several-fold in the last 50 years, both in women and in men. Currently, according to the American Cancer Society, men account for approximately 1per cent of all breast cancer cases. This subject is discussed in more detail in our womens health section.

There is also a third way by which SLS can potentially cause cancer. Carcinogenic nitrates can form in the manufacturing of Sodium Lauryl Sulfate or by its inter-reaction with other nitrogen bearing ingredients within a formulation utilizing this ingredient (many shampoos contain nitrate compounds). A single shampooing can produce more cancer-causing nitrates in the body than eating a pound of bacon, which is VERY high in nitrates!

Whether it is by these means or not, SLS in a known mutagen - it is capable of damaging the genetic material found every cell in your body. As mutagenicity has been strongly linked to cancer, this is a major concern.

IF you are worried about the effects of SLS on your health, please contact us and we will do our best to help and advise you on ways to limit both your exposure and your risk.

why is a dangerous chemical like sodium lauryl sulfate used in our soaps and shampoos?

The answer is simple - it is cheap. The sodium lauryl sulfate found in our soaps is exactly the same as you would find in a car wash or even a garage, where it is used to degrease car engines.

In the same way as it dissolves the grease on car engines, sodium lauryl sulfate also dissolves the oils on your skin, which can cause a drying effect.

It is also well documented that it denatures skin proteins, which causes not only irritation, but also allows environmental contaminants easier access to the lower, sensitive layers of the skin.

Perhaps most worryingly, SLS is also absorbed into the body from skin application. Once it has been absorbed, one of the main effects of sodium lauryl sulfate is to mimic the activity of the hormone Oestrogen. This has many health implications and may be responsible for a variety of health problems from PMS and Menopausal symptoms to dropping male fertility and increasing female cancers such as breast cancer, where oestrogen levels are known to be involved.

Products commonly found to contains Sodium Lauryl Sulfate or SLES

Soaps
Shampoos
Bubble-baths
Tooth paste
Washing-up liquid / dish soap
Laundry detergent

Childrens soaps / shampoos
Stain Remover
Carpet Cleaner
Fabric glue
Body wash
Shave cream
Mascara
Mouthwash
Skin cleanser
Moisture lotion / Moisturiser
Sun Cream

SLS-free Products and Recommended Product for Each Category

SLS-Free Soaps	Natural Soap
SLS-free Shampoos	Herbal shampoo 100% natural
SLS-Free Hair Colouring	Natural hair colouring
SLS-free Bubble-baths	Aromatherapy bath crystals
SLS-free Tooth paste	Natural Toothpaste
SLS-free Washing-up liquid / dish soap	Dish Soap
SLS-free Laundry detergent	Newbrite
SLS-free Childrens soaps / shampoos	Natural shampoo
SLS-free Body wash	Natural body wash
SLS-free Shave cream	His and hers shave soap
SLS-free Cosmetics	Chemical-free cosmetics
SLS-free Mouthwash	Eliminator mouthwash
SLS-free Skin cleanser	Clearskin face wash
SLS-free Moisture lotion / Moisturiser	Natural Hand and body cream
SLS-free Sun Cream/After Sun	Tanacity / Rebound

Do not believe that just because a product is labeled as "natural" it is free from SLS or SLES. Most common brands of "Natural" or "Herbal" shampoos and cleansers still use these harmful chemicals as their main active ingredient.

Propylene Glycol, Alcohol

A colourless, volatile, flammable liquid produced by fermentation of yeast and carbohydrates. Alcohol is frequently used as a solvent. As an ingredient in ingestible products, alcohol may cause body tissues to be more vulnerable to carcinogens.

Mouthwashes with an alcohol content over 25per cent have been implicated in mouth, tongue and throat cancers. Note, the forms of alcohol used in cosmetics are not necessarily those used in alcoholic drinks, and may be much more dangerous.

Some examples

Ethylene glycol, is commonly used in acrylic paints, brake fluid, antifreeze, tile grout, primer, sealant paste, floor polish, tyre sealant and shoe polish. Oh yes.......and it's also used in some "smoothing lotions" and "firming moisturizer"!

Known health effects: Throat irritation, headache, backache, kidney problems, oedema (swelling), necrosis (cell death). If swallowed, can cause drowsiness, and slurred speech, possibly stupor, vomiting, respiratory failure, coma, convulsions, and death

Found in: skin "firming" lotions.

Propylene Glycol

A cosmetic form of mineral oil found in automatic brake and hydraulic fluid and industrial antifreeze. In the skin and hair, propylene glycol works as a humescent, which causes retention of moisture content of skin or cosmetic products by preventing the escape of moisture or water. The Material Safety Data Sheet warns users to avoid skin contact with propylene glycol as this strong skin irritant can cause liver abnormalities and kidney damage.

Known health effects.

Eye irritation, skin irritation, skin drying, defatting. Ingestion has serious health effects similar to above.

Propylene glycol is Commonly found in:

- Makeup
- Shampoo
- Deodorant
- Detangler
- Styling mousse
- Cleansing cream
- Mascara
- Soap
- Skin cream
- Bubble bath
- Baby powder
- Conditioner
- Toner
- After shave
- Baby wipes

Also in:

- Tyre sealant
- Rubber cleaner
- De-icer
- Stain removers
- Fabric softener
- Degreaser
- Paint
- Adhesive
- Wallpaper stripper

Propylene Glycol-free Products and Recommended Product for Each Category	
Propylene Glycol-Free Soaps	Natural Soap
Propylene Glycol-free Shampoos	Herbal shampoo 100% natural
Propylene Glycol-Free Hair Colouring	Natural hair colouring
Propylene Glycol-free Bubble-baths	Aromatherapy bath crystals
Propylene Glycol-free Tooth paste	Natural Toothpaste
Propylene Glycol-free Washing-up liquid/dish soap	Dish Soap
Propylene Glycol-free Laundry detergent	Newbrite
Propylene Glycol-free Childrens soaps/shampoos	Natural shampoo
Propylene Glycol-free Body wash	Natural body wash
Propylene Glycol-free Shave cream	His and hers shave soap
Propylene Glycol-free Cosmetics	Chemical-free cosmetics
Propylene Glycol-free Mouthwash	Eliminator mouthwash
Propylene Glycol-free Skin cleanser	Clearskin face wash
Propylene Glycol-free Moisture lotion/Moisturiser	Natural Hand and body cream
Propylene Glycol-free Sun Cream	Tanacity / Rebound

1, 4 Dioxane

1, 4 Dioxane (not to be confused with dioxin) is a chemical contaminant and by-product of the production of ethoxylated surfactants (including SLS and SLES). Dioxane is a known carcinogen, and is included in the list of such compounds submitted to congress by Senator R Kennedy in 1997.

Nothing has yet been done about the presence of these dangerous chemicals in everyday personal care products and cosmetics. In fact, manufacturers of cosmetics can put in as much 1,4 dioxane, or any other chemical they like into your cosmetics, and they don't even have to tell you!

In 1992, John Bailey, of the FDA's division of Colours and Cosmetics raised the FDA's concerns about the presence of this dangerous chemical in shampoos and conditioners in "excessive levels". Nothing has been done to limit its presence since then. Most human exposure to this 1,4 dioxane comes from its contamination of shampoos and conditioners formulated with SLS or SLES. These are in themselves very nasty chemicals which are thought to be responsible for all sorts of problems from male infertility to PMS (PMT) and Menopause symptoms (due to their estrogen-like effects on the body).

HARMFUL CHEMICALS FOUND IN COSMETICS

BENTONITE, KAOLIN, LANOLIN

Bentonite (sodium bentonite, quaternium-18 bentonite) is a soft, porous clay that expands to many times its own volume as it absorbs water. Often found in cosmetic foundations, bentonite's thick consistency enables bentonite to clog pores and suffocate the skin, preventing normal oxygen transport. Known health effects of bentonite: Bentonite may block pores, suffocating the skin. Bentonite has also been linked to asthma and lung injury. Bentonite has also been linked to Eye irritation, skin irritation, skin drying, defatting. Ingestion of bentonite has serious health effects which may include death.

Bentonite is commonly found in:

- Foundation
- Anti-perspirant
- Shampoo

Bentonite can also be found in

- Cat litter
- Joint compound
- Putty

KAOLIN and Lanolin have very similar properties to Bentonite and can have similar effects. Many cosmetic manufacturers add bentonite to their products to attract water from the deeper layers of the skin to the surface. Whilst this bentonite can "puff up" superficial layers of the skin, it does so at the expense of dehydrating the lower, living layers of skin, damaging the cells in the process. It is through methods like this that bentonite and other chemicals added to "anti-wrinkle" products do more harm than good and eventually can lead to permanent skin damage.

GLYCERIN

Glycerin is a syrupy liquid that is chemically produced by combining water and fat. It is used as a solvent and a plasticiser. Unless the humidity of air is over 65per cent, glycerin draws moisture from the lower layers of the skin and holds it on the surface, drying the skin from the inside out.

Known health effects: Serious risks if ingested.

The vast majority of skin care, hair care and other products marketed as "moisturising", "hydrating" or "replenishing" use glycerin as one of their main active ingredients. It is clear from the data above, that this is a myth, resulting in an increased level of moisture at the surface of the skin at the expense of drying out the deep, sensitive, newly formed skin in the basal layers.

It doesn't take a rocket scientist to work out that if you keep drying out the skin from the inside out, it will get progressively worse!

Commonly found in:

- Hand cream
- Soap
- Moisturising lotion / bars (that's rich!)
- Eye colour
- Baby lotion!
- Shampoo
- Hydrating fluid
- After shave
- Toothpaste
- Cold cream
- Shave cream
- Insect repellant

- Sun cream
- Mascara

Also in

- Car glaze
- Leather cream
- copper ceaner
- Paint cleaner

COLLAGEN

Collagen is an insoluble, fibrous protein that is too large a molecule to penetrate the skin, i.e. no matter how much is present in a cosmetic, it will not get into your skin. The collagen found in most skin care products is derived from animal skins and ground-up chicken feet! Whilst it is a major body component, the form used in cosmetics is foreign and does not even get absorbed through the skin, much less help "rebuild it" as some products claim.

When applied to the skin, most collagen products form a film layer, which can stop oxygen transport, suffocating the skin. *Known health effects:* May block pores, suffocating the skin. Eye irritation, skin irritation, skin drying, defatting. Ingestion has serious health effects that can include death and coma.

Commonly found in:

- Shampoo
- Conditioner
- Moisturising lotion
- Toner

Elastin (elastin of high molecular weight)

Elastin is another skin component, similar to collagen, which, in its natural form is responsible for keeping skin supple. The elastin in skin products, like that of collagen, is derived mainly from animal sources. It's effect is similar to collagen.

TALC / TALCUM POWDER

Many people use talc after bathing or to soothe irritated or inflamed skin. What most people don't know is that there is an inextricable link between tacl and asbestos, which virtually everyone knows causes cancer. In fact, the two are virtually chemically identical. As a consequence, talc manufacturers have played down the link, whilst doing there best to ensure that their powders contain no asbestos fibres. Unfortunately, this isn't enough.

A 1993 US National Toxicology Programme report found that comsetic grade talc, even though it contained no asbestos-like fibres, caused tumours in animals. It would therefore appear that, whether asbestos-like fibres are present or not, cosmetic grade talcum powder is a carcinogen (causes cancer).

Ref: National toxicology programme. "Toxicology and carcinogenesis studies of talc (GAS No 14807-96-6) in F_{344}/N rats and B_6C_3F mice (inhalation studies). Technical report Series No 421, September 1993."

HARMFUL CHEMICALS FOUND IN SUNSCREENS

The British Medical Journal recently showed that sunbathers using some suntan lotions have a higher risk of developing malignant Skin Cancer, and a possible link with Oxybenzone. Oxybenzone is the Chemical used in many sun products with high Sun Protection Factors.

Oxybenzone's function is to 'filter' ultra violet light on the surface of the skin, converting it from light to heat, but it can also be absorbed through the skin. As yet we have not seen any research to indicate what happens when the oxybenzone is absorbed through the skin, but UV light causing cell damage is well known and readers may choose to avoid this form of sun protection. If light is converted to heat in the basal layers of the skin, damage to growing cells is very likely. Environmental + Health News May 98 reports that scientists at the Memorial Sloan Cancer centre N.Y., U.S.A. and the U.S Government Brookhaven Laboratory have found that sunscreens protected against sunburn and cancers like Carcinoma, but not against Melanoma which has increased twenty fold in Europe and U.S.A. since 1935.

Environmental + Health News also reports that Oxford University scientist John Knowland shows that PABA* seemed to damage DNA, thus increasing the risk of skin cancer. He recommends using non-PABA products.

*Para-amino-benzoic acid is very common and works by absorbing UV rays in much the same way as oxybenzone.One of the oldest tried and tested methods is to use tin oxide as a 'reflecting' coating to the skin. Tin oxide is widely used in wound dressings and considered safe, Applied as a cream it is visible in daylight. Although 'safe' it should be avoided by people with dry skin conditions as it has a drying action.

Many people will remember Calamine Lotion as both a sun protection and a soothing after sun. It is based on Zinc Oxide, is pink, visible in daylight and is easily washed off in water. It is likely that this, and other "reflective" sunscreens will, in the long run turn out to be much safer than "absorbing" lotions containing PABA and/or oxybenzone or benzophenone.

If you are in doubt about the safety of your sunscreen, e-mail us, or simply use non-oxybenzone sunscreens rather than "converting" ones.

4

Elements of Environmental Chemistry

UNDERSTANDING ENVIRONMENTAL CHEMISTRY

Environmental chemistry is the scientific study of the chemical and biochemical phenomena that occur in natural places. It should not be confused with green chemistry, which seeks to reduce potential pollution at its source. It can be defined as the study of the sources, reactions, transport, effects, and fates of chemical species in the air, soil, and water environments; and the effect of human activity on these. Environmental chemistry is an interdisciplinary science that includes atmospheric, aquatic and soil chemistry, as well as heavily relying on analytical chemistry and being related to environmental and other areas of science.

Environmental chemistry involves first understanding how the uncontaminated environment works, which chemicals in what concentrations are present naturally, and with what effects. Without this it would be impossible to accurately study the effects humans have on the environment through the release of chemicals.

Environmental chemists draw on a range of concepts from chemistry and various environmental sciences to assist in their study of what is happening to a chemical species in the environment. Important general concepts from chemistry include understanding chemical reactions and equations, solutions, units, sampling, and analytical techniques.

CONTAMINATION

A contaminant is a substance present in nature at a level higher than typical levels or that would not otherwise be there. This may be due to human activity. The term contaminant is often used interchangeably with pollutant, which is a substance that has a detrimental impact on the surrounding environment. Whilst a contaminant is sometimes defined as a substance present in the environment as a result of human activity, but without harmful effects, it is sometimes the case that toxic or harmful effects from contamination only become apparent at a later date. The "medium" (*e.g.* soil) or organism (*e.g.* fish) affected by the pollutant or contaminant is called a receptor, whilst a sink is a chemical medium or species that retains and interacts with the pollutant.

ENVIRONMENTAL INDICATORS

Chemical measures of water quality include dissolved oxygen (DO), chemical oxygen demand (COD), biochemical oxygen demand (BOD), total dissolved solids (TDS), pH, nutrients nitrates and phosphorus), heavy metals (including copper, zinc, cadmium, lead and mercury), and pesticides.

APPLICATIONS

Environmental chemistry is used by the Environment Agency, the Environmental Protection Agency the Association of Public Analysts, and other environmental agencies and research bodies around the world to detect and identify the nature and source of pollutants. These can include:

- Heavy metal contamination of land by industry. These can then be transported into water bodies and be taken up by living organisms.
- Nutrients leaching from agricultural land into water courses, which can lead to algal blooms and eutrophication.
- Urban run-off of pollutants washing off impervious surfaces (roads, parking lots, and rooftops) during rain storms. Typical pollutants include gasoline, motor oil and other hydrocarbon compounds, metals, nutrients and sediment (soil).
- Organometallic compounds.

METHODS

Quantitative chemical analysis is a key part of environmental chemistry, since it provides the data that frame most environmental studies. Common analytical techniques used for quantitative determinations in environmental chemistry include classical wet chemistry, such as gravimetric, titrimetric and electrochemical methods.

More sophisticated approaches are used in the determina-tion of trace metals and organic compounds. Metals are commonly measured by atomic spectroscopy and mass spectrometry: Atomic Absorption Spectrophotometry (AA) and Inductively Coupled Plasma Atomic Emission (ICP-AES) or Inductively Coupled Plasma Mass Spectrometric (ICP-MS) techniques. Organic compounds are commonly measured also using mass spectrometric methods, such as Gas chromatography -mass spectrometry (GC/MS) and Liquid chromatography-mass spectrometry (LC/MS).

Non-MS methods using GCs and LCs having universal or specific detectors are still staples in the arsenal of available analytical tools. Other parametres often measured in environmental chemistry are radiochemicals. These are pollutants which emit radioactive materials, such as alpha and beta particles, posing danger to human health and the environment. Particle counters and Scintillation counters are most commonly used for these measurements. Bioassays and immunoassays are utilized for toxicity evaluations of chemical effects on various organisms.

PUBLISHED ANALYTICAL METHODS

Peer-reviewed test methods have been published by government agencies and private research organizations. Approved published methods must be used when testing to demonstrate compliance with regulatory requirements.

ORIGIN AND EVOLUTION OF EARTH

Earth is an active place. Earthquakes rip along plate boundaries, volcanoes spew fountains of molten lava, and mountain ranges and seabed are constantly created and destroyed. Earth scientists have long been concerned with deciphering the history—and predicting the future—of this active planet. Over the past four decades, Earth scientists have made great strides in understanding Earth's workings.

Scientists have ever-improving tools to understand how Earth's internal processes shape the planet's surface, how life can be sustained over billions of years, and how geological, biological, atmospheric, and oceanic processes interact to produce climate—and climatic change. At the request of the U.S. Department of Energy, National Aeronautics and Space Administration, National Science Foundation, and U.S.

Geological Survey, the National Research Council assembled a committee to propose and explore grand questions in Earth science. This report, which is the result of the committee's deliberations and input solicited from the Earth science community, describes ten "big picture" Earth science issues being pursued today. Answers to these fundamental questions could profoundly improve under-standing of the planet on which we live and strategies for managing our environment.

ORIGINS

Formation Earth and Other Planets

The Solar System is composed of a set of radically different types of planets and moons— from the gas giants Jupiter, Saturn, Uranus, and Neptune to the rocky inner planets. Centuries of studying Earth, its neighbouring planets, and meteorites have enabled the development of models of the birth of the Solar System.

Astronomical observations from increasingly powerful telescopes have added a new dimension to these models, as have studies of asteroids, comets, and other planets via spacecraft, as well as geochemical studies of stardust and meteorites. While it is generally agreed that the Sun and planets all coalesced out of the same nebular cloud, little is known about how Earth obtained its particular chemical composition, or why the other planets ended up so different from Earth and from each other. For example, why has Earth, unlike every other planet, retained the unique properties—such as the presence of water—that allow it to support life? New measurements of Solar System bodies and extra solar planets and objects, will further advance understanding of the origin of Earth and the Solar System.

Earth's "Dark Age"

It is now believed that during Earth's formation, a Mars-sized planet collided with it, creating a huge cloud of debris that became Earth's Moon and releasing so much heat that the entire planet melted. But little is known about how the resulting molten rock evolved during the planet's infancy into the Earth we know today.

The first 500 million years of Earth's existence, known as the Hadean Eon, is a critical missing link in understanding how the planet's atmosphere, oceans, and differentiated layers of core, mantle, and outer crust developed. Scientists have almost no idea how fast the surface environment evolved, how the transition took place, or when conditions became hospitable enough to support life. Some clues from Earth's oldest minerals (zircons), as well as from Earth's Moon and other planets are allowing a clearer picture of the Hadean Eon to gradually emerge. The future is certain to provide additional breakthroughs. The amount of information that can be extracted from even the tiniest samples of old rocks and minerals is increasing rapidly, and with concerted effort, it is expected that many more ancient rocks and mineral samples will be found.

Beginning of Life

In The Origin of Species, Charles Darwin (1859) hypothesized that new species arise by the modification of existing ones—that the raw material of life is life. But somehow and somewhere, the tree of life had to take root from non-living precursors. When, where, and in what form did life first appear? The origin of life is one of the most intriguing, difficult, and enduring questions in science. Scientists have toiled to create life from sparks and gasses in the laboratory to illuminate how life first formed in Earth's early conditions.

But even pinning down what those early conditions were remains an elusive goal. From what materials did life originate? Did life, as Darwin speculated, originate in a "warm little pond," perhaps a tidal pool repeatedly dried and refreshed? Or might life be rooted among hydrothermal vents? Could life's origins even lie beyond Earth? Developing an accurate picture of the physical environments and the chemical building blocks available to early life is a critical Earth science challenge. Clues to shed light on these mysteries stem largely from investigations of Earth's ancient rocks and minerals—the only remaining evidence of the time when Earth's life first emerged.

EARTH'S INTERIOR

Earth's Interior Work and Its Effect on Surface

As planets age and cool off, their internal and surface processes gradually change. Manifestations of changes within Earth's interior—such as the development of mountains and volcanoes—have a huge influence on the nature of Earth's surface and atmosphere. Scientists know that much of the rock in the Earth's mantle (the thick layer between the core and crust), which is under extreme pressure and very high temperature, behaves like a viscous liquid. This vast interior, however, is largely inaccessible to direct study. For over a century, seismic wave, geomagnetic, and gravity measurements made at the surface have been improving understanding of Earth's internal structure. Despite continuing advances, however, scientists are only beginning to explore the connections between Earth's core, magnetic field, mantle, and surface and to investigate why Earth differs from other planets, or how it may change in the future.

Earth's Plate Tectonics and Continents

A major focus of Earth science has been on deciphering the nature of the continents—the features that make Earth habitable for land-dwelling life. How and when did the continents form? How have they changed? Why do the Atlantic coastlines of South America and Africa look like pieces of a puzzle? Plate tectonics—the description of Earth's outermost layers in terms of a small number of large rigid plates in relative motion—has offered many breakthrough insights since it became a central paradigm for geology nearly 40 years ago. The motion of tectonic plates and interactions at their boundaries are now known to be a driving force behind earthquakes, volcanic eruptions, the formation of mountains, and the slow drift of continents over Earth's surface. Although plate tectonic theory explains many of Earth's surface features, it is not known why Earth has plates or what the relationships are between plate tectonics and Earth's abundant water, continents, and the existence of life. Answering such questions will require improved models, studies of modern plate boundaries, and comparisons with other planets.

Earth Processes Controlled by Material Properties

It's the little things that make all the difference. Scientists now recognize that large-scale processes on Earth, such as plate tectonics, are driven by the nature of the materials that make up the planet—down to the smallest details of their atomic structures. The high pressures and temperatures of Earth's interior, the enormous size of the planet and its structures, the long expanse of geological time, and the vast diversity of materials present challenges to studying Earth materials.

Breakthroughs in this area are now at hand, however; new analytical tools and advanced computing capabilities are improving the study and simulation of Earth materials at the atomic level and promise to improve predictions of how these material properties will affect planetary processes.

EARTH'S CLIMATE AND HABITABILITY

Causes of Climate Change

It is widely recognized that Earth's mean global surface temperature has risen since the beginning of the industrial age, and that emissions of CO_2 and other greenhouse gases are at least partly responsible. The potentially serious consequences of global warming underscore the need to determine how much of the warming is caused by human activities and what can be done about it. Earth science has an important role in answering both questions.

The geological record has revealed the history of the planet's climate to be a peculiar combination of both variability and stability. Global climate conditions have been favourable for life and relatively stable for the past 10,000 years and suitable for life for over 3 billion years.

But geological evidence also shows that momentous changes in climate can occur in periods as short as decades or centuries.

How does Earth's climate remain relatively stable in the long term, even though it can change so abruptly? Understanding periods in which the planet was extremely cold, extremely hot, or changed especially quickly are leading to new insights about Earth's climate. Observations of ancient rocks could eventually improve prediction of the magnitude and consequences of climate changes.

Shaping Earth and Life

Scientists know that the composition of Earth's atmosphere, especially its high concentration of oxygen, is a consequence of the presence of life. At the microscopic scale, life is an invisible but powerful chemical force: organisms catalyze reactions that would not happen in their absence, and they accelerate or slow down other reactions.

These reactions, compounded over immense stretches of time by a large biomass, can generate changes of global consequence. Likewise, Earth's geologic evolution, as well as catastrophic events like meteorite impacts, has clearly affected the evolution of life. But even when extinctions and major evolutionary changes can be documented, the causes still remain a mystery.

To what extent were they caused by geological as opposed to biological processes? Exactly how geological events have affected evolution, and how much control life has had on climate, are still topics of debate. Understanding the interrelationships between life and the processes that shape the land presents a critical challenge.

LIVING WITH AN ACTIVE EARTH

Prediction of Earthquakes, Volcanic Eruptions and their Consequences

Earthquakes and volcanic eruptions are the most sudden and hazardous manifestations of the gradual movements of Earth's interior. The need to predict such events has escalated as human populations increasingly concentrate in volcano- and earthquake prone areas. Thanks largely to sensitive new instrumentation and better understanding of causes, geologists are moving towards predictive capabilities for volcanic eruptions. For earthquakes, progress has been made in estimating the probability of future events, but we may never be able to predict the exact time and place an earthquake will strike.

Continuing challenges include deciphering how fault ruptures start and stop, simulating how much shaking can be expected near large earthquakes, and increasing the warning time once a dangerous earthquake begins.

Human Environment: Fluid Flow and Transport

The presence and placement of many of Earth's most valuable resources are determined by processes of fluid flow and transport. The ability to assess and extract minerals, petroleum, natural gas, and groundwater and to safely dispose of wastes depends on understanding fluids, both at the surface and below ground. Some of the scientific questions associated with fluids also have application to earthquake prediction, climate prediction, the evolution of continents, the behaviour of volcanoes, and the properties of Earth materials.

New experimental tools and techniques, plus airborne and space borne measurements, are offering an unprecedented view of the processes that affect Earth's fluids. But many questions remain. The ultimate objective—to produce mathematical models that can predict the performance of natural systems far into the future—is still out of reach, but attaining it will be critical to making informed decisions about the future of the land and resources that support us.

THE ORIGIN OF LIFE AND EVOLUTION OF THE ATMOSPHERE

The history of the Earth describes the most important events and fundamental stages in the development of the planet Earth from its formation 4.6 billion years ago to the present day. Nearly all branches of natural science have contributed to the understanding of the main events of the Earth's past. The age of Earth is approximately one-third of the age of the universe. Immense geological and biological changes have occurred during that time span.

HADEAN AND ARCHAEAN

Starting with the Earth's formation by accretion from the solar nebula 4.54 billion years ago (4.54 Ga), the first eon in the Earth's history is called the

Hadean. It lasted until the Archaean eon, which began 3.8 Ga. The oldest rocks found on Earth date to about 4.0 Ga, and the oldest detrital zircon crystals in some rocks have been dated to about 4.4 Ga, close to the formation of the Earth's crust and the Earth itself.

Because not much material from this time is preserved, little is known about Hadean times, but scientists hypothesize at an estimated 4.53 Ga, shortly after formation of an initial crust, the proto-Earth was impacted by a smaller protoplanet, which ejected part of the mantle and crust into space and created the Moon. During the Hadean, the Earth's surface was under a continuous bombardment by meteorites, and volcanism must have been severe due to the large heat flow and geothermal gradient.

The detrital zircon crystals dated to 4.4 Ga show evidence of having undergone contact with liquid water, considered as proof that the planet already had oceans or seas at that time. From crater counts on other celestial bodies it is inferred that a period of intense meteorite impacts, called the "Late Heavy Bombardment", began about 4.1 Ga, and concluded around 3.8 Ga, at the end of the Hadean. By the beginning of the Archaean, the Earth had cooled significantly. It would have been impossible for most present day life forms to exist due to the composition of the Archaean atmosphere, which lacked oxygen and an ozone layer. Nevertheless it is believed that primordial life began to evolve by the early Archaean, with some possible fossil finds dated to around 3.5 Ga. Some researchers, however, speculate that life could have begun during the early Hadean, as far back as 4.4 Ga, surviving the possible Late Heavy Bombardment period in hydrothermal vents below the Earth's surface.

Origin of the Solar System

The Solar System (including the Earth) formed from a large, rotating cloud of interstellar dust and gas called the solar nebula, orbiting the Milky Way's galactic centre. It was composed of hydrogen and helium created shortly after the Big Bang 13.7 Ga and heavier elements ejected by supernovas. About 4.6 Ga, the solar nebula began to contract, possibly due to the shock wave of a nearby supernova. Such a shock wave would have also caused the nebula to rotate and gain angular momentum.

As the cloud began to accelerate its rotation, gravity and inertia flattened it into a protoplanetary disk oriented perpendicularly to its axis of rotation. Most of the mass concentrated in the middle and began to heat up, but small perturbations due to collisions and the angular momentum of other large debris created the means by which protoplanets up to several kilometres in length began to form, orbiting the nebular centre.

The infall of material, increase in rotational speed and the crush of gravity created an enormous amount of kinetic energy at the centre. Its inability to transfer that energy away through any other process at a rate capable of relieving the build-up resulted in the disk's centre heating up. Ultimately, nuclear fusion of hydrogen into helium began, and eventually, after contraction, a T Tauri star ignited to create the Sun. Meanwhile, as gravity caused matter to condense around the previously perturbed objects outside the gravitational grasp of the new sun, dust particles and the rest of the protoplanetary disk began separating into rings. Successively larger fragments collided with one another and became larger objects, ultimately becoming protoplanets. These included one collection about 150 million kilometres from the centre: Earth. The planet formed about 4.54 billion years ago (within an uncertainty of 1per cent) and was largely completed within 10–20 million years.

The solar wind of the newly formed T Tauri star cleared out most of the material in the disk that had not already condensed into larger bodies. Computer simulations have shown that planets with distances equal to the terrestrial planets in our solar system can be created from a protoplanetary disk. The now widely accepted nebular hypothesis suggests that the same process, which gave rise to the solar system's planets, produces accretion disks around virtually all newly forming stars in the universe, some of which yield planets.

Origin of the Earth's Core and First Atmosphere

The Proto-Earth grew by accretion, until the inner part of the protoplanet was hot enough to melt the heavy, siderophile metals. Such liquid metals, with now higher densities, began to sink to the Earth's centre of mass. This so called iron catastrophe resulted in the separation of a primitive mantle and a (metallic) core only 10 million years after the Earth began to form, producing the layered structure of Earth and setting up the formation of Earth's magnetic field. During the accretion of material to the protoplanet, a cloud of gaseous silica must have surrounded the Earth, to condense afterwards as solid rocks on the surface.

What was left surrounding the planet was an early atmosphere of light (atmophile) elements from the solar nebula, mostly hydrogen and helium, but the solar wind and Earth's heat would have driven off this atmosphere. This changed when Earth accreted to about 40per cent its present radius, and gravitational attraction retained an atmosphere which included water.

The Giant Impact Hypothesis

The Earth's relatively large natural satellite, the Moon, is unique. During the Apollo programme, rocks from the Moon's surface were brought to Earth. Radiometric dating of these rocks has shown the Moon to be 4527 ± 10 million years old, about 30 to 55 million years younger than other bodies in the solar system. (New evidence suggests the Moon formed even later, 4.48±0.02 Ga, or 70–110 Ma after the start of the Solar System.) Another notable feature is the relatively low density of the Moon, which must mean it does not have a large metallic core, like all other terrestrial bodies in the solar system.

The Moon has a bulk composition closely resembling the Earth's mantle and crust together, without the Earth's core. This has led to the giant impact hypothesis, the idea that the Moon was formed during a giant impact of the proto-Earth with another protoplanet by accretion of the material blown off the mantles of the proto-Earth and impactor. The impactor, sometimes named Theia, is thought to have been a little smaller than the current planet Mars. It could have formed by accretion of matter about 150 million kilometres from the Sun and Earth, at their fourth or fifth Lagrangian point.

Its orbit may have been stable at first, but destabilized as Theia's mass increased due to the accretion of matter. Theia oscillated in larger and larger orbits around the Lagrangian point until it finally collided with Earth about 4.533 Ga. Models reveal that when an impactor this size struck the proto-Earth at a low angle and relatively low speed (8–20 km/sec), much material from the mantles and crusts of the proto-Earth and the impactor was ejected into space, where much of it stayed in orbit around the Earth.

This material would eventually form the Moon. However, the metallic cores of the impactor would have sunk through the Earth's mantle to fuse with the Earth's core, depleting the Moon of metallic material. The giant impact hypothesis thus explains the Moon's abnormal composition. The ejecta in orbit around the Earth could have condensed into a single body within a couple of weeks. Under the influence of its own gravity, the ejected material became a more spherical body: the Moon.

The radiometric ages show the Earth existed already for at least 10 million years before the impact, enough time to allow for differentiation of the Earth's primitive mantle and core. Then, when the impact occurred, only material from the mantle was ejected, leaving the Earth's core of heavy siderophile elements untouched. The impact had some important consequences for the young Earth. It released an enormous amount of energy, causing both the Earth and Moon to be completely molten.

Immediately after the impact, the Earth's mantle was vigourously convecting, the surface was a large magma ocean. The planet's first atmosphere must have been completely blown away by the impact. The impact is also thought to have changed Earth's axis to produce the large 23.5° axial tilt that is responsible for Earth's seasons (a simple, ideal model of the planets' origins would have axial tilts of 0° with no recognizable seasons). It may also have sped up Earth's rotation.

Origin of the Oceans and Atmosphere

Because the Earth lacked an atmosphere immediately after the giant impact, cooling must have occurred quickly. Within 150 million years, a solid crust with a basaltic composition must have formed. The felsic continental crust of today did not yet exist. Within the Earth, further differentiation could only begin when the mantle had at least partly solidified again. Nevertheless, during the early Archaean (about 3.0 Ga) the mantle was still much hotter than today, probably around 1600°C.

This means the fraction of partially molten material was still much larger than today. Steam escaped from the crust, and more gases were released by volcanoes, completing the second atmosphere. Additional water was imported by bolide collisions, probably from asteroids ejected from the outer asteroid belt under the influence of Jupiter's gravity. The large amount of water on Earth can never have been produced by volcanism and degassing alone. It is assumed the water was derived from impacting comets that contained ice. Though most comets are today in orbits farther away from the Sun than Neptune, computer simulations show they were originally far more common in the inner parts of the solar system. However, most of the water on Earth was probably derived from small impacting protoplanets, objects comparable with today's small icy moons of the outer planets.

Impacts of these objects can have enriched the terrestrial planets (Mercury, Venus, the Earth and Mars) with water, carbon dioxide, methane, ammonia, nitrogen and other volatiles. If all water on Earth was derived from comets alone, millions of comet impacts would be required to support this theory. Computer simulations show that this is not an unreasonable number.

As the planet cooled, clouds formed. Rain created the oceans. Recent evidence suggests the oceans may have begun forming by 4.2 Ga, or as early as 4.4 Ga. In any event, by the start of the Archaean eon the Earth was already covered with oceans. The new atmosphere probably contained water vapour, carbon dioxide, nitrogen, and smaller amounts of other gases.

As the output of the Sun was only 70per cent of the current amount, significant amounts of greenhouse gas in the atmosphere most likely prevented the surface water from freezing. Free oxygen would have been bound by hydrogen or minerals on the surface. Volcanic activity was intense and, without an ozone layer to hinder its entry, ultraviolet radiation flooded the surface.

The First Continents

Mantle convection, the process that drives plate tectonics today, is a result of heat flow from the core to the Earth's surface. It involves the creation of rigid tectonic plates at mid-oceanic ridges. These plates are destroyed by subduction into the mantle at subduction zones. The inner Earth was warmer during the Hadean and Archaean eons, so convection in the mantle must have been faster. When a process similar to present day plate tectonics did occur, this would have gone faster too. Most geologists believe that during the Hadean and Archaean, subduction zones were more common, and therefore tectonic plates were smaller.

The initial crust, formed when the Earth's surface first solidified, totally disappeared from a combination of this fast Hadean plate tectonics and the intense impacts of the Late Heavy Bombardment. It is, however, assumed that this crust must have been basaltic in composition, like today's oceanic crust, because little crustal differentiation had yet taken place. The first larger pieces of continental crust, which is a product of differentiation of lighter elements during partial melting in the lower crust, appeared at the end of the Hadean, about 4.0 Ga. What is left of these first small continents are called cratons. These pieces of late Hadean and early Archaean crust form the cores around which today's continents grew.

The oldest rocks on Earth are found in the North American craton of Canada. They are tonalites from about 4.0 Ga. They show traces of metamorphism by high temperature, but also sedimentary grains that have been rounded by erosion during transport by water, showing rivers and seas existed then. Cratons consist primarily of two alternating types of terranes. The first are so called greenstone belts, consisting of low grade metamorphosed sedimentary rocks.

These "greenstones" are similar to the sediments today found in oceanic trenches, above subduction zones. For this reason, greenstones are sometimes seen as evidence for subduction during the Archaean. The second type is a complex of felsic magmatic rocks. These rocks are mostly tonalite, trondhjemite or granodiorite, types of rock similar in composition to granite (hence such terranes are called TTG-terranes).

TTG-complexes are seen as the relicts of the first continental crust, formed by partial melting in basalt. The alternation between greenstone belts and TTG-complexes is interpreted as a tectonic situation in which small proto-continents were separated by a thorough network of subduction zones.

Origin of Life

The details of the origin of life are unknown, but the basic principles have been established. There are two schools of thought about the origin of life. One suggests that organic components arrived on Earth from space, while the other argues that they originated on Earth. Nevertheless, both schools suggest similar mechanisms by which life initially arose. If life arose on Earth, the timing of this event is highly speculative—perhaps it arose around 4 Ga.

It is possible that, as a result of repeated formation and destruction of oceans during that time period caused by high energy asteroid bombardment, life may have arisen and been extinguished more than once. In the energetic chemistry of early Earth, a molecule gained the ability to make copies of itself—a replicator. (More accurately, it promoted the chemical reactions which produced a copy of itself.) The replication was not always accurate: some copies were slightly different from their parent.

If the change destroyed the copying ability of the molecule, the molecule did not produce any copies, and the line "died out". On the other hand, a few rare changes might have made the molecule replicate faster or better: those "strains" would become more numerous and "successful". This is an early example of evolution on abiotic material. The variations present in matter and molecules combined with the universal tendency for systems to move towards a lower energy state allowed for an early method of natural selection. As choice raw materials ("food") became depleted, strains which could utilize different materials, or perhaps halt the development of other strains and steal their resources, became more numerous.

The nature of the first replicator is unknown because its function was long since superseded by life's current replicator, DNA. Several models have been proposed explaining how a replicator might have developed. Different replicators have been posited, including organic chemicals such as modern proteins, nucleic acids, phospholipids, crystals, or even quantum systems. There is currently no way to determine whether any of these models closely fits the origin of life on Earth.

One of the older theories, one which has been worked out in some detail, will serve as an example of how this might occur. The high energy from volcanoes, lightning, and ultraviolet radiation could help drive chemical reactions producing more complex molecules from simple compounds such as methane and ammonia.

Among these were many of the simpler organic compounds, including nucleobases and amino acids, which are the building blocks of life. As the amount and concentration of this "organic soup" increased, different molecules reacted with one another. Sometimes more complex molecules would result—perhaps clay provided a framework to collect and concentrate organic material. Certain molecules could speed up a chemical reaction.

All this continued for a long time, with reactions occurring at random, until by chance it produced a replicator molecule. In any case, at some point, the function of the replicator was superseded by DNA; all known life (except some viruses and prions) use DNA as their replicator, in an almost identical manner. Modern life has its replicating material packaged inside a cellular membrane.

It is easier to understand the origin of the cell membrane than the origin of the replicator, because a cell membrane is made of phospholipid molecules, which often form a bilayer spontaneously when placed in water. Under certain conditions, many such spheres can be formed. The prevailing theory is that the membrane formed after the replicator, which perhaps by then was RNA (the RNA world hypothesis), along with its replicating apparatus and other biomolecules.

Initial protocells may have simply burst when they grew too large; the scattered contents may then have recolonized other "bubbles". Proteins that stabilized the membrane, or that later assisted in an orderly division, would have promoted the proliferation of those cell lines. RNA is a likely candidate for an early replicator, because it can both store genetic information and catalyze reactions.

At some point DNA took over the genetic storage role from RNA, and proteins known as enzymes took over the catalysis role, leaving RNA to transfer information, synthesize proteins and regulate the process. There is increasing belief that these early cells evolved in association with undersea volcanic vents known as black smokers or even hot, deep rocks. It is believed that of this multiplicity of protocells, only one line survived.

Current phylogentic evidence suggests that the last universal common ancestor lived during the early Archean eon, perhaps roughly 3.5 Ga or earlier. This "LUCA" cell is the ancestor of all life on Earth today. It was probably a prokaryote, possessing a cell membrane and probably ribosomes, but lacking a nucleus or membrane-bound organelles such as mitochondria or chloroplasts.

Like all modern cells, it used DNA as its genetic code, RNA for information transfer and protein synthesis, and enzymes to catalyze reactions. Some scientists believe that instead of a single organism being the last universal common ancestor, there were populations of organisms exchanging genes in lateral gene transfer.

CHEMISTRY OF THE STRATOSPHERE

STRATOSPHERIC OZONE–BACKGROUND

Why is Understanding Stratospheric Ozone Important?

Ozone is one of the most important trace species in the atmosphere. Ozone plays two critical roles: it removes most of the biologically harmful ultraviolet light before the light reaches the surface, and it plays an essential role in setting up the temperature structure and therefore the radiative heating/cooling balance in the atmosphere, especially the stratosphere.

Location of the Ozone Layer and Climatology

Ozone is mainly found in two regions of the atmosphere. Most of the ozone can be found in a layer between 10 and 60 km above the Earth's surface. This ozone region located in the stratosphere is known as the "ozone layer." Some ozone can also be found in the lower atmosphere in the region known as the troposphere. Although chemically identical to stratospheric ozone, tropospheric ozone is quite distinct and geophysically different from stratospheric ozone.

Ozone and UV—Biological Threat

Ozone is produced by the photolysis of molecular oxygen, O_2. The oxygen atom, O, produced by this photolysis recombines with O_2 to form ozone, O3. Ozone formation primarily occurs in the tropical upper stratosphere, where it is transported poleward and downward by the largescale Brewer-Dobson circulation. The 13-year average of the total ozone measurements taken by the Nimbus-7 Total Ozone Mapping Spectrometre instrument.

The formation of ozone by the photolysis of molecular oxygen removes most of the incident sunlight with wavelengths shorter than 200 nm. The wavelengths between 200 and 310 nm are removed by the photolysis of ozone itself. This photolysis of ozone in the stratosphere is the process by which most of the biologically damaging ultraviolet sunlight is filtered out. As this filtering process occurs, the stratosphere is heated. This heating is responsible for the temperature structure of the stratosphere, where the temperature increases as the altitude increases. Without this filtering, larger amounts of UV-B would reach the surface. Numerous studies have shown that excessive exposure to UV-B is harmful to plants, animals, and humans.

Ozone and Climate Change

If ozone in the stratosphere were to be removed, the stratosphere would cool. How a cooler stratosphere affects radiative balance in the rest of the atmosphere has been the subject of many detailed studies. These studies have been reanalysed and integrated into the latest Intergovernmental Panel on Climate Change report, "Climate Change 1994: Radiative Forcing of Climate Change". The result of that report is that stratospheric ozone loss leads to a "small but non-negligible offset to the total greenhouse forcing from CO_2, N_2O, CH4, CFCs...." It is ironic that the size of the negative radiative forcing from ozone loss is nearly equal to the positive radiative forcing from chlorofluorocarbons, the source of the stratospheric ozone loss. The size of the radiative forcing due to stratospheric ozone loss has also been shown to be very sensitive to the profile shape assumed for that loss.

Observed Ozone Changes

While the global amount of ozone is fairly constant, there are significant local, seasonal, and long-term changes. The seasonal ozone changes are basically determined by the winter-summer changes in the stratospheric circulation. Since ozone has a lifetime of weeks to months in the lower stratosphere, the amount of ozone can strongly vary due to transport by stratospheric wind systems.

Since weather conditions in the stratosphere, as in the troposphere, vary from year to year, there is also interannual variability in ozone amounts. Interseasonal changes in ozone are also linked to the 11-year solar cycle in UV output and the amount of volcanic aerosols in the stratosphere. Changes in ozone have also been linked to anthropogenic pollutants, especially the release of manmade chemicals containing chlorine. We describe the more-significant recent global changes in ozone observed by a variety of instruments.

Polar Ozone Changes

The first ozone measurements in the Antarctic were made during the 1950s. A Dobson instrument was installed at Halley Bay in late 1956 in preparation for the International Geophysical Year in 1957. One of the first discoveries made by this instrument was that the seasonal cycle of ozone in the south polar region is very different from that which had been observed in the north. Dobson which pointed out that its cause was a difference in the circulation patterns of the Antarctic relative to the Arctic.

In the Arctic, the total ozone amount grew rapidly in the late winter and early spring to about 500 Dobson Units. In contrast, the Antarctic early springtime amounts remained near 300 DU. The Dobson instrument at Halley Bay continued to make measurements each year. Farman showed that the springtime ozone amounts over Halley Bay had declined from nearly 300 DU in the early 1960s to about 180 DU in the early-to-mid 1980s.

This result has been confirmed at a number of other stations and shown using satellite data to occur over an area larger than the Antarctic continent. These large ozone changes implied that losses must be taking place in the lower stratosphere where most of the ozone exists. This was shown to be true in a series of ozonesonde measurements. More-recent sonde measurements have shown instances of near-zero concentrations of ozone over a 5-km altitude range. The 1996-1997 Northern Hemisphere winter experienced a significant ozone depletion over the Arctic and subsequent total ozone values achieved record low values in the spring. Long term records of the Total Ozone Mapping Spectrometre and ground based observations show a downward change over the past several years occurring mostly in February and March and confined to the lower stratosphere similar to the depletion over the Antarctic. Total ozone changes for both polar regions. Chlorine radicals have been conclusively identified as the causes of ozone depletion now in both hemispheres. Measurements of ClO by the UARS MLS instrument observed elevated levels in late February over Northern Hemisphere polar regions. The winter of 1996-1997, showed extremely low temperatures in the Stratosphere.

These cold temperatures led to the formation of polar stratospheric clouds whose particles shift chlorine gas away from its HCl reservoir to active ClO through heterogeneous chemistry. This is the primary mechanism producing the Antarctic ozone hole. Although the buildup of chlorine has occurred approximately uniformly in both hemisphere the unusually low temperatures reached in high northern latitudes mostly likely precipitated the concurrent ozone losses over the Arctic.

Midlatitude Ozone Loss

Midlatitude ozone loss estimates must be extracted from long time series using statistical models. The longest time series of total ozone data is from Arosa in Switzerland. This time series, which dates back to 1926. The Arosa data show a relatively constant amount of ozone for over 4 decades and a decrease in the last decade and a half. Analysis of a more-extensive network of 30+ stations which have been in operation for about 35 years shows negative trends over the last 1.5 decades, especially in the winter and early spring. High-quality global satellite data records began in November 1978 with the launch of the Nimbus-7 Solar Backscatter Ultraviolet radiometre and TOMS instruments.

These data show midlatitude trends in the Northern Hemisphere which are largest in the winter and early spring, peaking at about 6-8per cent per decade at 40°-50° N in February. These satellite trends are in the process of being updated with a version 7 algorithm for the TOMS and SBUV instruments. Changes in the profile of ozone with altitude can be deduced from sonde data or from the Stratospheric Aerosol and Gas Experiment satellite measurements. Analyses of sonde data show ozone decreases between the tropopause and about 24- km altitude. Analyses of SAGE data show larger decreases than those derived from sondes. SAGE results show negative ozone trends in the lower stratosphere in the tropics.

Column ozone changes deduced from SBUV and TOMS show only small downward trends. Hollandsworth used SBUV profile and total ozone trends to deduce that ozone in the tropics below 32 hPa has increased slightly over the last decade. The resolution of the uncertainty in the magnitude of lower stratospheric and upper tropospheric ozone trends is an important measurement and analysis issue for the coming years.

The Stratospheric Ozone Distribution

Chemical Processes

Ozone is being continuously created and destroyed by the action of ultraviolet sunlight. The overall amount of ozone in the global stratosphere is determined by the magnitude of the production and loss processes and by the rate at which air is transported from regions of net production to those of net loss. Production of ozone requires the breaking of an O_2 bond, with the extra or "odd" oxygen atom attaching to another O_2 to form O_3 This most frequently occurs via the photo dissociation of O_2 by solar ultraviolet radiation. In the lower stratosphere and troposphere ozone can also be produced by photochemical smog-like reactions. In these reactions H or CH_3 or higher hydrocarbon radicals attach to an O_2 forming HO_2 or CH_3O_2, etc., which then react with NO. This reaction breaks the O_2 bond by forming NO_2. When NO_2 is photolyzed an O atom is formed which reacts with O_2 to form O_3.

Loss of ozone occurs when an O atom reacts with O_3 to re-form the O_2 bond. More importantly, this loss process is catalyzed by the oxides of hydrogen, nitrogen, chlorine, and bromine. These oxides are produced in the stratosphere from longlived, unreactive molecules released at the surface of the Earth. The major source molecules for HOx are methane and water vapour. The main source of NOx is nitrous oxide.

The major sources of chlorine are industrially-produced CFCs and naturallyoccurring methyl chloride. The major sources of bromine are methyl bromide and the halons. These source molecules are transported to the stratosphere where they react or are photodissociated to produce the catalytically-active oxide radicals. The catalytic efficiency of hydrogen, nitrogen, chlorine, and bromine oxides is determined by a set of interlocking reactions which convert the active oxides to catalytically-inactive temporary reservoirs, such as HNO_3, HCl, $ClONO_2$, H_2O, HOCl, HOBr, and $BrONO_2$, and vice versa. In the lower stratosphere, the balance between catalytic oxides and temporary reservoirs is strongly affected by reactions on the surfaces of stratospheric aerosols.

The balance is even more profoundly affected in the polar winter by reactions on the surface of Polar Stratospheric Cloud particles. In the early spring, the chlorine balance is shifted to almost 100per cent ClOx. This shift in the chemical balance results in a large calculated chemical sensitivity of ozone towards chlorine perturbations and a relatively small calculated sensitivity of ozone towards nitrogen oxide perturbations.

Although the basic outline of the chemistry controlling stratospheric ozone is now known, many important aspects of the problem remain to be solved. The primary difference between the Northern and Southern Hemispheric polar ozone loss regions appears to be a result of the "denitrification" that occurs in the Antarctic winter. Denitrification means the removal of nitrogen oxides and HNO3 by large particles which fall into the troposphere. Denitrification takes place when temperatures are cold enough to form large stratospheric ice crystals. When springtime comes there are no nitrogen oxides to convert ClOx to $ClONO_2$ and slow down the rate of ozone depletion. There is some evidence for denitrification when temperatures are not cold enough to form ice crystals. Under those conditions the mechanism for denitrification is not completely understood.

Transport

Much of the currently-observed ozone interannual variability in the stratosphere is controlled by dynamical processes. In particular, this variability is driven by such processes as the quasi-biennial oscillation, El Niño-Southern Oscillation, tropospheric weather systems which extend into the stratosphere, and long-term fluctuations in planetary wave activity.

The annual cycle of total ozone is largely driven by transport effects. Relatively low values of ozone are observed in the tropics and high values are observed in the extratropics. These low tropical ozone values occur in spite of the large ozone production rates in the tropics. If ozone production were precisely balanced by ozone loss everywhere, total ozone would have extremely high values in the tropics.

The observed tropical low values result from vertical advection of low-ozone air from the tropical troposphere into the tropical stratosphere, and the subsequent transport of this air poleward and downward into the extratropics and polar regions. This advective circulation is known as the Brewer-Dobson circulation. The redistribution of ozone from the production region at low latitudes to extratropical latitudes is modulated by a variety of processes.

Foremost among these processes is the annual cycle in the circulation. It is now recognized that the Brewer-Dobson circulation is primarily controlled by large-scale waves in the winter stratosphere.

As these waves propagate through the westerly winds that dominate the winter stratosphere, they exert a westward zonal drag, which through the Coriolis force leads to a poleward and downward transport circulation, which in turn drives the temperatures away from radiative equilibrium. The large-scale waves breaking in the winter upper stratosphere also produce lifting in the tropics.

Since the lifetime of ozone increases with pressure, the poleward downward circulation causes ozone to accumulate in the lower stratosphere over the course of the winter. Since the large-scale waves are not present in the summer, the poleward and downward circulation is significantly weakened, and ozone amounts which have built up during winter begin to decrease due both to transport into the troposphere and to photochemistry. The exchange of mass between the troposphere and the stratosphere is the focus of considerable current research. Stratosphere-troposphere exchange is important for the budget of ozone in the lower stratosphere as well as the ozone budget in the troposphere. Upward transport occurs in the tropics, but the exact mechanism controlling the transport is not clear. Current research is focussing on the role of subvisible cirrus and the radiative impact of infrared heating of subvisible cirrus. Downward transport takes place in midlatitudes through jet stream folds—but the frequency and amount of mass irreversibly moving through these folds is still not understood.

Aerosols and Polar Stratospheric Clouds

It is now known that knowledge of stratospheric aerosols and PSCs is very important to our understanding of stratospheric ozone. The surfaces of aerosols and PSCs are sites for heterogeneous reactions which can convert chlorine from reservoir to radical forms. Likewise, radical nitrogen forms can be sequestered as nitric acid to shift the chemical loss process.

Aerosols

The long-term stratospheric aerosol record reveals at least three components: episodic volcanic enhancements, PSCs and clouds just above the tropical tropopause, and a background aerosol level. At normal stratospheric temperatures, aerosols are most likely super-cooled solution droplets of H_2SO4-H_2O, with an acid weight fraction of 55 to 80per cent. The primary source of stratospheric aerosols is volcanic eruptions that are strong enough to inject SO_2 buoyantly into the stratosphere.

Aerosol sizes range from hundredths of a micrometre to several micrometres. Although there is some variability, especially just after a volcanic eruption, a log-normal size distribution of spherical particles appears to aptly describe the aerosol. Just after an eruption, the size distribution becomes bimodal, and some particles are non-spherical because of the addition of crustal material. After an eruption, the SO_2 is converted to H_2SO4, which condenses to form stratospheric sulfuric acid aerosols, with a time scale of about 30 days. Subsequently, aerosol loading decreases due to a combination of sedimentation, subsidence, and exchange through tropopause folds. The loading decreases with an e-folding time of 9-to-12 months, although this appears quite variable with altitude and latitude.

The net effect of this post-volcanic dispersion and natural cleansing is a greatly enhanced aerosol concentration in the upper troposphere after a major eruption, especially poleward of about 30° latitude. Except immediately after an eruption, stratospheric aerosol droplets tend to be concentrated into 3 distinct latitudinal bands—one over the equatorial region and the other over each high-latitude region, 50° to 90° N and S. Following a low-latitude eruption, aerosol is dispersed into both hemispheres, whereas following a mid-to-high-latitude eruption, aerosols tend to stay primarily in the hemisphere of the eruption.

Potential sources of a background aerosol component include carbonyl sulfide from the oceans, low-level SO_2 emissions from volcanoes, and various anthropogenic sources, including industrial and aircraft emissions. Also, it is not clear whether there is an upward trend in this background aerosol, as has been hypothesized and linked to increasing aircraft emissions, since any increase may be due to incomplete removal of past volcanic aerosol. Stratospheric aerosol loading in 1979 was approximately 0.5, 10^{12}g, thought to be representative of background aerosol conditions.

The present status of the aerosol is one of enhancement due to the June 1991 eruption of Pinatubo which produced on the order of 30 ´ 10^{12}g of new aerosol in the stratosphere, about 3 times that of the 1982 eruption of El Chichón. This perturbation appears to be the largest of the century, perhaps the largest since the 1883 eruption of Krakatoa. By early 1993, stratospheric loading decreased to approximately 13 Mt, about equal to the peak loading values after El Chichón. Measurements in 1995 showed that the aerosol levels were approaching background levels.

Polar Stratospheric Clouds

The interannual variability in PSC sightings has been addressed by Poole and Pitts, who analysed more than a decade of data from the spaceborne Stratospheric Aerosol Measurement II sensor. They found noticeable variability in PSC sightings in the Antarctic from year to year, even though the southern polar vortex is typically quite stable and long-lived. This variability was found to occur late in the season and can be explained qualitatively by temperature differences. Poole and Pitts found even more year-to-year variability in SAM II Arctic PSC sighting probabilities.

This was expected since the characteristics and longevity of the northern polar vortex vary greatly from one year to the next. The year-to-year variability in Arctic sighting probabilities can also be explained qualitatively by differences in temperature, *e.g.*, zonal mean lower stratospheric temperatures in February 1988 were as much as 20 K colder than those one year earlier.

Solar Ultraviolet and Energetic Particles

Since ozone formation is fundamentally linked to the levels of ultraviolet radiation reaching the Earth, natural variations in that radiation must be understood in order to detect trends. The ultraviolet comprises only one-to-two per cent of the total solar radiation, but it displays considerably more variation than the longer wavelength visible radiation.

For example, from 1986 to 1990 the solar UV increased with onset of the 11-year solar cycle and resulted in an increase of global total ozone of almost 2per cent. This natural increase in ozone is comparable to the suspected anthropogenic decrease and needs to be understood in order to totally separate the anthropogenic decrease from this natural change. Studies of total ozone trends typically subtract solar cycle and other natural changes from the total ozone record in trend resolution.

Thus, more-quantitative knowledge of this natural solar-cycle-induced total ozone change would be especially valuable. Changes in energetic particle flux from the sun penetrate into the middle atmosphere and may also drive the natural ozone variations. A series of solar flares in 1989 spewed solar particles into the Earth's polar cap regions and led to polar ozone depletion. Further studies related to the very large solar particle events of October 1989 have predicted ozone depletions lasting for several months after the SPEs.

Although SPEs of this magnitude occur infrequently, they need to be understood more completely to be able to separate natural from anthropogenic ozone effects. Relativistic electron precipitations have been predicted to contribute substantially to the odd nitrogen budget of the stratosphere and, therefore, have been predicted to play a large role in controlling ozone in this region.

Another investigation has failed to find any REP-caused ozone depletion. Further work determined that REPs in May 1992, the largest measured relativistic electron flux precipitating in the atmosphere between October 1991 and July 1994, added only about 0.5 to 1per cent of the global annual source of odd nitrogen to the stratosphere and mesosphere. The actual importance of REPs in regulating ozone is thus not well understood nor characterized, and further work on REPs is required to thoroughly determine their importance regarding modulation of stratospheric ozone.

STEADY STATE APPROXIMATION IN CHEMICAL KINETICS

The steady state approximation, occasionally called the stationary-state approximation, involves setting the rate of change of a reaction intermediate in a reaction mechanism equal to zero. It is important to note that steady state approximation does not assume the reaction intermediate concentration to be constant, it assumes that the variation in the concentration of the intermediate is almost zero: the concentration of the intermediate is very low, so even a big relative variation in its concentration is small, if considered quantitatively.

Its use facilitates the resolution of the differential equations that arise from rate equations, which lack an analytical solution for most mechanisms beyond the most simple ones.

The steady state approximation is applied, for example in Michaelis-Menten kinetics. As an example, the steady state approximation will be applied to two consecutive, irreversible, homogeneous first order reactions in a closed system. This model corresponds, for example, to a series of nuclear decompositions like,

Validity

The analytical and approximated solutions should now be compared in order to decide when it is valid to use the steady state approximation. The analytical solution transforms into the approximate one whenbecause then and Therefore it is valid to apply the steady state approximation only if the second reaction is much faster than the first one because that means that the intermediate forms slowly and reacts readily so its concentration stays low.

The graphs show concentrations of A, B and C in two cases, calculated from the analytical solution:

- When the first reaction is faster it is not valid to assume that the variation of [B] is very small, because [B] is neither low or close to constant: first A transforms into B rapidly and B accumulates because it disappears slowly. As the concentration of A decreases its rate of transformation decreases, at the same time the rate of reaction of B into C increases as more B is formed, so a maximum is reached when From then on the concentration of B decreases.
- When the second reaction is faster, after a short induction period, concentration of B remains low because its rate of formation and disappearance are almost equal and the steady state approximation can be used.

The equilibrium approximation can be used sometimes in chemical kinetics to yield similar results as the steady state approximation: it consists in assuming that the intermediate is at chemical equilibrium.

Normally the requirements for applying the steady state approximation are laxer:

The concentration of the intermediate is only needed to be low and more or less constant but it is not needed to be at equilibrium, which is usually difficult to prove and involves heavier assumptions.

ORGANIC CONTAMINANTS IN SOIL

Identifying the sources of organic contaminants in the environment is of key importance to our understanding of pollution patterns and for making decisions concerning site remediation. In this study, soil/sediment samples from point sources of pollution were analysed for selected organic contaminants. These samples were used to determine the actual organic contaminants representative of that particular source.

The use of faecal sterols can indicate the source of faecal matter because they vary in type and quantity in the faeces of humans and animals and the microbiota in their digestive tract. The data sets obtained were subjected to various chemometric techniques including Cluster Analysis, Discriminant Analysis and Principal Component Analysis. Environmental data normally contain huge amounts of concentration values of chemicals, spread at distant geographical sites and during different time periods.

Moreover, the content of chemicals is also estimated at different environmental compartments. All these data values are difficult to cope and evaluate in a simple and fast way using simple univariate statistical tools due to their large number and due to their multivariate correlation. In order to discover relevant patterns within large multivariate data sets, the application of chemometric methods based on statistical multivariate data analysis is proposed.

Chemometric methods have often been used in exploratory data analysis tools for classification of samples or sampling stations and identification of pollution sources. In many other cases, the exploratory data analysis results will serve to achieve an insight into, *e.g.*, the contamination situation of a certain location and to make a plan for remediation or to prepare more focused sampling plans. Principal components analysis is an exploratory, multivariate, statistical technique that can be used to examine data variability. The principal components are ordered in such a way that the first PC explains most of the variance in the data, and each subsequent one accounts for the largest proportion of variability that has not been accounted for by its predecessors. Although the number of PCs equals the number of independent original variables, generally, most of the variation in the data set can be explained by the first few principal components that can be used to represent the original observations.

Multivariate techniques can consider a number of factors, which control data variability simultaneously and therefore offer significant advantages over univariate techniques. The pressurized liquid extraction technique was used to achieve faster extraction of organic contaminants from soil/sediment samples.

PHYTOREMEDIATION

Phytoremediation is a bioremediation process that uses various types of plants to remove, transfer, stabilize, and/or destroy contaminants in the soil and groundwater. There are several different types of phytoremediation mechanisms.

These are:

- Rhizosphere biodegradation. In this process, the plant releases natural substances through its roots, supplying nutrients to microorganisms in the soil. The microorganisms enhance biological degradation.
- Phyto-stabilization. In this process, chemical compounds produced by the plant immobilize contaminants, rather than degrade them.
- Phyto-accumulation. In this process, plant roots sorb the contaminants along with other nutrients and water. The contaminant mass is not destroyed but ends up in the plant shoots and leaves. This method is used primarily for wastes containing metals. At one demonstration site, water-soluble metals are taken up by plant species selected for their ability to take up large quantities of lead. The metals are stored in the plant's aerial shoots, which are harvested and either smelted for potential metal recycling/recovery or are disposed of as a hazardous waste. As a general rule, readily bioavailable metals for plant uptake include cadmium, nickel, zinc, arsenic, selenium, and copper. Moderately bioavailable metals are cobalt, manganese, and iron. Lead, chromium, and uranium are not very bioavailable. Lead can be made much more bioavailable by the addition of chelating agents to soils. Similarly, the availability of uranium and radio-cesium 137 can be enhanced using citric acid and ammonium nitrate, respectively.
- Hydroponic Systems for Treating Water Streams. Rhizofiltration is similar to phyto-accumulation, but the plants used for cleanup are raised in greenhouses with their roots in water. This system can be used for ex-situ groundwater treatment. That is, groundwater is pumped to the surface to irrigate these plants. Typically hydroponic systems utilize an artificial soil medium, such as sand mixed with perlite or vermiculite. As the roots become saturated with contaminants, they are harvested and disposed of.
- Phyto-volatilization. In this process, plants take up water containing organic contaminants and release the contaminants into the air through their leaves.
- Phyto-degradation. In this process, plants actually metabolize and destroy contaminants within plant tissues.
- Hydraulic Control. In this process, trees indirectly remediate by controlling groundwater movement. Trees act as natural pumps when their roots reach down towards the water table and establish a dense root mass that takes up large quantities of water. A poplar tree, for example, pulls out of the ground 30 gallons of water per day, and a cottonwood can absorb up to 350 gallons per day.

The plants most used and studied are poplar trees. The U.S. Air Force has used poplar trees to contain trichloroethylene in groundwater. In Iowa, EPA demonstrated that poplar trees acted as natural pumps to keep toxic herbicides, pesticides, and fertilizers out of the streams and groundwater. The US Army Corps of Engineers has experimented with wetland plants to destroy explosive compounds in the soil and groundwater.

Submersed and floating-leafed species decreased trinitrotoluene to 5per cent of original concentration. Submersed plants were able to decrease Royal Demolition Explosive levels by 40per cent, and when microbial degradation was added, RDX decreased by 80per cent. Sunflowers, using rhizofiltration, were used successfully to remove radioactive contaminants from pond water in a test at Chernobyl, Ukraine.

Limitations and Concerns

The toxicity and bioavailability of biodegradation products is not always known. Degradation by-products may be mobilized in groundwater or bio-accumulated in animals. Additional research is needed to determine the fate of various compounds in the plant metabolic cycle to ensure that plant droppings and products do not contribute toxic or harmful chemicals into the food chain.

Scientists need to establish whether contaminants that collect in the leaves and wood of trees are released when the leaves fall in the autumn or when firewood or mulch from the trees is used. Disposal of harvested plants can be a problem if they contain high levels of heavy metals. The depth of the contaminants limits treatment. The treatment zone is determined by plant root depth. In most cases, it is limited to shallow soils, streams, and groundwater.

Pumping the water out of the ground and using it to irrigate plantations of trees may treat contaminated groundwater that is too deep to be reached by plant roots. Where practical, deep tilling, to bring heavy metals that may have moved downward in the soil closer to the roots, may be necessary.

Generally, the use of phytoremediation is limited to sites with lower contaminant concentrations and contamination in shallow soils, streams, and groundwater. However, researchers are finding that the use of trees allows them to treat deeper contamination because tree roots penetrate more deeply into the ground. The success of phytoremediation may be seasonal, depending on location. Other climatic factors will also influence its effectiveness. The success of remediation depends in establishing a selected plant community.

Introducing new plant species can have widespread ecological ramifications. It should be studied beforehand and monitored. Additionally, the establishment of the plants may require several seasons of irrigation. It is important to consider extra mobilization of contaminants in the soil and groundwater during this start-up period. If contaminant concentrations are too high, plants may die. Some phytoremediation transfers contamination across media. Phytoremediation is not effective for strongly sorbed contaminants such as polychlorinated biphenyls. Phytoremediation requires a large surface area of land for remediation.

Applicability

Phytoremediation is used for the remediation of metals, radionuclides, pesticides, explosives, fuels, volatile organic compounds and semi-volatile organic compounds. Research is underway to understand the role of phytoremediation to remediate perchlorate, a contaminant that has been shown to be persistent in surface and groundwater systems. It may be used to cleanup contaminants found in soil and groundwater. For radioactive substances, chelating agents are sometimes used to make the contaminants amenable to plant uptake.

5

Stereoisomerism Chemistry in Organic Compounds

STEREOISOMERISM

The isomerism which arises due to different spatial arrangements of atoms or groups is called stereo isomerism.

It is broadly divided into:

- Geometrical isomerism,
- Optical isomerism,
- Conformational isomerism.

The stereo isomers have same structural formula *i.e.*, same connectivity order between atoms. However the atoms or groups are arranged differently in space.

REACTION STOICHIOMETRY

The idea that a chemical equation for a reaction is essentially a recipe. Our purpose in doing that was to ground the rather abstract notions of chemical formulas and equations to something concrete, with which almost everyone has experience. All of the important quantitative aspects of chemical equations—collectively known as reaction stoichiometry—can be understood in terms of the recipe idea. It is instructive to return to the fruit salad recipe for our first example.

Example: You intend to make 20 fruit salads for a large dinner party according to the following recipe:

$$1 \text{ apple} + 5 \text{ oranges} + 10 \text{ grapes} \rightarrow 1 \text{ fruit salad}$$

You go to the store to purchase starting materials, and come home with the 10 pounds of apples, 15 pounds of oranges, and 1 pound of grapes. Are these materials sufficient to produce 20 fruit salads?

Solution: We have a problem at the outset: the equation is expressed in terms of numbers of apples, oranges, and grapes; however, we know only the total amounts of each fruit by mass. Clearly we need more information, specifically, the average mass of an apple, an orange, and a grape. Suppose that we can weigh a typical apple, orange, and grape on a simple kitchen scale. Suppose further that the apples, oranges, and grapes are of unusually uniform size, so that the weight of one apple faithfully represents the weight of any apple. Our weighings give the following results:

1 apple-weight = 0.25 lb/apple

1 orange-weight = 0.33 lb/orange

1 grape-weight = 0.00033 lb/grape

These unit weights enable us to convert grocery store masses into numbers of apples, oranges, and grapes:

Number of apples = 10 lbs/(0.25 lb/apple) = 40 apples

Number of oranges = 10 lbs/(0.33 lb/orange)
= 30 oranges

Number of grapes = 1 lb/(0.00033 lb/grape)
= 3000 grapes

Now we are in a position to apply the recipe, which tells us that for each apple used, we must use 5 oranges and 10 grapes. To use 40 apples would require 200 oranges and 400 grapes. We have more than enough grapes to use up all of the apples, but we do not have enough oranges. We can conclude that the number of fruit salads will be limited by the number of oranges that we have.

Because we need 5 oranges per fruit salad, we can make only 6 fruit salads from 30 oranges. This will require 6 apples and 6 × 10 = 60 grapes. We will need to return to the store to get quite a few more oranges. Specifically, we will require enough for 14 more fruit salads, or a total of 14 × 5 = 70 oranges. Since each orange weighs a third of a pound, we need somewhat over 20 pounds more of oranges.

From this example we learn a few important ideas. First, the recipe is expressed in numbers of units of each ingredient, not in terms of their masses.

Similarly a chemical equation is expressed in terms of numbers of atoms, molecules, or formula units, not in terms of their masses. Second, the amounts of our fruit salad ingredients are expressed as masses when we buy them at the store. Similarly, when we buy chemical substances, the amounts are expressed as masses. Third, to apply the recipe, we must convert the masses to numbers of things using the mass per thing.

Having done this, we figure out how many fruit salads it is possible to make with the given amounts of ingredients. We must do the same thing with a chemical equation. Finally, one of the ingredients is found to limit the number of fruit salads that can be made. We will find that in many cases, one of the reactants limits the amount of product that can be formed in a chemical reaction. This will be the reactant that is used up first; it is called the limiting reactant. Let's now see how the recipe analogy helps us in doing stoichiometric calculations for chemical reactions.

Example: What mass of pure silicon can be obtained by reduction of $SiCl_4$ with 10.0 g of magnesium metal at elevated temperature? (This reaction is used in the process of extracting silicon from naturally occurring ores. Silicon is used to make semiconductors.) Assume that a large supply of $SiCl_4$ is on hand.

Solution: We require first a balanced chemical equation (recipe), which follows from the description of the process. We are told that silicon is produced by reaction of Mg and $SiCl_4$. We put Mg and $SiCl_4$ as reactants, and Si as a product. The reaction is carried out at elevated temperature, so magnesium is present as a liquid.

$$SiCl_4(l) + Mg(l) \rightarrow Si(s)$$

The equation is not and indeed can not be balanced as it is, because there are no magnesium and chlorine atoms in the products. We must first propose at least one other product for the reaction in order to balance it. The simplest choice is magnesium chloride, which can be shown experimentally to be correct. Thus

$$SiCl_4(l) + Mg(l) \rightarrow Si(s) + MgCl_2(s)$$

Next we must balance the equation to conform with the law of mass conservation. This is accomplished by putting coefficient 2 before $MgCl_2$ to balance chlorine; then coefficient 2 before Mg. Balancing is simple in this case.

$$SiCl_4(l) + 2Mg(l) \rightarrow Si(s) + 2MgCl_2(s)$$

At this point, we are ready to approach the problem.

The equation gives us relationships between the numbers of moles of reactants and products. Keeping in mind the recipe analogy, the following roadmap results:

Convert the given mass of Mg to moles Mg.

Use the coefficients in the equation to relate moles Mg (reactant) to moles Si (product).

Convert moles Si to grams of Si.

Mass Mg $\rightarrow$ moles Mg $\rightarrow$ moles Si $\rightarrow$ mass Si.

With the exception of the last step, this is exactly what we did in the recipe calculation in Example .

$$\text{Moles Mg} = 10.0 \text{ g Mg}/(24.305 \text{ g Mg/mole}) = 0.4114 \text{ moles Mg}$$

$$\text{Moles Si} = 0.4114 \text{ moles Mg} \times (1 \text{ mole Si}/2 \text{ moles Mg}) = 0.2057 \text{ moles Si}$$

The stoichiometric factor 1 mole of Si per 2 moles of Mg follows directly from the chemical equation, which tells us that 2 moles of magnesium is required to produce, and is therefore equivalent to, 1 mole of Si. Finally, we convert to mass of Si:

$$\text{g Si} = 0.2057 \text{ moles Si} \times (28.086 \text{ g Si/mole}) = 5.78 \text{ g Si}$$

We can therefore expect to obtain 5.78 g of pure silicon from 10 g of magnesium. Note that the amount of $SiCl_4$ does not play a role in the problem because we are told to assume that we have more than enough of it. This is equivalent to saying that Mg is the limiting reactant.

The situation can be made more realistic by specifying starting amounts for both $SiCl_4$ and Mg. This is done in the following example.

Example: How much pure silicon can be produced by reaction of 38.0 g of Mg and 26.7 g of $SiCl_4$(l)?

Solution: In order to use the chemical equation developed in the preceding example, we must determine the number of moles of each reactant:

moles Mg = 38.0 g Mg/(24.305 g/mole Mg) = 1.563 moles Mg

moles $SiCl_4$ = 26.7 g $SiCl_4$/(169.90 g/mole $SiCl_4$)
= 0.1572 moles $SiCl_4$

The chemical equation indicates that we require 2 moles of Mg for each mole of $SiCl_4$ that we desire to react; i.e., at the least, we must have twice as much Mg in moles as we have $SiCl_4$. The numbers above show that we have more than enough Mg to react with all of the $SiCl_4$ available. We conclude that $SiCl_4$ is the limiting reagent. The amount of Si product will be limited by the amount of $SiCl_4$ available.

moles Si = moles $SiCl_4$ * (1 mole Si)/(1 mole $SiCl_4$)

moles Si = 0.1572 moles

Now convert back to mass:

mass Si = 0.1572 moles Si * 28.085 g/mole = 4.41 g Si

ELECTRON TRANSFER REACTIONS

Before considering additional examples of reaction stoichiometry, we will take some time to extend our knowledge of electron transfer reactions. The reaction between Mg and $SiCl_4$ used in the preceding two examples is an electron transfer reaction. Magnesium begins as an element with uncharged atoms, but finishes as an ionic compound in which it has a charge of 2+. It has lost electrons, and has thus been oxidized.

Silicon begins as a component of $SiCl_4$, in which it has a charge of 4+; and ends as elemental silicon, with zero charge. Si has been reduced because it has gained electrons. For this reaction, the charge changes are fairly easy to see. In the reaction below, however, it may not be obvious whether or not electrons are transferred, and it is certainly not obvious how many have been transferred:

$$Cr_2O_7^{2-} + C_2H_6O \rightarrow Cr^{3+} + C_2H_4O_2$$

There is a simple set of rules that can be used to readily determine the charge on an atom in any compound.

The charge determined according to these rules is called the oxidation state of the atom.

This is generally different from the formal charge, because the rules for assigning electrons are different. Oxidation state, like formal charge, is an electron bookkeeping procedure. The oxidation state of an element in a compound can be determined using the following simple rules for oxidation state.

The Assignment of Oxidation State

- An atom of an element in its uncombined form has oxidation state zero. Thus, each oxygen atom in O_2 has oxidation state (O.S.) = 0. Each atom in a sample of iron metal, Fe(s), has O.S. = 0.
- For simple monatomic ions the oxidation state is the charge on the ion. Thus sodium is present in NaCl as Na+, and has O.S. = 1+; chlorine is present as Cl^- and has O.S. = –1.
- In compounds and polyatomic ions, oxygen has oxidation state –2. Because oxygen is the second-most electronegative element, it either gains two electrons to form the O^{2-} ion, or has the lion's share of electrons in a covalent compound. In Na_2O, the O.S. of oxygen is -2, and that of sodium is 1+. In SO_3, the oxidation state of oxygen is –2, and that of sulfur is 6+.
  There are two exceptions to this rule. The first is hydrogen peroxide, in which the two oxygen atoms are bonded together and must be considered to equally share the electrons of that bond. The O.S. of oxygen is –1 in peroxides. The other is the compound OF_2, in which the O.S. of oxygen is 2+.
- In compounds hydrogen has oxidation state 1+. The most common ion of hydrogen is the proton, H^+. With only a few exceptions, hydrogen may be considered to have O.S. = 1+ in its compounds.
- In most compounds, the oxidation states of elements not mentioned explicitly in these rules can be calculated from the overall charge on the molecule or ion, and the known oxidation states of the remaining elements.

Let's examine a few examples.

Example: Determine the oxidation state of each element in each molecule or ion:

$$NaBr$$
$$H_2SO_4$$
$$Na_2Cr_2O_7$$
$$HClO_4$$
$$C_2H_6O$$

Solution: NaBr. This is an ionic compound. The ions have their normal charges. Consequently,

$$O.S.(Na) = 1^+; O.S. (Br) = 1^-.$$

H_2SO_4. This is a covalent compound. We can use the rules for the oxidation states of oxygen and hydrogen to determine the oxidation state of sulfur. Each oxygen atom has an oxidation state of 2-. The total for four oxygens is then 8-. Each of the two hydrogens has O.S. 1+, for a total of 2+. The compound is electrically neutral overall, so the oxidation state of sulfur must be such that its sum with the total oxidation states of hydrogen and oxygen is zero:

$$2 + O.S.(S) - 8 = 0$$

This gives O.S.(S) = 6+. We can use the following schematic approach to determine unknown oxidation states.

- Write the formula: H_2SO_4
- Under the formula, write the oxidation state of each atom for which the oxidation state is known:

  H_2 S O_4
  1+ 2-
- to determine the total charge of each element, multiply the per-atom oxidation state by the number of atoms, and write the results above the formula:

  2+ 8-
  H_2 S O_4
  1+ 2-
- Determine the total charge of the element of unknown oxidation state using the known element charges and the known overall charge on the molecule or ion, and write the result above the element.

  2+ 6+ 8-
  H_2 S O_4
  1+ 2-
- Divide the total charge of the element of unknown O.S. by the number of atoms of the element, and write the result under the formula. This is the oxidation state of the element.

  2+ 6+ 8-
  H_2 S O_4
  1+ 6+ 2-

The oxidation state of sulfur in H_2SO_4 is 6+.

Using a similar approach gives 6+ for Cr in $Na_2Cr_2O_7$ and 7+ for Cl in $HClO_4$. We work through C_2H_6O using the shorthand system developed above to obtain 2- as the oxidation state for carbon:

4- 6+ 2-
C_2 H_6 O
2- 1+ 2-

We are now equipped to examine reaction for changes in oxidation states. Using our newly developed methods, we conclude that the oxidation state of chromium decreases from 6+ to 3+ as a result of the reaction, while that of carbon increases from 2- to 0. The reaction is clearly electron transfer.

$$: Cr_2O_7^{2-} + C_2H_6O \rightarrow Cr^{3+} + C_2H_4O_2$$

To develop a method for balancing the equations for redox reactions like that in reaction, we take advantage of the fact that in all electron transfer processes, the total number of electrons lost by the oxidized species is exactly the same as the total number of electrons gained by the reduced species. This is seen not to be true in reaction, because the two chromium atoms of $Cr_2O_7^{2-}$ gain a total of six electrons, whereas the two carbon atoms of C_2H_6O lose a total of only four. We have some balancing to do.

Balancing Electron Transfer Equations by the Method of Half Reactions

The method of half reactions is a systematic approach to the balancing of electron transfer reactions. We treat it as a series of steps.

Step: Divide the reaction into two half-reactions, one involving the oxidized element, the other involving the reduced element:

$$Cr_2O_7^{2-} \rightarrow Cr^{3+}$$
$$C_2H_6O \rightarrow C_2H_4O_2$$

Step: Balance each half reaction separately, first balancing atoms of the element that loses or gains electrons; then balancing oxygen and hydrogen; and finally, balancing electrical charge.

- Redox element balance. In the first half reaction, the element undergoing electron gain is chromium. To balance chromium, put coefficient 2 before Cr^{3+}. The second half reaction is already balanced in carbon.

$$Cr_2O_7^{2-} \rightarrow 2Cr^{3+}$$
$$C_2H_6O \rightarrow C_2H_4O_2$$

- Oxygen/hydrogen balance. Most redox reactions are carried out in aqueous solution, often in acidic solution, in which there are surpluses of water molecules and protons. Half reactions can be balanced for oxygen and hydrogen by adding H_2O and H^+ as needed. The chromium half-reaction is balanced by adding, first, 7 molecules of water to the right to balance oxygen; then 14 protons to the left to balance hydrogen. The carbon half reaction is balanced by adding one water molecule to the left, and 4 protons to the right.

$$Cr_2O_7^{2-} + 14H^+ \rightarrow 2Cr^{3+} + 7H_2O$$
$$C_2H_6O + H_2O \rightarrow C_2H_4O_2 + 4H^+$$

- Charge balance. Each half reaction is now balanced for charge by adding electrons to one side or the other so that the total charge on both sides of the half reaction is the same. For the chromium reaction, total charge after completion of step b is 12+ on the left, and 6+ on the right. Six electrons, each with a charge of 1-, are added to the left side to equalize charge. Similarly in the carbon half reaction, four electrons must be added to the right:

$$Cr_2O_7^{2-} + 14H^+ + 6e \rightarrow 2Cr^{3+} + 7H_2O$$
$$C_2H_6O + H_2O \rightarrow C_2H_4O_2 + 4H^+ + 4e$$

Step: Equalize electrons in the two half reactions. It is at this point that we make use of the key fact about redox reactions: that every electron produced in oxidation must be consumed in reduction. We equalize the electrons transferred if we multiply the chromium half reaction by two, and the carbon half reaction by 3:

$$2 * (Cr_2O_7^{2-} + 14H^+ + 6e \rightarrow 2Cr^{3+} + 7H_2O)$$
$$3 * (C_2H_6O + H_2O \rightarrow C_2H_4O_2 + 4H^+ + 4e)$$

Step: Recombine the balanced half reactions to give a balanced overall electron transfer reaction.

$$2Cr_2O_7^{2-} + 28H^+ + 3C_2H_6O + 3H_2O \rightarrow 4\,Cr^{3+}$$
$$+ 14H_2O + 3C_2H_4O_2 + 12H^+$$

Step: Simplify and check. Here we cancel common species, reduce coefficients to lowest whole numbers, and check to make sure that the same numbers of atoms of all types, and charges, occur on both sides. We obtain

$$2Cr_2O_7^{2-} + 16H^+ + 3C_2H_6O \rightarrow 4\,Cr^{3+} + 11H_2O + 3C_2H_4O_2$$

This checks for balance, with 4 Cr atoms, 17 O atoms, 34 H atoms, and 12+ charge on each side.

Example: What mass of acetic acid, $C_2H_4O_2$, is obtained from reaction of 146.2 g of potassium dichromate, $K_2Cr_2O_7$ with 84.6 g of ethyl alcohol, C_2H_6O? Discuss changes in molecular stereochemistry when C_2H_6O is converted to $C_2H_4O_2$.

Solution: The balanced equation is, First, we calculate moles of each reactant:

moles $K_2Cr_2O_7$ = 146.2 g/(294.181 g/mole) = 0.497 moles

moles C_2H_6O = 84.6 g/(46.069 g/mole) = 1.84 moles

The equation tells us that we need 3 moles of ethyl alcohol for each 2 moles of dichromate anion reacted. Because 1 mole of potassium dichromate contains one mole of dichromate anion, moles $Cr_2O_7^{2-}$ = 0.497. Using the 3 to 2 ratio

moles C_2H_6O = 0.497 moles $Cr_2O_7^{2-}$ * (3 moles ethanol/2 moles dichromate) = 0.746 moles.

This is all we need to use up all of the dichromate. Clearly we have available much more ethanol than we need. Dichromate is the limiting reactant and determines the quantity of acetic acid obtained:

moles $C_2H_4O_2$ = moles $Cr_2O_7^{2-}$ * (3 moles acetic acid/2 moles dichromate) = 0.746 moles

mass $C_2H_4O_2$ = 0.746 moles * 60.052 g/mole = 44.8 g of acetic acid

The NMR spectrum of ethanol shows an upfield triplet (intensity 3), a midfield quartet (intensity 2) and a midfield singlet (intensity 1), consistent with the structural formula. The spectrum for acetic acid shows an upfield singlet of intensity 3, and a far downfield singlet of intensity 1, consistent with the structural formula. In ethanol, both carbon atoms are surrounded by four electron groups, tetrahedrally distributed, and each group is used to attach an atom or group of atoms to carbon.

The stereochemistry is therefore tetrahedral at both carbon atoms. The oxygen atom also has four electron groups distributed tetrahedrally, but only two are used to attach atoms. The stereochemisty at oxygen is bent. The CH_3 carbon of acetic acid has the same stereochemistry as the CH_3 carbon of ethanol: tetrahedral. However, the second carbon atom is surrounded by only three electron groups, distributed in a trigonal planar arrangement. Since all three groups are used to attach atoms, the shape at this carbon is also trigonal planar. The stereochemistry at the singly bonded oxygen is bent.

ADDITIONAL ASPECTS OF REACTION STOICHIOMETRY

It is usually helpful in learning new concepts to have a visual representation of the çoncept.

Example.provides such a representation for a double displacement reaction.

Example: $AgNO_3$ and NaCl react in solution via double displacement according to

$$AgNO_3(aq) + NaCl(aq) \rightarrow NaNO_3(aq) + AgCl(s)$$

We react a fixed mass (1.500g) of $AgNO_3$ with varying masses of NaCl and weigh the amount of AgCl obtained as a solid in each reaction. We then plot mass AgCl obtained versus mass NaCl used.

- Why does the plot increase linearly in region 1?
- Why is the plot horizontal in region 2?
- What is the significance of the point where the region 1 and 2 lines meet (the breakpoint)?
- Suppose we were to plot mass NaCl/mass $AgNO_3$ on the x-axis. At what value of this quantity would the breakpoint occur?
- Suppose we were to plot moles NaCl/moles $AgNO_3$ on the x-axis. At what value of this quantity would the breakpoint occur?
- If we were to use KCl rather than NaCl, what would the various plots look like?

Solution:

- In this region, the amount of product goes up linearly with the amount of NaCl added, because there is sufficient $AgNO_3$ in solution to react with all of the added NaCl. Reaction runs until the limiting reagent, NaCl, is gone. Region 1 defines the range in which NaCl is the limiting reagent.
- In region 2, the same amount of product is obtained no matter how much NaCl is added. Now the amount of product is determined by the fixed amount of $AgNO_3$ present in the solution. The same amount

of product is always obtained because the amount of $AgNO_3$ is always the same. In region 2, $AgNO_3$ is the limiting reagent.

- The two lines intersect at the point where stoichiometrically equivalent amounts of NaCl and $AgNO_3$ are present. This gives the amount of NaCl that will exactly react with 1.500 g $AgNO_3$. When reaction is finished, both reactants will be gone. Neither reactant is limiting.
- What mass of NaCl is required to just use up 1.500 g $AgNO_3$? The two substances react in a 1:1 mole ratio. We convert 1.500 g $AgNO_3$ to moles, then calculate what mass of NaCl gives this same number of moles of NaCl:
  Moles $AgNO_3$ = 1.500g /(169.872 g/mole)
  = 8.830×10^{-3} moles
  Mass NaCl = 8.830×10^{-3} moles * 58.443 g/mole
  = 0.516 g
  The breakpoint would occur at a value of 0.516/1.500
  = 0.344.
- The breakpoint would occur at a mole ratio of 1, based on the stoichiometry of the chemical reaction.
- It would be necessary to add a larger mass of KCl to reach the breakpoint because it has a larger formula weight. The mass AgCl versus mass KCl plot would have a region 1 with smaller slope than region 1 for NaCl. However, the same limiting mass of AgCl would be reached. The plot of moles KCl/mole $AgNO_3$ would coincide with the mole/mole plot for NaCl, because the stoichiometry is the same for the two reactions.

The power of the mole concept is illustrated in the next example.

Example: In order to obtain copper metal, a sample of the ore, $CuFeS_2$, is taken through a process consisting of several steps, as detailed below:

$$2CuFeS_2(s) + 3O_2(g) \rightarrow 2CuS(s) + 2FeO(s) + 2SO_2(g)$$

$$2CuS(s) \rightarrow Cu_2S(l) + S(s)$$

$$Cu_2S(l) + O_2(g) \rightarrow 2Cu(l) + SO_2(g)$$

The copper is then solidified and refined by electrochemical methods.

What mass of copper can be obtained from 150.0 g of ore, $CuFeS_2$, by the above sequence of reactions, assuming each reaction runs at 100per cent efficiency?

Solution: It might at first seem necessary to work through the reactions one at a time to finally arrive at a mass of Cu. This is unnecessary, however. Seemingly complex problems of this type are really quite simple, if we take advantage of the conservation of mass. If we assume that all of the copper containing material produced in one step is used as input for the next, we can expect by the law of conservation of mass that all of the copper contained in the ore ends up as pure copper metal after the 3-step process. Thus moles Cu = moles Cu in $CuFeS_2$ = moles $CuFeS_2$ (since 1 mole of ore contains 1 mole of copper).

The calculation is simple:

Moles ore = 150.0 g ore/(183.51 g ore/mole ore) = 0.8174 moles

Moles Cu in ore = 0.8174 moles

Moles Cu(s) recovered = 0.8174 moles g Cu(s) = 0.8174 moles Cu * 63.5 g Cu/mole = 51.90 g Cu

It is interesting to note that the detailed stoichiometries of the three reactions are not used at all in solving this problem. You might want to try to classify each of the reactions as electron transfer, proton transfer, or double displacement.

Stoichiometric calculations are crucially important in the methods used by chemists to determine the relative amounts of elements in a new compound. Elemental analysis is the chemist's name for the process by which the identities and amounts of the elements constituting a particular compound are determined. For the numerous compounds containing carbon and hydrogen, combustion analysis is useful. In this process, a precisely known mass of compound is burned in an excess of oxygen.

The carbon in the compound is completely converted to carbon dioxide, CO_2, and the hydrogen to water, H_2O. The CO_2 and H_2O are collected separately and weighed. From the masses of CO_2 and H_2O produced, the masses of carbon and hydrogen in the original mass of compound can be obtained. From this information, their relative numbers of moles can be calculated. Masses of other elements present can be determined by separate methods, leading to an empirical formula.

Example: When 0.617 grams of a compound of carbon, hydrogen, and nitrogen is burned in excess oxygen, 1.716 g CO_2 and 0.3513 g H_2O are collected. What is the empirical formula of the compound?

Solution: The masses of carbon dioxide and water can be converted to moles.

From this information, the moles and consequently the masses of carbon and hydrogen in the original mass of compound can be obtained. The mass of nitrogen is then obtained by difference. Once the masses of all elements in the original mass of compound are known, conversion to the relative numbers of moles gives the formula.

$$\text{Moles } CO_2 = 1.716 \text{ g} * (1 \text{ mole}/44.009 \text{ g}) = 3.899 \times 10^{-2} \text{ moles}$$

$$\text{Moles C} = \text{moles } CO_2 = 3.899 \times 10^{-2} \text{ moles}$$

$$\text{Mass C} = 3.899 \times 10^{-2} * (12.011 \text{ g/mole}) = 0.468 \text{ g C}$$

$$\text{moles } H_2O = 0.3513 \text{ g} * (1 \text{ mole}/18.015 \text{ g}) = 1.95 \times 10^{-2} \text{ moles}$$

$$\text{Moles H} = 2 * \text{moles } H_2O = 3.90 \times 10^{-2} \text{ moles}$$

$$\text{Mass H} = 3.90 \times 10^{-2} \text{ moles} * (1.008 \text{ g/mole}) = 0.0393 \text{ g H}$$

Now the mass of nitrogen in the original mass of compound can be obtained:
Mass N = 0.617 0.468 0.039 = 0.110 g

Moles N = 0.00785 moles

We now have

Moles C = 0.03899

Moles H = 0.0390

Moles N = 0.00785

Dividing all 3 by moles N gives the formula: C_5H_5N

Example: A 32.1 mg sample of a compound containing carbon, hydrogen, and oxygen was burned in oxygen to produce 78.4 mg of CO_2 and 32.0 mg of H_2O. ^{1}H-NMR and IR data for the compound are given below:

NMR	IR
triplet, D = 1.05 (3H)	Strong band near 1720 cm^{-1}
singlet, D = 2.13 (3H)	
quartet, D = 2.47 (2H)	

What is the structural formula for the compound?

Solution: The combustion analysis data allow us to determine empirical formula. Then the NMR and IR data can be used to deduce the arrangement of atoms—the structural formula.

millimoles CO_2 = 78.4 mg/(44.009 mg / m mole)

= 1.781 mmole CO_2

m mole C = mmole CO_2 = 1.781 mmole C

mg C = 1.781 mmole * 12.011 mg/mmole = 21.4 mg

(Here we have used the fact that the molar mass is the same in units of mg/mmole as in g/mole.)

m mole H_2O = 32.0 mg/(18.015 mg/mmole)

= 1.776 mmole H_2O

mmole H = 2 * mmole H_2O = 3.553 mmole H

mg H = 3.553 mmole * 1.008 mg/mmole = 3.58 mg

Therefore, mg O = 32.1 – 21.4 – 3.58 = 7.12 mg

Convert oxygen mass to mmoles:

mmoles O = 7.12 mg/(15.999 mg/mmole) = .445 mmole

Divide mmoles of each element by the smallest value, 0.445:

relative mmoles C = 1.781/0.445 = 4.00

relative mmoles H = 3.553/0.445 = 7.98

relative mmoles O = 0.445/0.445 = 1

The empirical formula is C_4H_8O. The infrared data suggest that a C = O group is present, so we use the NMR data to build a structure based around this group. There are two 3-proton signals in the NMR spectrum, suggestive of CH_3 groups. One CH_3 group produces a singlet signal, so there are no protons on the atom attached to the CH_3 carbon atom. This CH_3 group is probably attached to the carbon atom of the C=O group. So far, we have the following arrangement:

$$- C(= O) - CH_3$$

The second CH_3 signal is split to a triplet, suggesting that there are two protons on the neighboring atom. Consistent with this, the two proton signal is split to a quartet, suggesting coupling to a CH_3 group. It appears that the structure is as shown below:

$$H_3C - CH_2 - C(= O) - CH_3$$

This compound is called methyl-ethyl ketone.

REACTIONS INVOLVING GASES

The properties of gases extensively, learning that a very simple relationship describes the interrelationship of volume, pressure, temperature, and moles of gas. Gases are frequently reactants, products, or both in chemical reactions, and often it is convenient to determine the extent of reaction by measuring the amount of gas produced or consumed. The following example illustrates this application.

Example: 17.24 g of liquid C_6H_{14} (hexane) is enclosed with 3.80 moles of $O_2(g)$ in a cylinder fitted with a piston. The initial temperature of the mixture is 27°C,and the external (outside) pressure on the piston is 1.00 atm. The hexane and oxygen are then caused to react according to the following equation:

$$C_6H_{14}(l) + O_2(g) \rightarrow CO_2(g) + H_2O(l)$$

All of the hexane is used up. The heat produced by the reaction causes the temperature to rise to 77°C. The external pressure is maintained at 1.00 atm.

- Balance the equation
- Calculate the initial volume of reactants in the cylinder (assume that liquid hexane occupies a negligible volume.)
- Calculate the final volume of products and left-over reactants in the cylinder (assume that liquid water occupies a negligible volume.)

Solution: This problem will give us a workout on gas and stoichiometry concepts.

- The equation is balanced by placing 6 before CO_2, 7 before H_2O, and 19/2 before O_2. Fractions are then cleared via multiplication by 2. The result is

$$2C_6H_{14}(l) + 19O_2(g) \rightarrow 12CO_2(g) + 14H_2O(l)$$

- Initial volume of gaseous reactants in the cylinder. We initially have only liquid hexane and gaseous O_2 in the cylinder. We can ignore the very small volume occupied by the hexane. We have 3.80 moles $O_2(g)$ at 27 °C and a pressure of 1.00 atm. The ideal gas law gives us the volume:

  V = nRT/P = (3.80 moles)(0.08206 L-atm/K mole)

  (300 K)/(1.00 atm) = 93.55 L

- Initial amounts of both reactants are specified; we must determine which is limiting. Oxygen is already expressed in moles, so all we need do is compute moles hexane:

moles hexane = 17.24 g/ (86.178 g/mole)

= 0.200 moles hexane

The balanced equation indicates that 19 moles of oxygen are needed for each 2 moles of hexane. For 0.200 moles hexane,

moles O_2 = 0.200 moles hexane × (19 moles O_2/2 moles hexane) = 1.900 moles oxygen

Much more than this is available; hexane is limiting. Conclusions thus far: all hexane is used up

1.900 moles O_2 is used up; therefore 1.900 moles O_2 is left over CO_2 and H_2O are formed; volume of H_2O can be ignored.

To calculate the moles of CO_2 formed is simple:

moles CO_2 = moles hexane * (12 moles CO_2/ 2 moles hexane)

= 0.200 × 6 = 1.200 moles CO_2

The total amount of gas at the end of reaction is 1.200 moles CO_2 + 1.900 moles O_2 = 3.100 moles gas. The volume can be calculated from the ideal gas law, using T = 77 + 273, and P = 1 atm:

V = nRT/P = (3.1)(0.08206)(350)/1 = 89.0 L

The piston moves slightly in during reaction.

SOLUTIONS AND CONCENTRATION

A solution is a mixture of pure substances that is homogeneous at the molecular level. That is, the molecules of the substances are intimately intermixed. Usually, one of the substances is a liquid, called the solvent. A relatively small amount of a second substance, usually a solid, is then dissolved in the liquid.

The substance that dissolves is called the solute. A solution is obtained when table sugar is dissolved in water. Water is the solvent, and sugar is the solute. Solutions are extremely important in chemistry. Most chemical reactions are carried out by dissolving the reactants in a solvent, and some substances are sold commercially as solutions because they are unstable in pure form.

The amount of solute per unit amount of solvent is called the concentration of the solution. There are several ways of expressing this, the most common of which is called the molarity. Molarity, symbol M, is defined as the number of moles of solute per liter of solution. You will be expected to use and understand molarity in the laboratory; we introduce it here to initiate this process.

Example: A solution of K_2SO_4 (potassium sulfate) in water is prepared by the following stepwise procedure:

- 26.1743 g K_2SO_4 is transferred to a 500-mL volumetric flask.
- About 250 mL distilled water is added to the flask, which is then swirled to dissolve the salt. When the salt has completely dissolved,
- More distilled water is added until the water level coincides with the mark on the neck of the flask. This mark indicates a volume of exactly 500.0 mL.
- The solution is vigorously shaken to insure uniformity.

What is the molarity of the solution?

Solution: We calculate the moles of K_2SO_4 and divide by the total volume of solution prepared:

moles K_2SO_4 = 26.1743 g/(174.253 g/mole) = 0.1502 moles

Molarity = M = 0.1502 moles solute/0.5000 L solution = 0.3004 M

Molarity is a useful concept for dealing with the stoichiometry of reactions in solution. We consider two examples here, one involving double displacement and the other involving proton transfer.

Example: 34.6 mL of the 0.3004 M K_2SO_4 solution from Example is mixed with 52.3 mL of a 0.0954 M solution of $BaCl_2$. How much $BaSO_4$(s) forms?

Solution: Aqueous potassium sulfate and barium chloride undergo double displacement in aqueous solution to produce barium sulfate as a solid and leaving potassium chloride dissolved in the water.

$$K_2SO_4(aq) + BaCl_2(aq) \rightarrow BaSO_4(s) + 2KCl(aq)$$

This equation relates moles of reactants and products, which can be obtained easily from molarities.

$$\text{Moles } K_2SO_4 = 0.0346 \text{ L} * 0.3004 \text{ moles/L}$$
$$= 10.40 \times 10^{-3} \text{ moles } K_2SO_4$$
$$\text{Moles } BaCl_2 = 0.0523 \text{ L} * 0.0954 \text{ moles/L}$$
$$= 4.99 \times 10^{-3} \text{ moles } BaCl_2$$

Because 1 mole of K_2SO_4 is required per mole of $BaCl_2$, $BaCl_2$ is the limiting reagent. It follows that 4.99×10^{-3} moles $BaSO_4$ are produced. Converting to grams,

$$\text{Mass } BaSO_4 = 4.99 \times 10^{-3} \text{ moles} * 233.386 \text{ g/mole}$$
$$= 1.165 \text{ g } BaSO_4.$$

Example: 31.05 mL of a 0.151 M solution of the acid, H_2SeO_4, is mixed with 58.74 mL of a 0.201 M solution of the base, NaOH. The acid and base react via proton transfer. Will the final solution be acidic (i.e., have an excess of H^+) or basic (i.e., have an excess of OH^-)?

Solution: The proton transfer reaction involving H_2SeO_4 and NaOH is given in reaction.

$$H_2SeO_4 + 2NaOH \rightarrow Na_2SeO_4 + 2H_2O$$

As in any reaction stoichiometry calculation, we must determine the numbers of moles of reactants that are initially present.

mmoles H_2SeO_4 = 31.05 mL * 0.151 m mole/mL
= 4.689 m moles H_2SeO_4

mmoles NaOH = 58.74 mL * 0.201 m mole/mL
= 11.81 m moles NaOH

(Here we have recognized that because a millimole is 1/1000 of a mole, and a mL is 1/1000 of a liter, molarity can be expressed either as moles/L or mmoles/mL.). The equation indicates that 2 mmoles NaOH are required to react with each mmole H_2SeO_4. There is more than enough NaOH available. We conclude that H_2SeO_4 is the limiting reactant, and there will be some NaOH left over when reaction is complete. The final solution will contain an excess of OH-; it will be basic. Although we have answered the question at this point, we can take things one step further by determining the amount of NaOH remaining after reaction. This is

Amount unreacted NaOH = total NaOH – 2 * moles H_2SeO_4
= 11.81 mmole – 2 * (4.689 mmole)
= 2.432 mmoles

This quantity of NaOH is present in a total solution volume of 31.05 + 58.74 = 89.79 mL. The molarity of NaOH remaining is thus 2.432 mmole/89.79 mL = 0.0271 M

STOICHIOMETRY OF INCOMPLETE REACTIONS

We have made an implicit and drastic assumption about all of the chemical reactions that we have worked with thus far in the text: that they proceed until the limiting reactant is completely used up. When this happens, chemists say that the reaction has “proceeded to completion.” In practice, many reactions do not proceed to completion.

In most cases this is because the reaction reaches a state of chemical equilibrium in which both reactants and products are present.

Example: 5.25 moles of nitrogen and 12.00 moles of hydrogen are mixed and allowed to react according to equation. Calculate the amount of ammonia produced, assuming that the reaction proceeds only 20per cent of the way to completion, based on the limiting reactant.

$$N_2(g) + 3H_2(g) \rightarrow 2NH_3(g)$$

Solution: We must first determine the limiting reactant, based on the given starting amounts and the reaction stoichiometry. The extent of reaction does not enter in to this determination, which is done in the usual way.

In this problem, determination of limiting reagent is simple because the amounts are given in moles. The equation requires 3 moles of H_2 for each mole of N_2 used, but only 12.00/5.25 = 2.29 moles of H_2 are available per mole of N_2. We conclude that H_2 is limiting. At this point we bring in the extent of reaction. Based on hydrogen, reaction proceeds only 20per cent. This means that 20per cent of the hydrogen initially present will be converted to ammonia, leaving 80per cent unconverted.

Moles H_2 that react = 0.20 × 12.00 = 2.40 moles

Moles H_2 left unreacted = 12.00 – 2.40 = 9.60 moles

The equation dictates that for each 3 moles of H_2 reacted, 2 moles NH_3 must form:

moles NH_3 formed = 2.40 moles H_2 *

(2 moles NH_3/3 moles H_2) = 1.60 moles NH_3.

Incomplete reactions can be nicely handled in terms of a formalism that we describe as follows.

Step: Under the chemical equation, write the initial amounts of reactants and products present:

N_2 + $3H_2$→ $2NH_3$

initial 5.25 12.00 0

Zero is written under ammonia, because there is no ammonia initially present.

Step: Under the initial amounts, write the amounts of substances that are used (reactants) and formed (products). These numbers should be negative for reactants (because their amounts decrease during reaction), positive for products (because their amounts increase during reaction), and should be in accord with the stoichiometry of the reaction.

N_2 + $3H_2$ → $2NH_3$

initial 5.25 12.00 0

reacts/forms –2.40

The number, -2.40, under H_2 is 20per cent of the initial amount; this is the amount of hydrogen that is used up. The number is negative to indicate this. What numbers should be placed under N_2 and NH_3? This is determined from reaction stoichiometry:

moles N_2 used = moles H_2 used * (1 mole N_2/3 moles H_2)

= –2.40 * (1/3) = –0.80 moles N_2

moles NH_3 formed = moles H_2 used * (2 moles NH_3/3 moles H_2) = 2.40 * (2/3) = 1.60 moles NH_3

Enter these numbers in the "reacts/forms" line:

	N_2	$+ 3H_2 \rightarrow$	$2NH_3$
initial	5.25	12.00	0
reacts/forms	–0.80	–2.40	1.6

Step: Calculate the final amounts of the three substances by adding the amount reacted or formed to the initial amount:

	N_2	$+ 3H_2 \rightarrow$	$2NH_3$
initial	5.25	12.00	0
reacts/forms	–0.80	–2.40	1.6
final	4.45	9.60	1.60

This gives us a straightforward system for dealing with incomplete reactions.

OXIDATION STATES FROM LEWIS STRUCTURES

In an earlier section we presented some simple rules for the determination of oxidation state. In this section we show that it is possible to determine oxidation states of the atoms in covalent species (molecules and ions) directly from the Lewis structure of the species, using a procedure very similar to that used to obtain formal charge. Again, the procedure is systematic. It is carried through using the thiosulfate anion as an example.

Step: Develop an acceptable Lewis structure for the species, $S_2O_3^{2-}$. For $S_2O_3^{2-}$.

Step: Assign electrons to atoms according to these rules:

- Each atom is assigned all electrons of its lone pairs
- The bond electrons are assigned to the more electronegative atom of the bond. If both atoms are of the same element, the bond electrons are divided in half.

The three oxygen atoms of $S_2O_3^{2-}$ are bonded similarly to the central sulfur atom. Each is assigned 6 electrons from lone pairs, and both electrons of its bond to S. Each oxygen is assigned 8 electrons. The terminal sulfur atom is assigned the 6 electrons of its lone pairs, and one half of the bond electrons for a total of 7. The central sulfur has no lone pairs, and is assigned only one bonding electron from the bond with the terminal sulfur.

Step: Compare the number of electrons assigned with the number of valence electrons that the atom normally has. The oxidation state is given as the difference between the normal number and the number assigned in the Lewis structure.

Subtracting: from the normal number, 6, gives 2– for the oxidation state of each oxygen atom.

Subtracting: from the normal number, 6, gives the oxidation state of the terminal sulfur atom as 1^-.

Subtracting: from the normal number, 6, gives the central sulfur an oxidation state of 6 1 = 5+.

Note that these differ from formal charges, which would be 1- for each oxygen; 1- for the terminal sulfur; and 2+ for the central sulfur.

Applying the rules in section to the thiosulfate anion gives oxidation states 2- for oxygen and 2+ for sulfur. The sulfur value is the average of the values obtained from the Lewis structure.

The advantage of the Lewis structure method is that it produces individual atom oxidation states, rather than averages. You might try assigning oxidation states to the carbon atoms of acetic acid, C_2H_4O, using the Lewis structure approach. You should find that the central carbon has an oxidation state of 3+, and the other carbon has oxidation state 3-. These average to zero.

Supplement: Job's Method for Determination of Stoichiometry

Job's Method, also called the Method of Continuous Variation, is a simple and effective approach to the determination of chemical reaction stoichiometry. We will discuss it in the context of generic reaction,

$aA + bB \Leftrightarrow dD$

which can be rewritten in the form of by dividing all coefficients by "a".

$A + kB \Leftrightarrow mD$

where $k = b/a$ and $m = \Delta/a$. Job's method is based on the following fact: if a series of solutions is prepared, each containing the same total number of moles of A and B, but a different ratio, R, of moles B to moles A, the maximum amount of product, Δ, is obtained in the solution in which R = k (the stoichiometric ratio). To implement Job's Method experimentally, one prepares a series of solutions containing a fixed total number of moles of A and B, but in which the R is systematically varied from large to small, and measures the amount of product obtained in each solution. One then plots amount of product versus R, and obtains a maximum at the initially-unknown value of k.

That the maximum amount of product should occur at the stoichiometric ratio can be justified both intuitively and mathematically. The intuitive justification runs as follows. When R is greater than k, there is an excess of reagent B, so reagent A is the limiting reagent. As R is systematically decreased towards k (i.e., as moles A increases and moles B decreases such that moles A + moles B stays constant) the amount of product increases with the amount of limiting reagent, A, until R becomes equal to k. In contrast, when R is less than k, there is an excess of reagent A, and B is limiting. As R is systematically increased towards k (i.e., as moles B increases and moles A decreases), the amount of product increases with the amount of limiting reagent B, until R becomes = k. Putting this all together, we see that as R is varied over the range from zero to the maximum value investigated, the amount of product obtained increases until R = k, then decreases as R becomes larger than k. This demonstrates Job's Method intuitively.

The mathematical justification is also quite simple. We use variable "x" to represent the moles of A in a particular solution, and assume that the total moles of A and B is to be kept at 1.0 throughout the series of solutions. Then in each solution it will be true that

$$x = \text{moles A}$$
$$1 - x = \text{moles B}$$

Our goal is to show that the maximum amount of product is obtained when R = moles B/moles A = $(1 - x)/x$ is equal to k. We approach this by finding the value of x that maximizes product. According to equation, if x is less than the stoichiometrically correct amount of A, then A is limiting and moles product = mx.

A plot of moles product versus x over a series of solutions should be linear, with slope m. Similarly, if x exceeds the stoichiometrically correct amount of A, then B is limiting and moles product = m$(1 - x)/k$. A plot of moles product versus x over a series of solutions should also be linear, with slope = –m/k. The first plot will proceed up to the right as x increases. The second plot will proceed down to the right as x increases.

At some point then, the two straight lines will intersect. At the intersection, they have a point in common. The value of x corresponding to this point is obtained by equating the ordinate values and solving for x:

$$mx = m(1 - x)/k$$

Solving gives $x = 1/(1 + k)$.

Substituting in the expression for R, $(1 - x)/x$, we find that the two lines intersect when

$$R = (1 - x)/x = \{1 - [1/(1 + k)]\}/[1/(1 + k)] = k.$$

Because the amount of product increases as k is approached from either direction, the point of intersection of the lines occurs at the maximum amount of product obtainable. We have therefore shown that maximum product is obtained when $R = k$. This is what we set out to demonstrate.

Example: A and B are known to react to form Δ, but the stoichiometry is uncertain. A Job's Method study yields the following data. Plot quantity of product versus moles A to determine the stoichiometry.

moles A	*moles B*	*grams product*
0.2	1.8	2.5
0.3	1.7	3.75
0.4	1.6	5.0
0.6	1.4	4.38
0.8	1.2	3.75
1.0	1.0	3.12

Solution: The value of k is clearly 4, because at the maximum, moles B/ moles A = 1.6/0.4 = 4.

A Detailed Example: Stoichiometry of Copper Amino Acid Complexes by Job's Method

We now describe the application of Job's method to the determination of the stoichiometry of amino acid complexes with the Cu^{2+} ion.

Job's Method may be applied to stoichiometric detemination in either of 2 ways. For clarity, we discuss it in terms of the general reaction:

$$A + xB \rightarrow C$$

Table. The Standard Approach

Solution Number	Volume 0.1 M A solution, mL	Volume 0.1 M B solution, mL	Volume additional solvent, mL	Ratio [B]/[A]	mmoles Product
1	0	0.9	1.1	infinite	0
2	0.1	0.8	1.1	8	0.01
3	0.2	0.7	1.1	3.5	0.0175
4	0.3	0.6	1.1	2	0.015
5	0.4	0.5	1.1	1.25	0.0125
6	0.5	0.4	1.1	0.8	0.01
7	0.6	0.3	1.1	0.5	0.0075
8	0.7	0.2	1.1	0.28	0.005
9	0.8	0.1	1.1	0.125	0.0025
10	0.9	0	1.1	0	0

The Standard Method. A series of solutions is prepared in which the sum of the number of moles of A and the number of moles of B is kept constant, but in which the relative amounts of A and B are systematically varied. Thus moles B/moles .

A varied from 0 to some value certain to be larger than x. For example, the following series of solutions might be prepared from two stock solutions, one 0.1 M in A and the other 0.1 M in B:

Once the solutions are prepared, the amount of product formed is measured in each solution, and a plot of the amount of product formed versus moles of B in the solution is constructed.

The crux of Job's Method is that the maximum amount of product will form when A and B are present in the correct stoichiometric ratio.

Normally the plot looks like an inverted "V", with the vertex skewed towards one side or the other.

Straight lines are drawn through the points on both sides of the vertex, and the moles of B corresponding to their intersection point allows the stoichiometry to be determined. The value of x is then calculated from equation:

$$x = \text{(moles of B at intersection)/[total moles}$$

$$A + B \quad \text{moles of B at intersection]}$$

Because the intersection in the plot occurs at moles B = 0.072/(0.09 – 0.072) = 4, the coefficient of reactant B in equation is 4.

Table. The Limiting Reagent Approach

Solution Number	Volume 0.1 M A solution, mL	Volume 0.1 M B solution, mL	Volume additional solvent, mL	Ratio [B]/[A]	mmoles Product
1	0.15	0	1.85	0	0
2	0.15	0.05	1.8	0.33	0.00125
3	0.15	0.1	1.75	.667	0.0025
4	0.15	0.15	1.7	1	0.00375
5	0.15	0.2	1.65	1.33	0.005
6	0.15	0.25	1.6	1.67	0.00625
7	0.15	0.3	1.55	2	0.0075
8	0.15	0.45	1.4	3	0.01125
9	0.15	0.60	1.25	4	0.015
10	0.15	0.75	1.1	5	0.015
11	0.15	0.90	0.95	6	0.015

The challenging aspect of the application of Job's Method is in finding a way to measure the amount of product formed in each solution. If the product precipitates, the product in each solution may be recovered by filtration and weighed. If the product absorbs light at a known wavelength, then each solution can be placed in a spectrometer and the absorbance at that wavelength determined.

There are many other possibilities. The primary restraint, particularly when using a spectrometric method, is that only the amount of product is measured. If, for example, one of the reactants also has significant absorbance at the wavelength being used to monitor the amount of product, the shape of the Job's method plot will be affected, sometimes to an extent that prevents determination of the stoichiometry in a straightforward way. It is important to be alert for this problem.

The Limiting Reagent Method. In this modification of Job's Method, a series of solutions is prepared containing a fixed number of moles of reactant A and varying amounts of reactant B, such that the ratio, moles B/moles A runs the gamut from 0 to a value known to be larger than x. The amount of product in each solution is then measured. Once the amount of B exceeds the stoichiometrically required amount, A is the limiting reagent, and the amount of product formed will remain constant. Thus by noting the ratio [B]/[A] at which the amount of product formed levels off, the stoichiometry can be determined. A plot of amount of product versus [B]/[A] will reveal this point visually. Table provides data for a series of solutions.

GEOMETRICAL ISOMERISM

The geometrical isomerism arises when atoms or groups are arranged differently in space due to restricted rotation of a bond or bonds in a molecule.

- Two different spatial arrangements of methyl groups about a double bond in 2-butene give rise to the following geometrical isomers.
  i.e., cis-2-butene and trans-2-butene

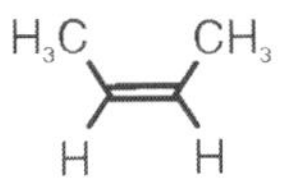

Cis-2-butene 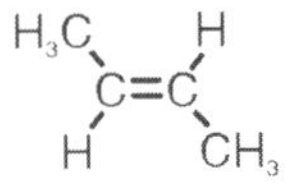 Trans-2-butene

Above two forms are not inter convertible due to restricted rotation of double bond. In the cis isomer, the two methyl groups are arranged on the same side of a double bond. Whereas in the trans isomer, they are on the opposite side.

- There are two geometrical isomers (cis & trans) possible in case of 1,4–dimethylcyclohexane as shown below:

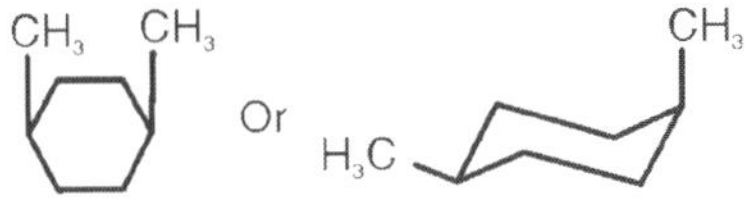

Cis-1,4-dimethylcyclohexane

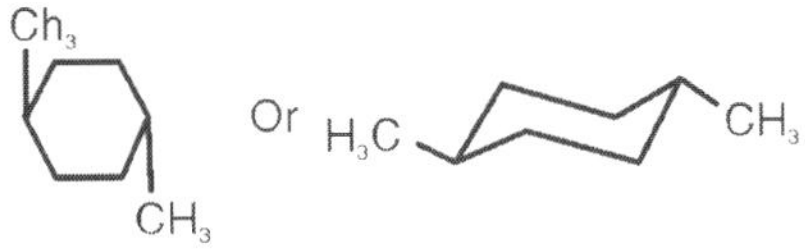

Trans-1,4-dimethylcohexane

In the above geometrical isomers, the methyl groups are arranged differently about the plane of the cyclohexane ring. These isomers are not inter convertible since it is not possible to rotate the bonds in the cyclohexane ring.

- The geometrical isomers often show different physical and chemical properties. The difference in their physical properties is more significant when there is more difference in their polarity.
- Usually the dipole moment of cis-isomers is greater than that of trans isomers. Hence the cis isomers usually have more solubility in polar solvents.
- In general, the trans isomers are more stable than cis isomers.

E-Z Notation: The simple convention of denoting the geometrical isomers by cis/trans descriptors is not sufficient when there are more than two different substituents on a double bond. To differentiate the stereochemistry in them, a new system of nomenclature known as the E-Z notation method is to be adopted.

According to this method, if the groups with higher priorities are present on the opposite sides of the double bond, that isomer is denoted by E.

Where E = Entgegen (the German word for 'opposite')

However, if the groups with higher priorities are on the same side of the double bond, that isomer is denoted by Z.

Where Z = Zusammen (the German word for 'together')

The letters E and Z are represented within parentheses and are separated from the rest of the name with a hyphen.

Step by Step Procedure to Determine the E-Z Configuration: The following procedure is to be adopted to denote the geometrical isomers by E and Z descriptors:

- First determine the higher priority group on each end of the double bond.
- If the higher priority groups are on the opposite sides of double bond, the isomer is denoted by the descriptor, E.
- Otherwise if they are on the same side of double bond, the Z descriptor must be used.

The priorities are assigned by following Cahn-Ingold-Prelog sequence rules (CAN rules) described below:

- Rank the atoms directly attached to the olefinic carbon according to their atomic number. High priority is given to the atom with higher atomic number.
- If isotopes of same element are present, the higher priority is given to the isotope with higher atomic mass.

The Deuterium isotope (H^2 or D) has more priority than protium (H^1 or H).

The C^{13} isotope has more priority than C^{12}.

- If the atoms are still identical, examine the next atoms along the chain until a "first point of difference" is found. This is done by making a list of atoms linked directly to the first atom. Each list is arranged in order of decreasing atomic number. Then the higher priority is given to the list which contain atom with higher atomic number at first point of difference.

Examine the lists of atoms directly linked to the highlighted carbons in the following compound, (2Z)-2-tert-Butyl-3-methylpent-2-en-1-ol.

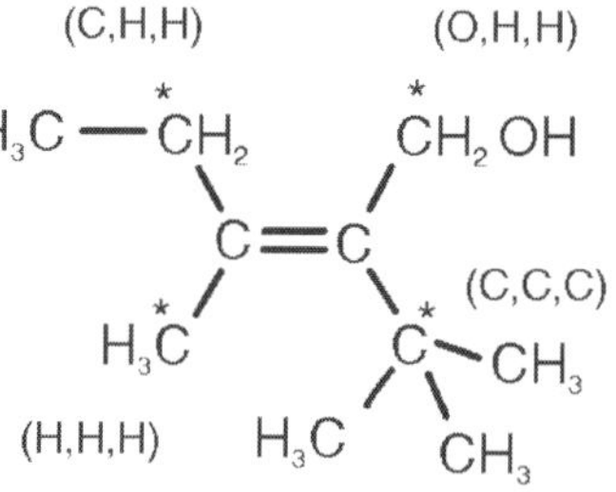

(C,H,H) (O,H,H)

$H_3C-\overset{*}{C}H_2$ $\overset{*}{C}H_2OH$

C=C

(C,C,C)

$H_3\overset{*}{C}$ $\overset{*}{C}-CH_3$

(H,H,H) H_3C CH_3

The (C,H,H) list has more priority over (H,H,H). Whereas, the (O,H,H) has more priority over (C,C,C).

Since the groups with highest priorities are on the same side of the double bond, the descriptor, Z is used to represent the stereochemistry of groups at double bond.

- The multiple bonds are counted as multiples of that same atom *i.e.*, each σ bond is treated as if it were another π bond to that type of atom.

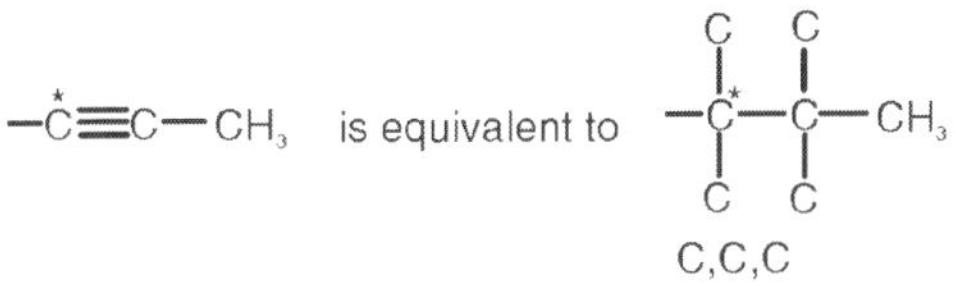

–C≡C–CH3 is equivalent to (C,C,C)

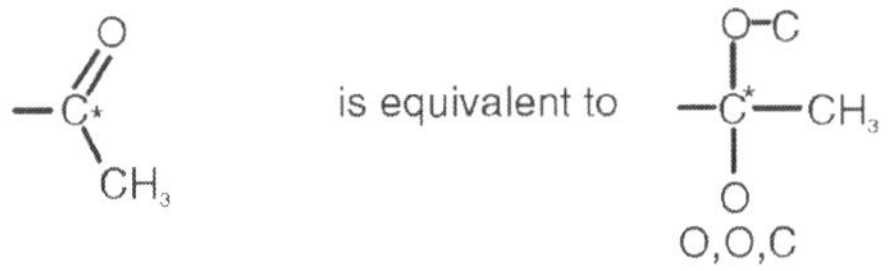

–C(=O)–CH3 is equivalent to (O,O,C)

Illustrations:

- The cis and trans 2-butenes can also be represented as Z and E isomers respectively.

(2Z)-but-2-ene
cis-2-butene

(2E)-but-2-ene
trans-2-butene

Since the double bond is at 2nd carbon of the butene, the number 2 is used before the descriptors *i.e.*, '2Z' or '2E' .

- There are four geometrical isomers possible with 3-methylhexa-2,4-diene as shown below. The geometries at each double bond are clearly mentioned by the numbers: 2 and 4.

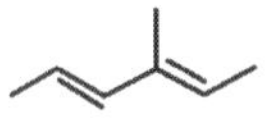

(2E,4E)-3-methylhexa-2,4-diene

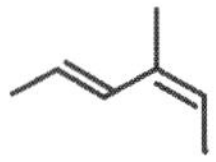

(2Z,4E)-3-methylhexa-2,4-diene

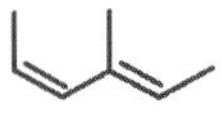

(2E,4Z)-3-methylhexa-2,4-diene

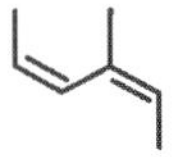

(2Z,4Z)-3-methylhexa-2,4-diene

- If the double bond has two similar groups on the same carbon, then there is no need of descriptor. For example, in the following compound, 2,3,4-trimethylhexa-2,4-diene, the double bond on 1st carbon has two methyl groups. Hence there is no need of the descriptor for this double bond.

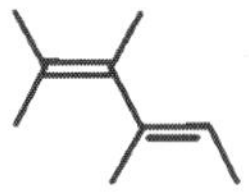

(4E)-2,3,4-trimethylhexa-2,4-diene

GEOMETRICAL ISOMERISM IN OXIMES

The oximes are formed when carbonyl compounds are treated with hydroxyl amine.

These are represented as:

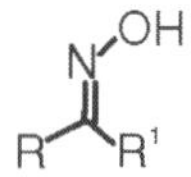

Oxime

Where R and R^1 are hydrogens; or alkyl or aryl groups.

The oximes are of two types:

1) *Aldoximes*: These are derived from aldehydes. In this case, at least either R or R^1 is hydrogen.
2) *Ketoximes*: These are derived from ketones. In this case, both R or $R^1$are alkyl or aryl groups only.

The oximes show geometrical isomerism due to restricted rotation of C=N bond. Two geometrical forms are possible for the oximes as shown below.

The descriptors, *syn* and *anti* are used to distinguish them.

- In case of aldoximes, the *syn* form is the one in which both the hydrogen and the hydroxyl (–OH) group are on the same side of the C=N. Whereas in the *anti* form, they are on the opposite side.

The syn and anti forms of acetaldoxime are shown below.

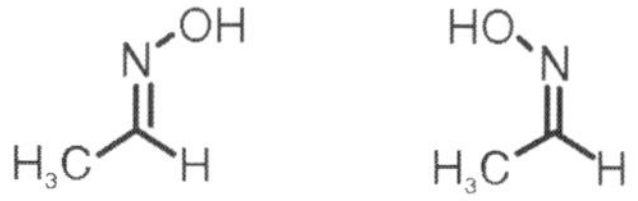

syn oxime anti oxime

- However with ketoximes, the *syn* and *anti* descriptors indicate the spatial relationship between the first group cited in the name and the hydroxyl group. For example, the following ketoxime of butanone can be named as either syn methyl ethyl ketoxime or anti ethyl methyl ketoxime.

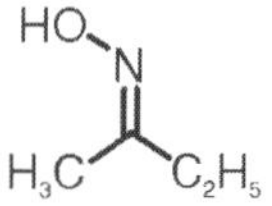

E-Z Notation of Oximes: The geometrical isomers are better differentiated by using E-Z notations. The Z oxime has hydroxyl group and the group with higher priority on the same side of C=N. However in the E oxime, they are arranged on the opposite sides of the C=N. The syn acetaldoxime is named as (E)-acetaldoxime, since the hydroxy group and the group with higher priority *i.e.*, methyl group are on the different sides of the C=N. Whereas the anti form is named as (Z)-acetaldoxime.

IUPAC System of Nomenclature of Oximes

- In the IUPAC system of nomenclature, the oximes are named as: N-hydroxyalkanimines, where the C=N group is represented by the suffix, imine and the –OH group on nitrogen is represented by the prefix, N-hydroxy.

The *syn* and *anti* forms of acetaldoxime are named as follows:

syn form	anti form
N–OH	HO–N
H_3C H	H_3C H
(1E)-N-hydroxyethanimine	(1Z)-N-hydroxyethanimine

STRUCTURAL ISOMERISM IN ORGANIC COMPOUNDS

The structural isomerism or constitutional isomerism arises when atoms within a molecule are arranged in different orders. The structural isomers have the same molecular formula but different structural formulae. The structural isomers usually show different physical and chemical properties.

The structural isomerism is further divided into:

- Chain isomerism
- Positional isomerism
- Functional isomerism
- Metamerism
- Tautomerism

Chain Isomerism: The chain isomerism arises due to different arrangements of carbon atoms leading to linear and branched chains. The chain isomers have same molecular formula but different types of chains i.e., linear and branched.

The chain isomers have almost similar chemical properties but different physical properties. For example, the branched chain isomers have lower boiling points than that of their linear counterparts. It is because, the linear ones have more surface area of contact and hence the intermolecular forces of attraction are maximum.

Two chain isomers possible with the molecular formula, C_4H_{12}.

i.e.,

- n-butane: a linear chain isomer.
- isobutane (or 2-methylpropane): a branched isomer

Linear	Branched
$H_3C-CH_2-CH_2-CH_3$	$H_3C-CH(CH_3)-CH_3$
Butane	Isobutane or 2-methylpropane

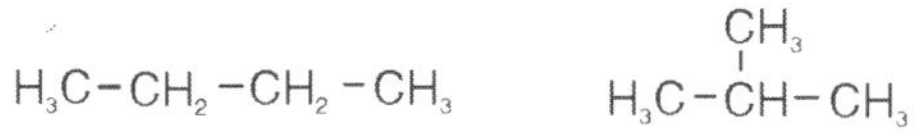

Positional Isomerism: The positional isomerism arises due to different positions of side chains, substituents, functional groups, double bonds, triple bonds etc., on the parent chain.

- Propyl chloride and isopropyl chloride are the positional isomers with the molecular formula, C_3H_7Cl. These isomers arise due to difference in the position of the chloro group on the main chain.

$$H_3C-CH_2-CH_2-Cl \qquad H_3C-\underset{\displaystyle Cl}{\underset{|}{C}H}-CH_3$$

Propyl chloride	Isopropyl chloride
or	or
1-chloropropane	2-chloropropane

- In the following positional isomers i.e., but-1-ene and but-2-ene the position of double is different but they have same molecular formula, C_4H_8.

$$\underset{\text{but-1-ene}}{H_2C=CH-CH_2-CH_3} \quad \underset{\text{but-2-ene}}{H_3C-CH=CH-CH_3}$$

- The functional group, -OH is at different positions on the main chain of the following alcohols. They have the same molecular formula, C_3H_7OH.

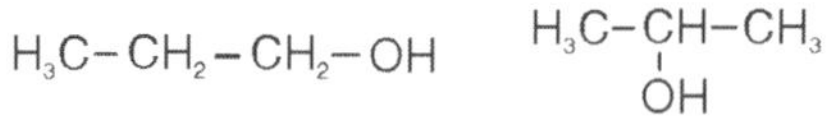

$$H_3C-CH_2-CH_2-OH \qquad H_3C-\underset{\displaystyle OH}{\underset{|}{C}H}-CH_3$$

Propyl alcohol	isopropyl alcohol
or	or
propan-1-ol	propan-2-ol

- In the following positional isomers i.e., ortho xylene, meta xylene and para xylene, the relative positions of methyl groups on the benzene ring are different

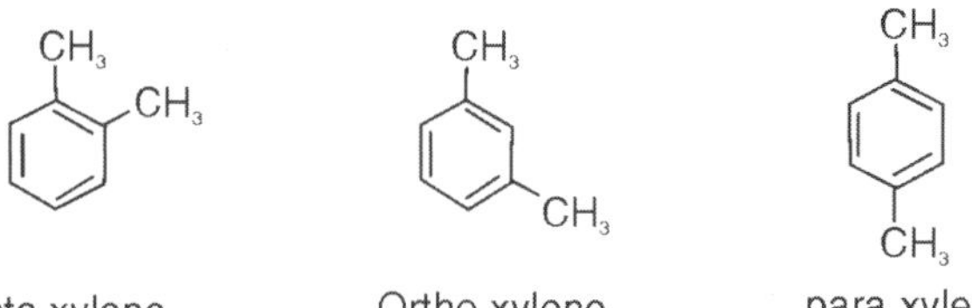

Meta xylene	Ortho xylene	para xylene
or	or	or
1,2-dimethylbenzene	1,3-dimethylbenzene	1,4-dimethylbenzene

Functional Isomerism: The functional isomerism arises due to presence of different functional groups. The functional isomers have the same molecular formula but possess different functional groups.

The functional isomers show different physical as well as chemical properties:

- The dimethyl ether is the functional isomer of ethyl alcohol. Both have the same molecular formula, C_2H_6O. However they have different functional groups.

$H_2C - CH_2 - OH$ $H_3C - O - CH_3$

ethyl alcohol dimethyl ether

- Acetaldehyde is a functional isomer of acetone. They have the same molecular formula, C_3H_6O but different functional groups. Acetaldehyde contains aldehyde, -CHO functional group, whereas acetone has ketone, >C=O functional group.

$H_3C{-}CH_2{-}CHO$ $H_3C{-}C({=}O){-}CH_3$

acetaldehyde or propanal — Acetone or propan-2-one

- Both Acetic acid and methyl formate are also functional isomers. They have the same molecular formula, $C_2H_4O_2$. Acetic acid is a carboxylic acid with -COOH group whereas methyl formate is an ester with -$COOCH_3$ group.

$H_3C{-}C({=}O){-}OH$ $H{-}C({=}O){-}OCH_3$

acetic acid — methyl formate

Metamerism: The metamerism arises when different alkyl groups are attached to same functional group:

- The following metamers contain the ether functional group. However they differ by the nature of alkyl groups attached to the oxygen atom.

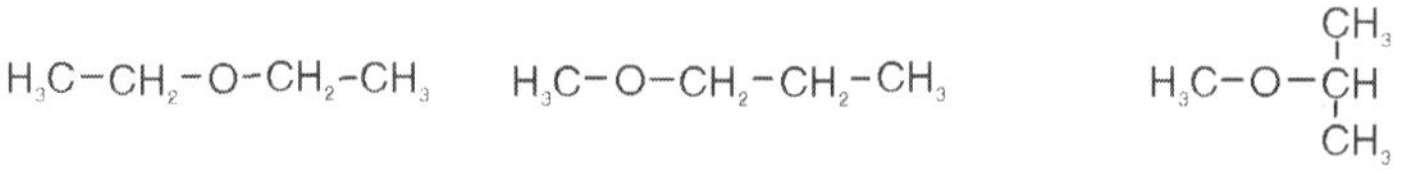

$H_3C{-}CH_2{-}O{-}CH_2{-}CH_3$ $H_3C{-}O{-}CH_2{-}CH_2{-}CH_3$ $H_3C{-}O{-}CH(CH_3)_2$

diethyl ether or ethoxyethane — methyl propyl ether or 1-methoxypropane — isopropyl methyl ether or 2-methoxypropane

- The metamerism is also possible in amines as shown below.

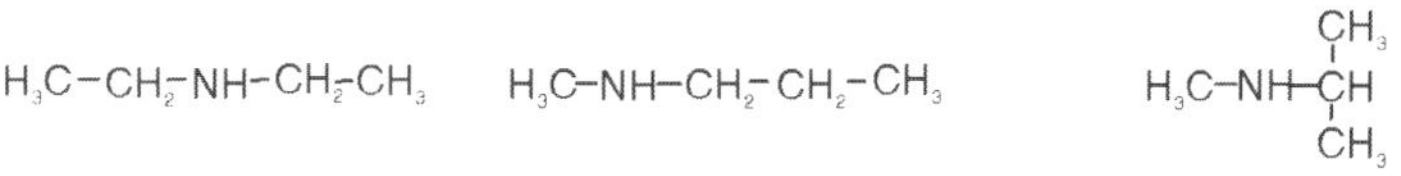

$H_3C{-}CH_2{-}NH{-}CH_2{-}CH_3$ $H_3C{-}NH{-}CH_2{-}CH_2{-}CH_3$ $H_3C{-}NH{-}CH(CH_3)_2$

diethyl amine or N-ethylethanamine — methyl propyl amine or N-methylpropan-1-amine — isopropyl methyl amine or N-methylpropan-2-amine

Tautomerism: Tautomerism refers to the dynamic equilibrium between two compounds with same molecular formula. It is also called as desmotropism or kryptotropism or prototropy or allelotropism.

It is most often a special case of functional group isomerism. In general, the tautomers have different functional groups and exist in dynamic equilibrium with each other due to a rapid interconversion from one form to another.

It is very important to note that the tautomers are not the resonance structures of same compound.

- *Keto-enol Tautomerism:* The carbonyl compounds containing at least one á-hydrogen atom exhibit keto-enol tautomerism. The carbonyl group can be converted to enol form due to transfer of one of the á-hydrogen onto the oxygen atom.

  Note: An alcoholic group on C=C is called enol. It is an alkene alcohol.

The per cent composition of keto enol tautomeric mixture depends on the relative stabilities of these two forms. In general the keto form is the low energy form and is more stable than the enol form. However the enol form is also stable in certain cases due to other stability factors.

The dynamic equilibrium between a keto form and an enol form exhibited by acetone is shown below. The tautomers are formed due to complete transfer of hydrogen. However the relatively stable keto form exists in higher percentage (about 99per cent).

$$H_3C-\overset{O}{\overset{\|}{C}}-CH_3 \rightleftharpoons H_3C-\overset{OH}{\overset{|}{C}}=CH_2$$

Acetone keto form — Prop-1-en-2-ol enol form

However enol form is also appreciably stable in case of 1,3-dicarbonyl compounds like acetoacetic ester (ethyl aceto acetic ester) as depicted below.

keto enol tautomerism in acetoacetic ester

$$H_3C-\overset{O}{\overset{\|}{C}}-CH_2-\overset{O}{\overset{\|}{C}}-O-C_2H_5 \rightleftharpoons H_3C-\overset{OH}{\overset{|}{C}}=CH-\overset{O}{\overset{\|}{C}}-O-C_2H_5$$

keto form 92.4% — enol form 7.6%

The enol form is more stable than the keto form in case of phenols due to resonance stabilization of aromatic ring.

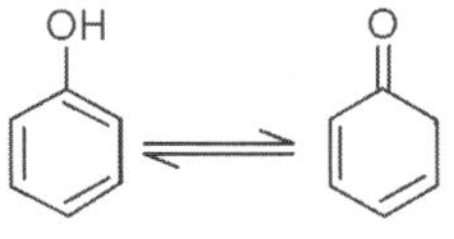

OH (phenol) ⇌ O (cyclohexadienone)

more stable enol form — less stable keto form

- The nitro-aci tautomerism: It is exhibited by nitro compounds containing atleast one á-hydrogen. In this case also, the hydrogen atom is transferred completely from one atom to another during the conversion of nitro form to aci form and vice versa.

The nitro-aci tautomerism exhibited by nitromethane is shown below.

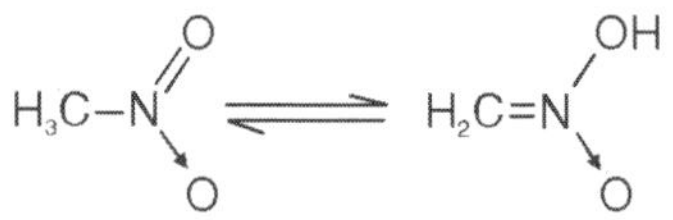

$$H_3C-N(=O)\rightarrow O \rightleftharpoons H_2C=N(-OH)\rightarrow O$$

nitro form — aci form

CLASSIFICATION OF ORGANIC COMPOUNDS

Functional Groups

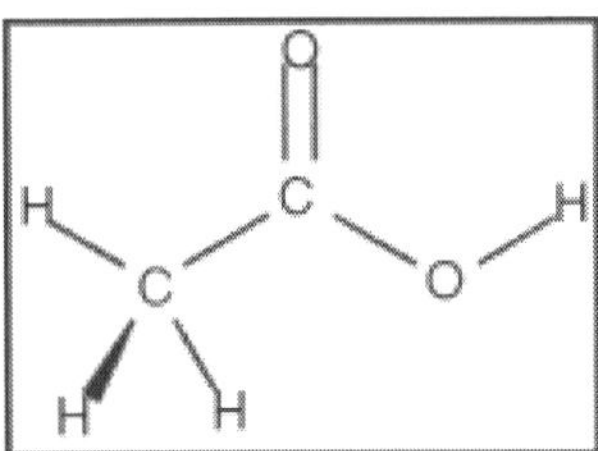

Fig. The Family of Carboxylic Acids Contains a Carboxyl (-COOH) Functional Group. Acetic Acid is an Example.

The concept of functional groups is central in organic chemistry, both as a means to classify structures and for predicting properties. A functional group is a molecular module, and the reactivity of that functional group is assumed, within limits, to be the same in a variety of molecules.

Functional groups can have have decisive influence on the chemical and physical properties of organic compounds. Molecules are classified on the basis of their functional groups. Alcohols, for example, all have the subunit C-O-H. All alcohols tend to be somewhat hydrophilic, usually form esters, and usually can be converted to the corresponding halides. Most functional groups feature heteroatoms (atoms other than C and H). Organic compounds are classified according to funcational groups, alcohols, carboxylic acids, amines, etc.

Aliphatic Compounds

The aliphatic hydrocarbons are subdivided into three groups of homologous series according to their state of saturation:

- Paraffins, which are alkanes without any double or triple bonds,
- Olefins or alkenes which contain one or more single double bonds, i.e di-olefins (dienes) or poly-olefins.
- Alkynes, which have one or more triple bonds.

The rest of the group is classed according to the functional groups present. Such compounds can be "straight-chain," branched-chain or cyclic. The degree of branching affects characteristics, such as the octane number or cetane number in petroleum chemistry. Both saturated (alicyclic compounds and unsaturated compounds exist as cyclic derivatives. The most stable rings contain five or six carbon atoms, but large rings (macrocycles) and smaller rings are common.

The smallest cycloalkane family is the three-membered cyclopropane ($(CH_2)_3$). Saturated cyclic compounds contain single bonds only, whereas aromatic rings have an alternating (or conjugated) double bond. Cycloalkanes do not contain multiple bonds, whereas the cycloalkenes and the cycloalkynes do.

Aromatic Compounds

Aromatic hydrocarbons contain conjugated double bonds. The most important example is benzene, the structure of which was formulated by Kekulé who first proposed the delocalization or resonance principle for explaining its structure. For "conventional" cyclic compounds, aromaticity is conferred by the presence of 4n + 2 delocalized pi electrons, where n is an integer. Particular instability (antiaromaticity) is conferred by the presence of 4n conjugated pi electrons.

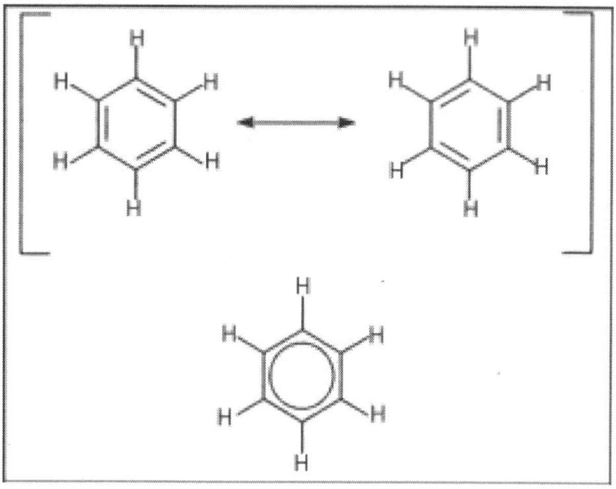

Fig. Benzene is One of the Best-known Aromatic Compounds as it is one of the Simplest Aromatics.

Heterocyclic Compounds

The characteristics of the cyclic hydrocarbons are again altered if heteroatoms are present, which can exist as either substituents attached externally to the ring (exocyclic) or as a member of the ring itself (endocyclic). In the case of the latter, the ring is termed a heterocycle. Pyridine and furan are examples of aromatic heterocycles while piperidine and tetrahydrofuran are the corresponding alicyclic heterocycles.

The heteroatom of heterocyclic molecules is generally oxygen, sulfur, or nitrogen, with the latter being particularly common in biochemical systems.Examples of groups among the heterocyclics are the aniline dyes, the great majority of the compounds discussed in biochemistry such as alkaloids, many compounds related to vitamins, steroids, nucleic acids (e.g. DNA, RNA) and also numerous medicines.

Heterocyclics with relatively simple strucures are pyrrole (5-membered) and indole (6-membered carbon ring). Rings can fuse with other rings on an edge to give polycyclic compounds. The purine nucleoside bases are notable polycyclic aromatic heterocycles..Rings can also fuse on a "corner" such that one atom (almost always carbon) has two bonds going to one ring and two to another. Such compounds are termed spiro and are important in a number of natural products.

Polymers

One important property of carbon in organic chemistry is that it can form certain compounds, the individual molecules of which are capable of attaching themselves to one another, thereby forming a chain or a network. The process is called polymerization and the chains or networks polymers, while the source compound is a monomer. Two main groups of polymers exist: those artificially manufactured are referred to as industrial polymers or synthetic polymers and those naturally occurring as biopolymers.Since the invention of the first artificial polymer, bakelite, the family has quickly grown with the invention of others.

Common synthetic organic polymers are polyethylene or polythene, polypropylene, nylon, teflon or PTFE, polystyrene, polyesters, polymethylmethacrylate (commonly known as perspex or plexiglas) polyvinylchloride or PVC, and polyisobutylene an important artificial or synthetic rubber also the polymerised butadiene, a rubber component.

The examples are generic terms, and many varieties of each of these may exist, with their physical characteristics fine tuned for a specific use. Changing the conditions of polymerisation changes the chemical composition of the product by altering chain length, or branching, or the tacticity. With a single monomer as a start the product is a homopolymer. Further, secondary component(s) may be added to create a heteropolymer (co-polymer) and the degree of clustering of the different components can also be controlled. Physical characteristics, such as hardness, density, mechanical or tensile strength, abrasion resistance, heat resistance, transparency, colour, etc. will depend on the final composition.

Biomolecules

Biomolecular chemistry is a major category within organic chemistry which is frequently studied by biochemists. Many complex multi-functional group molecules are important in living organisms. Some are long-chain biopolymers, and these include proteins, DNA, RNA and the polysaccharides such as starches in animals and celluloses in plants.

The other main classes are amino acids (monomer building blocks of proteins), carbohydrates (which includes the polysaccharides), the nucleic acids (which include DNA and RNA as polymers), and the lipids. In addition, animal biochemistry contains many small molecule intermediates which assist in energy production through the Krebs cycle, and produces isoprene, the most common hydrocarbon in animals.

Isoprenes in animals form the important steroid structural (cholesterol) and steroid hormone compounds; and in plants form terpenes, terpenoids, some alkaloids, and a unique set of structural hydrocarbons called biopolymer polyisoprenoids present in latex sap which is the basis for making rubber.

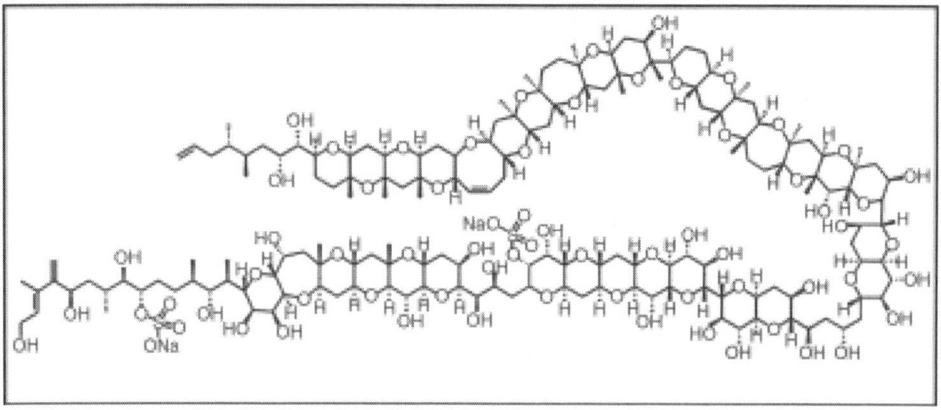

Fig. Maitotoxin, a Complex Organic Biological Toxin.

Small Molecules

In pharmacology, an important group of organic compounds is small molecules, also referred to as 'small organic compounds'. In this context, a small molecule is a small or ganic compound that is biologically active, but is not a polymer. In practice, small molecules have a molar mass less than approximately 1000 g/mol.

Fullerenes

Fullerenes are among the types of compounds engineered by organic chemists that have generated the most interest. The discovery of their unique electronic properties due to their spherical structure has stimulated new research into related fields such as carbon nanotubes.

Others

Organic compounds containing bonds of carbon to nitrogen, oxygen and the halogens are not normally grouped separately. Others are sometimes put into major groups within organic chemistry and discussed under titles such as organosulfur chemistry, organometallic chemistry, organophosphorus chemistry and organosilicon chemistry.

ORGANIC REACTIONS

Organic reactions are chemical reactions involving organic compounds. While pure hydrocarbons undergo certain limited classes of reactions, many more reactions which organic compounds undergo are largely determined by functional groups. The general theory of these reactions involves careful analysis of such properties as the electron affinity of key atoms, bond strengths and steric hindrance. These issues can determine the relative stability of short-lived reactive intermediates, which usually directly determine the path of the reaction.

The basic reaction types are: addition reactions, elimination reactions, substitution reactions, pericyclic reactions, rearrangement reactions and redox reactions. An example of a common reaction is a substitution reaction written as:

$$Nu^- + C\text{–}X \rightarrow C\text{–}Nu + X^-$$

where X is some functional group and Nu is a nucleophile.

The number of possible organic reactions is basically infinite. However, certain general patterns are observed that can be used to describe many common or useful reactions.

Each reaction has a stepwise reaction mechanism that explains how it happens in sequence—although the detailed description of steps is not always clear from a list of reactants alone.

ORGANIC SYNTHESIS

Synthetic organic chemistry is an applied science as it borders engineering, the "design, analysis, and/or construction of works for practical purposes". Organic synthesis of a novel compound is a problem solving task, where a synthesis is designed for a target molecule by selecting optimal reactions from optimal starting materials.

Complex compounds can have tens of reaction steps that sequentially build the desired molecule. The synthesis proceeds by utilizing the reactivity of the functional groups in the molecule.

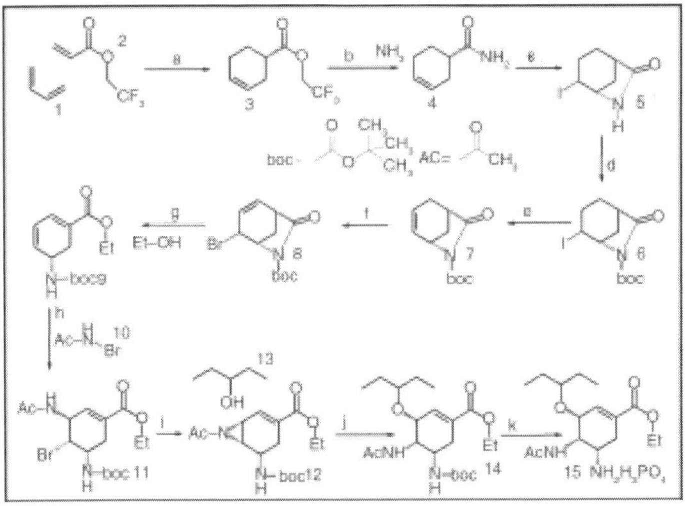

Fig. A Synthesis Designed.this Synthesis has 11 Distinct Reactions.

For example, a carbonyl compound can be used as a nucleophile by converting it into an enolate, or as an electrophile; the combination of the two is called the aldol reaction. Designing practically useful syntheses always requires conducting the actual synthesis in the laboratory. The scientific practice of creating novel synthetic routes for complex molecules is called total synthesis. There are several strategies to design a synthesis. The modern method of retrosynthesis, developed by E.J. Corey, starts with the target molecule and splices it to pieces according to known reactions.

The pieces, or the proposed precursors, receive the same treatment, until available and ideally inexpensive starting materials are reached. Then, the retrosynthesis is written in the opposite direction to give the synthesis. A "synthetic tree" can be constructed, because each compound and also each precursor has multiple syntheses.

CHARACTERIZATION

Since organic compounds often exist as mixtures, a variety of techniques have also been developed to assess purity, especially important being chromatography techniques such as HPLC and gas chromatography. Traditional methods of separation include distillation, crystallization, and solvent extraction.

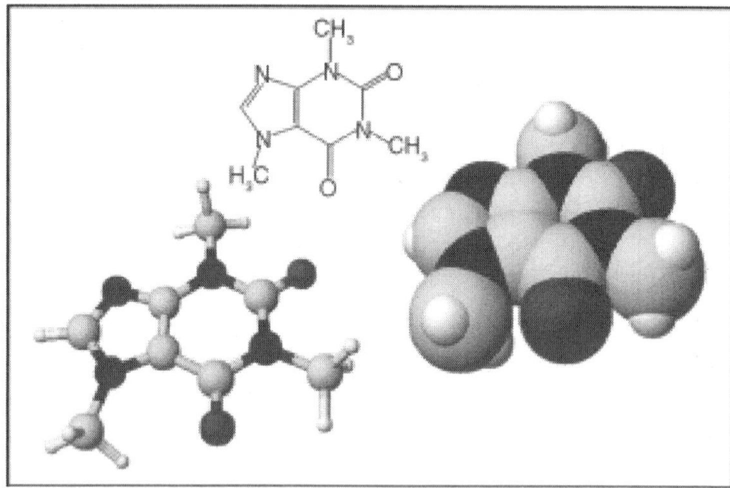

Fig. Molecular Models of Caffeine.

Organic compounds were traditionally characterized by a variety of chemical tests, called "wet methods," but such tests have been largely displaced by spectroscopic or other computer-intensive methods of analysis. Listed in approximate order of utility, the chief analytical methods are:

- Nuclear magnetic resonance (NMR) spectroscopy is the most commonly used technique, often permitting complete assignment of atom connectivity and even stereochemistry using correlation spectroscopy. The principle constituent atoms of organic chemistry - hydrogen and carbon - exist naturally with NMR-responsive isotopes, respectively ^{1}H and ^{13}C.
- *Elemental analysis*: A destructive method used to determine the elemental composition of a molecule.

- Mass spectrometry indicates the molecular weight of a compound and, from the fragmentation patterns, its structure. High resolution mass spectrometry can usually identify the exact formula of a compound and is used in lieu of elemental analysis. In former times, mass spectrometry was restricted to neutral molecules exhibiting some volatility, but advanced ionization techniques allows one to obtain the "mass spec" of virtually any organic compound.
- Crystallography is an unambiguous method for determining molecular geometry, the proviso being that single crystals of the material must be available and the crystal must be representative of the sample. Highly automated software allowing a structure to be determined within hours of obtaining a suitable crystal.

6

Chemical Fertilizers in Chemistry

FERTILIZERS IN ENVIRONMENT

Positive effects of fertilizer use on the environment are often overlooked and only the negative aspects are brought into focus.

Mineral and organic fertilizers are accused of accumulation of dangerous or even toxic substances in soil from fertilizer constituents, for example cadmium from mineral phosphate fertilizers or from town or industrial waste products; eutrophication of surface water, with its negative effect on oxygen supply (damaging to fish and other forms of animal life); nitrate accumulation in ground water, diminishing the quality of drinking water; unwanted enrichment of the atmosphere with ammonia from organic manures and mineral fertilizers, and with N_2O from denitrification of excessive or wrongly placed N fertilizer.

Regarding contamination of soils with toxic heavy metals, it can easily be shown that mineral fertilizers make only a small contribution in comparison with town wastes, for example. However, as soil fertility must be considered in the very long term and not only in decades or centuries, the annual addition should be kept at such a low level that the enrichment is negligible. Industrial waste products should always be carefully checked to determine whether they contain potentially toxic substances, and appropriate critical limits should be established.

Nutrient losses from the soil into surface and ground water (mainly nitrate by leaching and phosphate by erosion) occur even when fertilizers are not used, but they are increased slightly but unavoidably even by correct fertilizer use and are increased substantially by excessive or unbalanced use.

Considerable leaching of nitrate is caused, for example, by: excess application of organic liquid manure; intensively fertilized speciality crops; ploughing of grassland; fertilizer application for over-optimistic yield expectations which fail to materialize; part of the correctly estimated N requirement remaining unused because of other limiting factors not being taken into account - deficiencies of secondary or micronutrients, for example.

In other words, N losses are mainly due to mistakes in fertilizer use or crop management, not fertilizer use itself. Moreover, counter-measures can be taken to prevent loss of nitrate residues after harvesting (soil must not be left bare over winter) and to prevent soil erosion.

Nitrogen loss by leaching seems to range from 10kg/ha N to more than 100kg/ha N; in extreme cases more than 150kg/ha N depending on the usage of fertilizer and preventive methods. In Germany the average appears to be far below the officially (but wrongly) discussed figure of 100kg/ha N, but for most soils rather in the range of 30-60kg/ha N.

In any case, exaggerated overall averages do great injustice to farmers who apply fertilizers accurately and spend much effort in preventing excess leaching. From a scientific point of view, much more attention needs to be given to the enigma of N balance sheets before drawing premature conclusions on N losses. Loss of phosphate by leaching (<1kg/ha P) is negligible, while loss by erosion is due to bad soil management rather than fertilizer use.

Atmospheric pollution by ammonia is mainly due to primitive methods of storing and spreading organic manure. N immission (involuntary intake from the air) ranges, in Central Europe, from 10 to 15kg/ha N, with over 40kg/ha N recorded in the vicinity of intensive animal husbandry.

Of the mineral fertilizers, only urea and ammonium sulphate might cause significant NH_3^- volatilization losses, especially if not incorporated (grassland, topdressing of cereals, for instance). To minimize these losses, incorporation into the soil or application before rain or irrigation is recommended.

The contention that agriculture contributes considerably to N_2O production via denitrification, as a result of excessive or wrongly applied N fertilizer, is a serious problem. This gas contributes to the destruction of the ozone layer in the stratosphere which protects the world against ultra-violet radiation.

Official estimates, derived mainly under artificial conditions or by the difference method, showing losses of approximately 15 per cent or more of the applied N, are not really substantiated. Total denitrification losses in the range of 5-10 per cent of the applied N, of which only about 10 per cent is as N_2O, seem to be more realistic, especially for soils under normal moisture conditions.

Since pollution of the environment should be minimized, governments are trying to control the avoidable negative influences by special laws.

Under conditions of food shortage, the major goal of fertilizer use is a high crop yield giving a lower priority to food quality and possible negative influences on the environment. However, when production efforts have resulted in meeting the food demand or even in a surplus, the quality aspect and the potential pollution effects on soil, water and air receive the same or more importance as the crop yield itself.

FERTILIZERS CAN INFLUENCE QUALITY

They do so indirectly, by improving plant health, especially resistance to adverse climatic factors, diseases and pests, or directly, by increasing the content of essential and beneficial organic and mineral nutrients in human food and animal feed. They can have negative impacts, through their incorrect or imbalanced use or by the involuntary addition of toxic substances.

Examples of the improved resistance of well nourished plants to adverse climatic factors:

(1) *Resistance to drought:* A better supply of K improves their waterholding capacity, and P encourages early root growth and so ensures better survival in dry spells;

(2) *Resistance to frost and cold:* Increased by a better supply of K, P and some micronutrients (for example,Mn, Cu);

(4) *Resistance to ultra-violet radiation:* A good supply of Zn counteracts radiation-induced destruction of growth regulators.

Clearly, the damaging effects of plant diseases and pests cannot be completely eliminated simply by supplying abundant and balanced plant nutrition, but in many cases they can be contained and reduced to a lower and sometimes negligible level.

Examples of such instance include:

(a) Better resistance to some insect pests resulting from a good supply of K, as a result of better mechanical protection and a decrease in cell constituents attractive to insects;

(b) Better resistance to fungal attack resulting from a good supply of boron;

(c) Improved soil fertility also seems to result in soil fungi producing a better supply of antibiotics which protect plants therapeutically against some bacterial diseases.

Further research is certainly needed in this border area between plant nutrition and plant protection with the aim of minimizing the need for protective sprays.

There are two separate aspects of food and fodder quality:

1. *Market value:* Depending on easily recognizable external characteristics such as cleanness and absence of decay; furthermore, on the content of protein, sugar, etc, for the processing industry.
2. *Nutritional value:* Comprising palatability (taste and smell, difficult to categorize), content of the many important organic and mineral nutritional constituents, and absence of undesirable or even dangerous toxic substances.

Nutrient Supply Affects Food Quality

Better supply of nitrogen increases amounts of total and pure protein, protein quality (more of essential amino-acids), and some vitamins, especially B_1; excessive supply tends to increase amide content, resulting in bad flavour after cooking, or to raise nitrate content unacceptably.

Better supply of phosphorous improves protein quality and increases the content of some vitamins and of mineral phosphate, which is an important mineral nutrient; slightly increased radioactivity due to uranium present as natural impurity in P fertilizer seems to be of no importance whatsoever. Better supply of Potassium increases carbohydrates, and especially vitamin C; as with P, slightly increased radioactivity coming from the naturally occurring K_{40} isotope is of no importance.

Other nutrients: The advantage of having an optimum supply of all nutrients is obvious. Individual nutrients which may adversely affect quality when supplied in excess include the heavy metals such as Zn and Cu. Unwanted contamination with the toxic heavy metal Cd may arise from the use of town wastes or, less significantly, from P fertilizers.

DEFICIENCY DISEASES

Although the fertilizer-induced increase in the content of essential food constituents does not necessarily signify that fertilizers improve "health", it seems nevertheless to be so. Before the advent of fertilizer use, deficiency diseases in farm animals and humans were widespread, for example bone weaknesses due to lack of P, vitamin deficiencies due to inadequate plant nutrition, diseases in grazing livestock due to deficiencies of Cu and Co.

Furthermore, some virus and bacterial diseases seem to have diminished in their infective capacity as a result of improved nutrition. The considerable increase in human life expectancy must also be attributed in part at least to having more and better food, stemming in turn from fertilizers and such. Even so, it has to be admitted that a significant proportion of the benefits to food quality are lost in processing, for instance, in the production of white bread, and may even be lost during cooking, as with some heat-sensitive vitamins.

In view of the established generally positive effect of fertilizer use on food quality, it is surprising that certain groups of consumers in the developed countries are requesting so-called "natural" food, in the sense of food produced not only without chemical plant protection, but also without the use of synthetic mineral fertilizers (quite apart from food additives such as preservatives and colourants).

A special market has been developed for such products of "organic farming" using either organic manure alone or together with "natural" mineral fertilizer such as rock phosphate. This is fully acceptable so long as scientific principles are observed and no unfounded claims of superior quality are made.

PLANT NUTRITION AND THE SOIL-PLANT SYSTEM

The key-role of fertilizers and their judicious use in crop husbandry is well understood, when one is familiar with the general facts about plant nutrition. It is now known that at least 16 plant-food elements are necessary for the growth of green plants. These plant-nutrients are called essential elements. In the absence of any one of these essential elements, a plant fails to complete its life cycle, though the disorder caused can, however, be corrected by the addition of that element.

These 16 elements are Carbon(C), hydrogen(H), oxygen(O), nitrogen(N), phosphorous(P), sulphur(S), potassium(K), calsium(Ca), magnesium(Mg), iron(Fe), manganese(Mn), zinc(Zn), copper(Cu), molybdenum(Mb), boron(B) and chlorine(Cl). Green plants obtain carbon from carbon-di-oxide from the air; oxygen and hydrogen from water, whereas the remaining elements are taken from the soil.

Based on their relative amounts, normally found in plants, the plant nutrients are termed as macronutrients, if large amounts are involved, and micronutrients, if only traces are involved. The micronutrients essential for plant growth are iron, manganese, copper, zinc, boron, molybdenum, and chlorine. Most of the plant nutrients, besides carbon, hydrogen and oxygen, originate from the soil. The soil system is viewed by the soil scientists as a triple-phased system of solid, liquid and a gaseous phases. These phases are physically seperable. The plant nutrients are based in the solid phase and their usual pathway to the plant system is through the surrounding liquid phase, the soil solution and then to the plant root and plant cells. This pathway may be written in the form of an equation as

M(Solid)→M(Solution)→N(Plant root)→(Plant top) where'M'is the plant nutrient element in continual movement through the soil-plant system.

The operation of the above system is dependent on the solar energy through photosynthesis and metabolic activities. This is however, an oversimplified statement for gaining a physical concept of the natural phenomenon, but one should bear in mind that there are many physico and physico-chemical processes influencing the reactions in the pathway. The actual transfer in nature takes place through the charged ions, the usual form in which plant-food elements occur in solutions(liquid phase of the system).

Plant roots take up plant-food elements elements from the soil in these ionic forms. The positively charged ions are called'cations'which include potassium(K^+), Calcium(Ca^{++}), magnesium(Mg^{++}), iron(Fe^{+++}), zinc(Zn^{++}), and so on. The negatively charged ions are called anions and the important plant nutrients taken in this form include nitrogen(NO-3), phosphorous($H_2PO^-_4$), sulphur(SO^-_4), Chlorine(Cl), etc.

The process of nutrient uptake by plants refers to the transfer of the nutrient ions across the soil root interfaces into the plant cell. The energy for the process is provided by the metabolic activity of the plant and in its absence no absorption of nutrients take place. Nutrient absorption involves the phenomenon of ion exchange. The root surface, like soil, carries a negative charge and exhibits cation-exchange property. The most efficient absorption of the plant nutrients takes place on the younger tissues of the roots, capable of growth and elongation.

In this respect, root-systems are known to vary from crop to crop. Hence their feeding power differs. The extent and the spread of the effective root-system determines the soil volume trapped in the feeding-zone of the crop plant. This is indeed an important information in a given soil-plant system which helps us to choose fertilizers and fertilizer-use practices. The absorption mechanisms of the crop plants are fairly known now. There are three mechanisms in operation in the soil-water-plant systems. They are

- The contact exchange and root interception,
- The mass flow or convection,
- Diffusion.

In the case of contact exchange and root interception, the exchangeable nutrients ions from the clay-humus colloids migrates directly to the root surface through contact exchange when plant roots come into contact with the soil solids. Nutrient absorption through this mechanism is, however, insignificant as most of the plant nutrients occur in the soil solutions. Scientists have found that plant roots actually grow to come into contact with only 3 percent of the soil volume exploited by the root mass, and the nutrient uptake through root interception is even still less.

The second mechanism is mass flow or convection, which is considered to be the important mode of nutrient uptake. This mechanism relates to nutrient mobility with the movement of soil water towards the root surface where absorption through the roots takes place along with water. Some are called mobile nutrients. Others which move only a few millimetres are called immobile nutrients. Nutrient ions such as nitrate, chloride and sulphate, are not absorbed by the soil colloids and are mainly in solution.

Such nutrient ions are absorbed by the roots along with soil water. The nutrient uptake through this mechanism is directly related to the amount of water used by the plants (transpiration). It may, however, be mentioned that the exchangeable nutrient cations and anions other than nitrate, chloride and sulphate, which are absorbed on soil colloids are in equilibrium with the soil solution do not move freely with water when it is absorbed by the plant roots.

These considerations, therefore, bring out that there are large differences in the transport and root absorption of various ion through the mechanism of mass flow. Mass flow is, however, responsible for supplying the root with much of the plant needs for nitrogen, calcium and magnesium, when present in high concentrations in the soil solution, but does not do so in the case of phosphorous or potassium. The nutrient uptake through mass flow is largely dependent on the moisture status of the soil and is highly influenced by the soil physical properties controlling the movement of soil water. The third mechanism is diffusion. It is an important phenomenon by which ions in the soil medium move from a point of higher concentration to a point of lower concentration. in other words, the mechanism enables the movement of the nutrients ion without the movement of water.

The amount of nutrient-ion movement in this case is dependent on the ion-concentration gradient and transport pathways which, in turn, are highly influenced by the content of soil water. This mechanism is predomionant in supplying most of the phosphorous and potassium to plant roots. It is important to note that the rhizophere volume of soil in the immidiate neighbourhood of the effective plant root receives plant nutrients continously to be delivered to the roots by diffusion. However, when the nutrient concentration builds up far excess of the plant in the reverse direction. These are some of the choice of fertilizers and fertilizer practices for practising scientific agriculture.

The relationship in the soil-plant system stated in the simple equation give in the earlier paragraph reflects the highly dynamic nature of the soil solution. One knows that the roots of the growing plants continuously remove nutrient ions from the soil solutions. At the same time, the breakdown of the soil minerals and the generating of more exchangeable cations, the biological activity and the additions made to the anions, e.g. nitrates, continuously change the composition of the soil solution. At a given point of time, therefore, the available plant nutrients in the soil solution may range from a tiny amount to larger quantities. Under favourable conditions, crop plants, in general, require larger amounts of plant nutrients than the quality found in soil solution at any given time. Hence, the situation of nutrients supply to plants becomes a limiting factor, specially, at the critical stages of plant growth and low crop yeilds result in recognition, therefore, fertilizers application and the use of suitable fertilizers are recommended for higher crop yeilds in productive farming. The knowledge of the specific role of each essential element in the growth of crop plants and their amounts required for efficient crop production is considered necessary in adopting scientific fertilizers use.

PLANT NUTRIENTS AND THEIR FUNCTIONS

The plants require, the following essential nutrients for their normal development

Carbon	Nitrogen	Calcium
Hydrogen	Phosphorous	Magnesium
Oxygen	Potassium	Sulphur
Iron	Zinc	Chlorine
Manganese	Boron	—
Copper	Molybdenum	—

Carbon is obtained from carbon dioxide of the air; Oxygen from air and water; Hydrogen from water; Nitrogen from air and soil or both, and all other nutrients from the soil. Soil is a the most important source of plant food. Nitrogen, Phosphorous and Potassium are known as primary plant nutrients; Calcium, Magnesium and Sulphur are secondary nutrients; Iron, Manganes, Copper, Zinc, Boron, Molybdenum and Chlorine as trace elements or micronutrients. The primary nutrients and secondary nutrients elements are known as major elements. This classification is based on their relative abundance, and not their relative importance. the micronutrients are required in small quantities, but they are as important as the major elements in plant nutrients.

Air is the primary source of Nitrogen for plant nutrient. Only leguminous crops can directly use this free Nitrogen with the help of symbiotic bacteria of the genus Rhizobium. Other plant derive from soil their Nitrogen in the form of Nitrogen and Ammonium. Nitrogen and Ammonium are produced in the soil by action of micro-organisms on the soil organic matter. Non symbiotic micro-organisms can fix free Nitrogen of the air and make it available to plant in Ammonium and Nitrates forms.

Nitrogen encourages the vegetative development of plants by importing a healthy green colour to the leaves. It also controls, to some extent the efficient utilization of phosphorous and Potassium. Its dependency retards growth and root development, turns the foilage yellowish or pale green, histens maturity, causes the shrivelling of grains and lowers crop yield. The older leaves are affected first. An excess of Nitrogen produces leathery(sometimes crinkled), dark-green leaves and succulent growth. It also delays the maturation of plants, impairs the quality of crops like barley, potato, tobacco, sugarcane, and fruits; increases susceptibility to diseases and causes'lodging'of cereal crops by inducing an undue lengthoning of the stem internodes.

Phosphorous influences the vigour of plants and improves the quality of crops. It encourages the formation of new cells, promotes root growth(particularly the development of fibrous roots), and hastens leaf development through emergence of ears, the formation of grains, and the maturation of crops. It also increases resistence to diseases and strengthens the stems of cereal plants, thus reducing their tendency to lodge. It offsets the harmful effects of excess nitrogen in the plant. When applied to leguminous crops, it hastens and encourages the development of nitrogen-fixing nodule bacteria. If phosphorous is deficient in the soil, plants fail to make a quick start, do not develop a satisfactory root-system, remain stunted and sometimes develop a tendency to show a reddish or purplish discolouration of the stem and foilage owing to an abnormal increase in the sugar content and the formation of anthoscyanin.

However, the deficiency of this element is not easily recognised as that of nitrogen. It has also been observed that cattle feeding on the produce of deficient soils become dwarfed, develop stiff joints and lose the velvetty feel of the skin. Such animals show an abnormal craving for eating bones and even soil itself.

Potassium enhances the ability of plants to resist diseases, insect attacks, and cold and other adverse conditions. It plays an essential part in the formation of starch and in the production and translocation of sugars, and is thus of special value to carbohydrate-rich crops, e.g. sugarcane, potato and sugar-beet. The increased production of starch and sugar in legumes fertilized with potash benefits the symbiotic bacteria and thus enhances the fixation of nitrogen. It also improves the quality of tobacco, citrus etc. With an adequate supply of potash, cereals produce plump grains and strong straws. But an excess of element tends to delay maturity, though, not to be the same extent as nitrogen.

Plants can make up and store potassium in much larger for correcting zinc deficiency. The symptoms of zinc deficiency appear generally in younger leaves, starting with intervienal chlorosis leading to a reduction in shoot growth and the shortening of internodes. Mottle leaf, little leaf, etc. in the case of trees are symptoms of zinc difficiency. The buds of several defficient maize plants become white; in citrous interveinal chlorosis and mottled leaf occur. In calcareous soils and in soils with very high phosphorus content, zinc defficiency is commonly expected to occur. The principal function of zinc in plants is as a metal activator of enzymes.

In highly weathered coarse textured soils zinc deficiency appears under an intensive cropping programme. The availability of zinc is least between pH 5.5 and 7, but its availability increases at a lower pH. At a higher pH above 7, zinc availability becomes a complex problem, as the positively charged zinc ion gets converted into a negatively charged zincate complex whose availability tends to be reduced in alkaline soils. When the calcium ion is predominent, the highly insoluble calcium zincate is formed and zinc availability gets seriously limited. The application of soluble zinc salts or zinc chelates to the soil is generally recommended to correct its defficiency.

Foliar sprays are advocated, especially for orchard trees for amending zinc defficiency. About 5 to 50 kg of zinc sulphate per hectare is used for such purposes.

The symptoms of boron defficiency vary with the kind and age of the plant, the conditions of growth and the severity of the deffiency. Each crop produces its characteristic growth abnormalities associated with boron defficiency, such as yellows and rosetting in lucerne, snakehead in wallnuts, die-back and corking of fruits in apple, corking and pitting of fruits in tomatoes, hollow stem and the bronzing of curd in cauliflower, the brown-heart diseases in table-beets, turnips, etc. Molybdenum deficiency produces whip-tail in cauliflower, broccoli and other Brassica crops.

The deficiency of this element reduces the activity of the symbiotic and non-symbiotic nitrogen-fixing micro-organisms. It was in 1954 that chlorine was proved to be an essential micronutrient. Its defficiency under field conditions has not been reported so far. In water-culture solutions, the leaves of chlorosis, necrosis and an unusual bronze discolouration on tomatoes.

Sodium is not an essential element for plant growth. But some crops, such as beet, celery, cabbage, kale, knol-khol, radish, rape and turnip, benefit greatly by application of soluble sodium salts, specially if the soil is deficient in potassium. Sodium is also of direct benefit to plants indigneous to the sea-shore or to irrigated arid regions. Salts of this element are said to release more of potassium from the exchange complex and to help to maintain phosphorus in a more available form. They also serve as a partial substitute for potassium in the case of potatoes and cotton.

MAINTENANCE OF SOIL FERTILITY

No two soils are alike either in respect of their nature or in respect of quantities of plant nutrients they contain. Under a given situation, the system of farming, soil management and manuring practices, etc., influence the productiovity of soils and crop yields obtained from them. The quantities of the three primary nutrients- N, P_2O_5 and K_2O removed from a hectare of land by some of the important crop.

It is estimated that the different agricultural crops in India remove about 4.27 million tonnes of nitrogen, 2.13 million tonnes of phosphoric acid, 7.42 million tonnes of potash and 4.88 million tonnes of lime per year.

The production of larger yields through improved varities of crops and intensive cultivation will increase the depletion of nutrients still further. But erosion and leaching cause additional losses. The present production of synthetic nitrogenous fertilizers in the country reached 1.5 million tonnes of nitrogen in 1975-76, and the bulky organic manures might supply another 1.5 million tonnes of total nitrogen.

The amounts of phosphorous, potash, etc., added to the soil are very small. It is thus obvious that the current huge drain on nutrient supplies will continue to impoverish the soils unless these supplies are replenished by natural or by artificial means, the principal methods of supplementing natural recuperation and for improving the productive capacity of the soils are

- To add organic matter to the soil, so that through decay, it may furnish a more or less continuous supply of nuttrients for crops,
- To restore or increase the amount of defficient nutrients by the application of fertilizers.

The urgent need of the constantly expanding agriculture production to meet the requirements of the continually increasing human and cattle populations in India makes the supply of additional plant nutrients through fertilizers and organic manures, a problem of supreme importance.

Table The Quantities of Plant Nutrients Removed from soil by Different Crops (Kg/ha)

Crop	Yield (grain)kg/ha	N	P_2O^5	K_2O
Rice	2,240	34	22	67
Wheat	1,568	56	24	67
Jawar	1,792	56	15	146
Bajra	1,120	36	22	66
Maize	2,016	36	20	39
Barley	1,120	41	20	35
Sugarcane	67,200	90	17	202
Groundnut	1,904	78	22	45
Mustard	672	22	11	28
Linseed	1,008	19	12	33

Cotton	448	30	17	45
Jute	1,568	67	34	67
Tea	896	45	13	28
Coffee	896	34	11	34
Tobacco	1,456	94	57	91

TYPES OF MANURES AND FERTILIZERS

Indian soils are usually very poor in organin matter as well as in nitrogen. Pohsphate deficiency is less wide-spread and potash deficiency generally occurs in com-pact areas. In acid soils the addition of lome steps up production.

Materials which are commonly used to maintain and improve soil fertility may be classified as follows

- *Manures:* These are relatively bulky materials, such as animal or green manures, which are added mainly to improve the physical condition of the soil, to replenish and keep up its humus status, to maintain the optimum conditions for the activities of soil micro-organisms and make good a small part of the plant nutrients removed by crops or otherwise lost through leaching and soil erosion. They, thus, supply practically all the elements of fertility which crops require, though not in adequate proportions. The plant-food elements contained in a manure are released in an available form after it is applied to the soil and is decomposed by soil micro-organisms. Similarly, the green manures add not only substantial amounts of organic matter but also nitrogen.
- *Fertilizers:* Fertilizers are inorganic materials of a concentrated nature; they are applied mainly to increase the supply of one or more of the essential nutrients, e.g. nitrogen, phosphorous and potash. Fertilizers contain these elements in the form of soluble of readily available chemical compounds. This distinction is, however, not very rigid. In common parlance the fertilizers are sometimes called' chemical', 'artificial'or' inorganic' manures.
- *Concentrated organic manures:* Some of the concentrated materials, such as oil-cakes, bone-meal, urine and blood are of organic origin. The use of manures and fertilizers is complementary and not as a substitute for each other.
- *Bulky Organic Manures:* The properties and role of organic matter and humus in the soil have been explained already. The average nutrient contents of manures and other organic raw materials which may be used to maintain the humus content of the soil.

FARMYARD MANURE

Good-quality farmyard manure is perhaps the most valuable organic matter applied to a soil. It is the most commonly used organic manure in India. It consists of a mixture of cattle dung, the bedding, used in the stable and of any ramnants of straw and plant stalks fed to cattle. Though its crop-increasing value has been recognised from time immemorial, more than 50 per cent of the cattle dung produced in the country today is burnt as fuel and is thus lost to agriculture. Not only this tremendous waste, but also the tradition method of preparing and storing the farmyard manure is generally faulty.

The cattle-dung, together with stable-waste and house sweeping, is first collected in the open backyard, and when a cartload has been collected, it is removed to another heap or to an uncovered pit in a common plot outside the village. The loose heaps lie exposed to the sun, with the result that the raw organic matter dries up quickly and does not rot properly. Very often, a part of the dry dung is blown off by wind or washed away by rain.

Cattle urine is either not conserved or is stored in a defective manner. American studies on the distribution of soil derived elements between urine and faeces of dry cows have shown that 95 per cent pf Potassium, 63 per cent of nitrogen and 50 per cent of sulphur are contained in the urine. The wastage of nitrogen-rich urine, the loss of nitrogen(in the form of ammonia) due to the fermentation of exposed cattle dung, and the washing away of soluble mineral elements be leaching reduce its manurial value in India to a great extent.

About half of this nitrogen, one-sixth of phosphorous and more than half of potash are readily soluble and subject to dissipation. However, the loss of nitrogen and minerals elements caused by careless handeling can be reduced greatly by using absorbent bedding for cattle, storing dung in stone or brick-line pits, mixing large quantioties of straw and other vegitable matter with cattle dung, and keeping the heap compact and moist. Thus, if urine is properly conserved, the loss of soluble mineral elements through seepage is prevented, bacterial decomposition of raw organic matter is encouraged, plant nutrients are made soluble, and nitrogen losses are minimised.

If urine is not conserved in the bedding used for cattle it must be collected in covered pucca cistern and then, added to the dung in the manure pit. Nitrogen in the urine is mainly in the form of urea, which readily changes into the highly volatile ammonium carbonate through bacterial action, and quickly loses ammonia thereafter by evapouration. This loss can be reduced to a great deal if the manure and the urine-soaked absorptive itter for bedding are kept compacted in a pit. The pit may be 1 m in depth, 1.3 to 1.5 m in width and 4.5 to 6 m in length, depending upon the no. of cattle on a farm.

The filling of the pit should be'sectional'and when each section of three or 1.3 m in length is filled to about 45 cm above the ground level, it should be clustered with 2.5 cm layer of a mixture of mud and dung in equal proportions. Before plastering, 4 to 5 buckets of water should be added to the manure in the pit. Plastering conserves moisture and nitrogen and also prevents housefly nuisance. The manure becomes ready for use in about 4 to 5 months after plastering.

The quality of manure is also improved by the concentrated feeds given to cattle. Cotton-seed, cotton-seed cake, linseed-meal, wheat bran, grain husk, groundnut cake, gram, horse-gram, etc. are rich in nitrogen, phosphorous, potassium, magnesium and sulphur. It has been found that in the case of adult working-cattle about 80 per cent of nitrogen and the other mineral elements contained in the feed is recovered in urine, faeces and other animal by-products. Accordingly, manure from cattle fed on cereal straws and grass hay is much less valuable than that from animals fed on legume hays, grains and concentrates.

In foreign countries, considerable attention has been given to the use of preservatives on manure. Calcium sulphate or gypsum and superphosphate have proved most promising in preventing the escape of ammomnia. Gypsum has been found specially effective as an ammonia-absorbing agent. Superphosphate, besides absorbing ammonia, supplies additional phosphorous and, thus, improves the crop producing capacity of the manure.

Partially rotten farmyard manure should generally be applied to the soil about three to four weeks before the sowing of a crop. In case there is sufficient moisture in the soil, there sill be enough time for its decomposition and for improving the soil structure. Its application too long before sowing a crop will either cause a drying up of the rotted manure or too quick decomppostion, depending on the incidence of rains. But in each case, there will be a serious loss of ammonia and nitrogen. If the manure is already well rotted, it is advisable to apply it just before sowing to a crop.

This procedure is perticularly essential in the case of light soils. In any case, after the manure is carted to the field, it should be evenly spread and worked into the soil soon to avoid the loss of nitrogen. The existing practice of leaving the manure in small heaps scattered in the field for several days, before spreading it in the field and incorporating it into the soil, results in the serious deterioration of its quality, particularly if strong winds blow. In vagitable and fruit cultivation, the application of well-rotten manure, in conjunction with fertilizers to young plants individually, has been founnd to give the best results. In Egypt, even the cotton crop, which is invariably grown on ridges, is given a top-dressing a handful of the rotten manure being applied to the soil at the base of each plant before an irrigation, and is worked into the soil with a hand hoe.

The practice of penning cattle, sheep and goats in the fields in summer is common in some parts of the country. Folding 7,000 sheep for one night is said to add the equivalent of 149.3 quintals(14.93 tonnes) of cattle dung. The fresh dung left in the field in such cases rapidly dries up. This drying checks ammonification and loss of nitrogen. With the first fall of rain, the dung is worked into the soil. It, therefore, does not lose much of its fertilizing value. Further more, its beneficia effects on the physical condition of the soil are undeniable.

However, sheep-folding is said to make the land more weedy. There must be adequate moisture in the soil for the proper decomposition of organic matter. Farmyard manure can, therefore, be applied to all crops grown in the rainy season or grown under irrigation.

The quantity of manure to be applied to unirrigated crops varies from 1.5 to 2 cartloads per hectare in areas of heavier rainfall. If sufficient farmy manure is not available, it may be applied at the usual rate to a part of the land,say, to one-third or one-forth of the area, in rotation every year, so that all parts of the field receive the manure regularly once in three or four years. For irrigated field crops, the rate varies from 10 to 20 cartloads. Sugarcane, maize and garden crops, such as potatoes, turmeric, ginger, vegetables and fruits receive still higher doses, amounting sometimes to 15 to 25 cartloads. A cartload of manure, measuring 9 cubic metres, weighs about half a tonne. It must be stressed that the value of farmyard manure in soil improvement is due to its content of principal nutritive elements and its ability to

- Improve the soil tilth and aeration,
- Increase the water-holding capacity of the soil,
- Stimulate the activity of micro-organisms that make the plant-food elements in the soil readily available to crops.

The supply of organic matter, which is later converted into humus is a property of farmyard manure. One tonne of carttle dung can supply only 2.95 kg of nitrogen, 1,59 kg of phsophoric acid and 2,95 kg of potash. The use of farmyard manure alone causes an imbalance in nutirtion owing to its relatively low content of phosphoric acid. Therefore, to keep the soils well supplied with all the essential elements of plant food in a readily available form, and also to keep them in good'heart', it is advisable to use the bulky organic manures in conjuction with superphosphate and such other artificial fertilizers as contain the particular plant food or foods in which a soil may be deficient or which the crop to be grown may specially require.

COMPOSTED MANURE

Another method of augmenting the supplies of organic is the preparation of compost from farmhouse, and cattle-shed wastes of all types. Composting has been advocated and adopted extensively during the past 25 years. Composting is the process of reducing vegetable and animal refuse(rural or urban) to a quickly utilizable condition for improving the maintaining soil fertility. Research conducted in India and abroad has shown that good organic manure similar in appearance and fertilizing value to cattle manure similar in appearance and fertilizing value to cattle manure can be produced from waste materials of various kinds, such as cereal straws, crop stubble, cotton stalks, groundnut husk, farm weed and grasses, leaves, leaf-mould, house refuse, wood ashes, litter, urine-soaked earth from cattle-sheds and other similar substances. These raw vegetables materials are rich in cellulose and other readily decomposable carbohydrates and have a carbon-nitrogen ratio of 40 or more to 1. The direct application of such undecomposed, low nitrogen organic matter as manure brings about a temporary deficiency of mineral nutrients(specially nitrates and ammonium compounds) in the soil by stimulating the growth of micro-organisms, which in turn, compete with crop plants for available nitrogen, phosphorous and other elements. Hence, before using them as manure, it is necessary to compost or partially decompose them. This process lowers the carbon-nitrogen ratio to about 10 or 12 to 1.

Two methods of composting waste organic materials are usually recommended. One depends on aerobic and other one unaerobic decomposition. In both cases, the farm wastes have to be used as a bedding for cattle in order to absorb a large of the animal urine. In the aerobic process, the used bedding, the sweepings from cattle sheds and some urine-soaked earth from the stable floar are removed everyday, mixed with a little cattle dung and 2 or 3 handfuls of wood ashes are deposited on a well drained site to gradually builed up a lowpile about 30 or 45 cm in height, about 5 m in width and of any convenient length.

The pile is built up before to start of rainy season. After the first heavy shower, the waited material in 1.2 strip on each side of the long heap is turn with a rake on to 2.4 m wide stripe in the middle, so raising the height of the heap to nearly 1 m. This process prevents loss of misture and ensures a quick start of decomposition. When the heap sinks appreciably and such a sinking takes about 3 to 4 weeks, it is given a turning and made into fresh heap, thus mixing the outside material with that from inside. After about a month or more, depending on incidence of rains the heap is given a final turning on a cloudy or moderately rainy day and rebuilt the vacant part of the original position. The composed becomes ready for use in about 4 months. This method is eminently suited for composting in the rainy season.

In Tamil Nadu, the application of 90 kg of partially fermented dung in the form of a thin suspension, with 22.50 kg of bone-meal per tonne of dry material to be composted is recommended. If urine-soaked earth and cattle-dung are not available, the raw organic materials can still be composted, provided more than one-third of the residues is soft and finely broken, such as fallen leaves, leaf-mould, kitchen wastes, grass clippings, green and succulent weeds, plant trimming and clopped wheat, barley and soft cereal straw. The use of ordinary soil and wood ashes or lime is, however, essential.

In the anaerobic process, the mixed farm residues are collected in pits of a convenient size, say, 4.5 m*1.5 m*1 m. Each day's collection is spread in a thin layer, sprinkled with a mixture of fresh cowdung(4.5 kg), ashes(140 to 170 g) and water (18 to 22 litres) and compacted. The pit is filled till the raw material stands 38 to 46 cm above its edge, and then is then plastered with a 2-5 cm layer of a mixture of mud cowdung. Under such conditions, decompostion is anaerobic and high temperatures do not develop.

Insoluble nitrogen compunds gradually become soluble and the carbonaceous matter is broken down into carbon dioxide and water. The loss of ammonia is negligible, because in high concentrations of carbon dioxide, ammonium carbonate is stable. The plastered pit also prevents the fly nuisance. The well-made compost contains 0.8 to 1 per cent of nitrogen and has all the good properties of farmyard manure. It can be used in the same way as the latter. The anaerobic process is particularly suited for use by gardeners in or near cities and towns.

COMPOSTING METHODS AND Raw Materials

The materials are needed are mixed plant residues, animal dung and urine,earth, wood ash and water. All vegetables wastes available on a farm like weeds,stalks,stems,fallen purnings,chaff,fodder remnants,green matter and so on, are collected and stacked in a pile. Hard woody material like cotton or pigeon-pad stalks and stubble are first spread on the farm road and crushed under Vehicles viz. tractors or bullock carts before being piled. Such hard materials should in any case not exceed 10 percent of the total plants residues.

Green material which are soft and sufficient and allowed to wilt for 2 or 3 days to remove access moistures before stacking; they tend to pack closely, if they are stacked in the fresh state. While stacking, each material is spread in layers of about 15cm thickness until the heap is about 1m and 50cm high.

The mixture of different kinds of vegetable residuesensures a more efficient decomposition. The heap is then cut into vertical slices and about 20-25 kg is put under the feet of cattle in the shed as bedding for the night. The next morning the bedding, along with the dung and the urine, and urine earth, is taken to the pits where the composting is to be done.

PIT METHOD

This method includes following steps

Site and Size of Pit

The site selected for the compost pit should be near to cattle shed and water source at high level so that no rain water gets in during the mansoon season. A temporary shed amy be constructed over it to protect the compost from heavy raifall. The pit should be about 1 m deep * 1.5-2.0 m wide and of any suitable length.

Filling the Pit

The material brought from the cattle shed is spread and on each layer is spread a slurry of dung made with 4.5 kg urine earth and 4.5 kg of inoculum taken from a 15-day-old composting pit. A sufficient quantity of water (nearly 90per cent) is sprinkled over the material in the pit to the wet it. The pit is filled in this way layer-by-layer and it should not take longer than 1 week to fill. Care should be taken to avoid compacting the material in any way.

Turning

The material is turned 3 times during the whole period of composting (i) after 15 days from filling the pit, (ii) another 15 days, (iii) after another 30 days. At each turning the material is mixed throughly, moistened with water and replaced with the pit.

HEAP METHOD

During rainy seasons or in regions with heavy rainfall the compost may be prepared in heaps above ground. when sufficient nitrogeneous material is not available a green manure or leguminous crop like sunhemp is grown on the fermenting heap by sowing seeds after the first turning. The green mater is then turned in at the second mixing.

Dimensions

The basic Indore pile is about 2 m wide at the base, 1.5 m high 2 m long. The sides are tapered so that the top is about 0.5 m narrower in width than the base. A small bund is sometimes built around the pile to protect it from wind which tends to dry the heap.

Forming the Heap

The heap isusually commenced with a 20 cm layer of carbonceous material such as leaves, hay-straw, sawdust, wood chips and chopped corn stalks. This is then covered with the 10 cm of nitrogeneous material such as fresh grass, weeds or garden plant residues, garbage, fresh or dry manure or digested sewage sludge. The pattern of 20 cm carbonaceous material and 10 cm nitrogeneous material is followed until the pile is 1.5 m high and they are normally wetted so that they feel damp but not soggy. The pile is sometimes covered with soil or hay to retain heat and is turned at 6- and 12-weeks-interval. In the republic of korea, heaps are covered with thin plasic sheets to retain heat and it has also resulted in the death of insects.

If materials are limited, the alternate layres can be added as they become available. Also, all materials may be mixed together in the pile, if one is careful to maintain the proper proportions. Shredding the material speeds up deconposition considerably; most materials can be shredded by running over it several times with a rotary-mower.

Advantages and Limitations

Preparation on a large scale can be done through community composting. there is lack of protection from rain and wind. A considerable amount of water is needed ans so the heap method is not suitale in near areas of scanty rainfall. The intense aerobic decomposition to which the material is subjected no doubt shortens the period of composting but it leads to heavy losses of organic matter and nitrogen. Therefore the C N ratio should be maintained between 30 and 40 to reduce such losses.

BANGALORE METHOD

Preparation of the Pit

Trenches or pits 1m deep are dug; the breadth and length of the trenches can be made depending on the availability of land and the type of material to be composted. The selection of site for 1 pit is made as mentioned in the Indore method. The trenches should preferably have sloping walls and a floor of 90 cm slope to prevent waterlogging.

Filling the Pit

Organic residues and night soil are put in alternate layers and after filling, the pit is covered with a 15-20 cm thick layer of refuse. the materials are allowed to remain in the pit without turning and watering for 90 days. during this period the material settles down due to reduction in volume of the biomass and additional nightsoil and refuse in alternate layers are placed on top and plastered or covered with mud or earth to prevent loss of moisture and breeding of files. the material undergoes anaerobic decomposition at a vey slow rate and it takes about 180-240 days to abtain the finished product.

ADVANTAGES AND LIMITATIONS

The recovery of the finished product is greater as compared to aerobic composting but loss of nitrogen is negligible. Labour requirements is less than for the Indore method as turning of material is not done; labour is needed only for digging and filling the pits. The methods requires along time to produce a finished compost and so takes up more land use. A uniform high temperature is not assured in the biomass. Problems of odour and fly breeding need to be attended too.

SYNTHETIC COMPOST

In the preparation of synthetic compost, the organic nitrogen as dung, required by micro-organism, can be completely substituted by inorganic nitrogen compounds like ammonium sulphate and urea which are utilized equally effectively for decomposition of carbonaceous materials into compost. The Adco process of preparations of synthetic compost developed by Hutuchinson and richards is based on this principle.

This facilitates the utilization of large quantities of various organic waste material where supplies of dung are either short of the requirement or not available at all, as on mechanized farms.

The basic principle of C N ratio in manure preparation can be applied to add nitrogenous fertilizers in sufficient quantity to decompose. The material to be composted is moistened. This is sprinkled with the fertilizer solution and then with lime. Superphosphate may be added to fortify the phosphorous content of the manure. The treatment is continued layer-wise until the heap or pit is filled to the size and allowed to ferment. The manure becomes ready for application in about 120-180 days and resembles with farmyard manure in its action on soil and plant growth.

LEAF COMPOST

Leaf composting, can be achieved by heap or ditch composting or by windrow composting. windrow are preferred as they allow efficient handling of materials Provide good aeration, allow sufficient of water and are easy to be performed. It is suggested the formation of uniform-shaped windrow from 2.40 m-3.60 m at the base and 2.40m-3.00m high and of any convenient lenght. Windrow built too high will have excessive compaction at the base resulting in anaerobic conditions. Windrows built too low will not allow sufficient insulation to sustain thermophilic temperatures during cold weather. To ensure proper aeration, it is important to break apart tightly compacted leaves.

Though reduction in size may aid in rapid decomposition, it is not desirable in leaves because it increases the compaction making more frequent aeration necessary. When the incoming leaves are not adequately moist, it is desirable to add water to maintain a proper moisture regime of 40 to 60per cent.

The C N ratio of leaves is relatively high. It can be as high as 80, and needs to be amended with nitrogen. Sewage sludge, urea and grass clippings are good sources of nitrogen. If a nitrogen source is to be added, caution should be exercised to distribute it uniformly throughtout the windrow, lest it may result in undesirable anaerobic conditions and uneven decomposition. Proper aeration can be maintained by periodical mixing of the material.

Under optimum environmental composting conditions, leaf compost will be ready between 180 and 270 days. Leaf mould will have final PH range of 6-7. It is suggested the use of finished compost (leaf mould) as converting material (10-15cm) in the subsequent preparation of leaf compost to supply a heavy inoculum of micro-organisms.

ENRICHMENT WITH PHOSPHOROUS

Phosphorous enriched compost is prepared by adding 5per cent superphosphate at the filling of the compost pits. Other sources of phosphorous for this purpose are powered rock phosphate, preferably of low grade (less than 11per cent P), can be used with profit. Besides, phosphorous it is a source of calcium and micronutrients. Bonemeal will provide nitrogen as well as phosphorous; it contain 9-11per cent P and 2-4per cent N. Steamed bonemeal is more easily ground than the fresh material. It contains a little less nitrogen but more phosphorous than the raw material. Basic slag provides calcium, magnesium and trace nutrients and small amount of phosphorous. Banana residues contains about 1-1.5per cent phosphorous on an ash basis.

ENRICHMENT WITH POTASSIUM

Granite dust of powdered potassium-containing minerals like feldspars can be added to enrich compost. Potassium and other deficient elements can be added to compost by including plant materials which contain apperciable amounts of those elements.

Water hyacinth for example, is a rich source of potassium and of many other elements required by plants. Banana skin and stalks contain 34-42per cent potassiumon an ash basis, seaweeds are rich in iodine, boron, copper, magnesium, calcium and phosphorous. Leaves are also a good source of trace elements and should form a parta part of every compost heap. Potato peel is rich in trace elements and dry potato vines contain 1per cent potassium, 4per cent calcium and 1per cent magnesium.

VERMI COMPOSTING

VERMI-COMPOSTING is the use of earthworms for composting of organic residues. Earthworms can consume practically all kinds of organic matter. One worm, weighs about 0.5 to 0.6 g, eates wastes as their own body weight perday and produces cast of the same weight per day.

It is estimated that 1000 tonnes of moist organic matter can be converted by earthworms into 300 tonnes of compost. Organnic materials undergo comples biochemical changes in the intestines and vermi-composting is an appropriate technique for disposal of non-toxicsolid and liquid organic wastes. It helps in cost effective and efficient recyclicing of animal wastes (poultry, equine,piggery excreta and cattle dung) agriculture residue and industrial wastes using low energy, excreta together with their cocoons and undigested feed make up vermicasting. The castings of earthworms are rich in nutrients (N, P, K, Ca and Mg), and also in bacterial and actinomycetes population.

The actinomycetes population in worm cast is over 6 times more than in the original soil. A mosist compost heap (30-40per cent moisture level) of 2.4 m * 1.2 m * 0.6 m high, can support a population of mare than 50,000 worms. The temperature is the culture bed should be within the range of 200-300C. The introduction of worms into compost heap has been found to mix the materials, aerate the heaps and hasten decomposition.

Turning the heaps is not necessary, if earthworms are present to do the mixing and aeration. Besides rural and urban wastes, effluents from agro-industries viz dairies, tanneries, pulp and paper mills, distilleres etc.can be treated by using earthworms.

Benefits of Earthworms

Earthworms help in the preparation of compost maintaining soil health as follows:

- Improvement in fertility of soil.
- Arnelioration of physical condition of soil.
- Mixing of sub-soil and top soil.
- Correction of undetermined deficiences in plants.
- Use of earthworms in recycling of city and rural wastes, sewage waste waters and suldge, and industrial wastes eg. paper, food and wood industries.
- Supplementing traditional feeds

SPECIES FOR VERMI-COMPOSTING

Earthworm can be divided as surface living (epigeic) and burrowing (epianecic) worms. Epigeic or compost worms are found on surface and are reddish brown eg. Lumbricus rubellus (red worms). Of many species of earthworms tested for mass culture all over the world, Eisenia fetida, Eudrilus eugeniae and Perionyx excavatus come in the above order of preference for their ability to compost organic wastes. The shapes of cocoons of Eisenia frtida and Eudrilus eugeniae are dissimilar.

REARING OF EARTHWORMS

The worms are reared and multiplied from a commercially obatined breeder stock in shallow wooden boxes of 45 cm *60 cm, provided with drainage holes and stored on shelves and tiers. A bedding material is compounded from miscellaneous organic residues saw dust, cereal straw, rice husks, sugarcane trash, bagasse, paper, cardboard, coir waste, grasses etc. and is moistened well with water.

The wet mixture is stored for 30 days covered with a damp sack and is throughly mixed several times. When fermentation is complete, chicken manure and green matter eg. Leucaena leaves or water hyacinth is added. The material is placed in the boxes in the ans sufficiently loose for the worms to burrow and should be able to retain moisture. the The proportion of the different materials will vary according to the nature of the material but a final nitrogen content of about 2.4per cent should be amied at.

A pH value as near neutral as possible is necessary and the boxes should be kept at temperatures between 200C and 270C. At higher temperatures the worms will aestivate and at lower temperatures they hibernate. for each 0.1 m2 of surface area 100 g of breeder worms are added to the boxes. Inspite of their being able to eat the bedding material, the worm at this stage are regularly fed @ 1 kg of feed a day for every kg of worms. the feed stuffs used are again various types of organic matter and include partially digested cowdung, chicken manure, Leucaena leaves, vegetable waste and water hyacinth. Some form of protection against predators like birds, vats, ants, frogs, leeches and centipeded is provided to the worms.

VERMI-COMPOSTING IN PITS

A number of pits 2m *1m with sloping sides aredug having suitable dimension. Vermi-composting is done in pits and in vitro. Both of these are discussed here.

Bamboo poles are laid in paralled row on the pit's. Its floor is with a lattice of wood strips. Necessary drainage is provided because worms can not survive in a waterlogged condition. Alternatively to this and sand can be placed in the bottom of the pit to facilitate proper drainage. Above this a thick layer (15-20 cm) of good loamy soil should be spread. The pit can now be filled with available organic residues such as animal manure, leaves and green weeds, crop residues etc. Moisture levels of the contents of pit is maintained through addition of required amount of water. The worms from breeding boxes are introduced in the organic refuse, the worms immediately burrow down into the damp soil.

The compost pit is left for 60 days. It should be shaded from hot sunshine and it must be kept moist. Within 60 days about 10 kg of castinge would have been producer per kg of worms. The pit is then excavated to an extent of about two-thirds to three-quarters and the bulk of the worms removed by hand or by seving. This leaves sufficient worms in the pit for further composting and the pit can be refilled with fresh organic residues and continued.

IN-VITRO VERMI-COMPOSTING

This is also called as bioconversion in soil. This involves the application of the basal dose (5 tonnes/ha) of vermicastings and covering with 2.5 cm. Layer of organic mater (cowdung or pressmud) followed by 10 cm layer of sugarcane trash, crop residues or city swastes. The worms hatch out within 10 days.

VERMI-COMPOSTING OF AGRICULTURAL WASTES

With the aim of vermi-composting of agricultural wastes an experiment was conducted at the Indian Agricultural Research Institute, New Delhi. The usefulness and efficiency of earthworms in composting of agricultural residues was studied.

Two trials were conducted with mixed organic materials. Trial A, had more of green materials like grasses and Leucanena leaves mixed with soil and paper. Experient B had 4 kg of composting materials consisting of 2 kg of paddy straw, 1 kg soil 500 g twigs, and 500 g shredded paper and was laid in layers per pit. The bottom most layer was laid with twigs to allow for percolation of excess moisture. The compost pit was kept moist and care was taken to avoid waterlogging. After 10 days of initial decomposition, worms (Eisenia fetida) obtain from M/s Biogenic Ltd., Mumbai were introduced @ 100 worms/pit. Afterthe worms burrowed into the deeper layers, the pit was coveredwith a thin layer of soil. The results of both of the trials show that introduction of worms in the organic matter piles was found helpful in composting. Organic carbon was reduced at different intervals of composting. In the inoculated compost, nitrogen content wwas appreciably augmented and the C N ratio was narrowed to a desirable level.

The worms were also active in decomposition of high C N ratio of compost was brought to 51 only whereas with worms it was narrowed down to 21. The use of worms also increased the available phosphorous content of compost. Conjoint use of cellulolytic fungi and earthworms showed better result then other of them alone.

TOWN COMPOST

In recent years, large-scale composting of town refuse and night-soil in properly constructed trenches away from human habitations has been taken up successfully by the municipalities of many large and small towns. Trenches, 1 to 1.2 m wide, 75 cm deep and of convenient length, are filled with successive layers of night-soil, town refuse and earth, in this order. The compost gets ready in about three months. The following figures of volume-weight conversion will be found useful in the preperation of town compost.

Volume	Weight in kg or q
1 cu.m of refuse	= 318 kg or 3.18 q
1 litre of night-soil	= 0.991kg
1 cu. m of compost	= 636 kg or 6.36 q
1 cartload of refuse(0.849 cu. m),/font>	= 95.40 kg

Social prejudice against the use of this valuable compost has disappeared and town-composting is almost being rapidly adopted in other localities all over India. With a suitable modifications, such as the provision of trench latrines, it can be taken up in villages too. Particulars of the process may may be obtained from the National Extension Service Officers.

SEWAGE AND SLUDGE

The liquid waste, like sulage and sewage contain large quantities of plant nutrients and are used for growing of sugarcane, vegetables and fodder crops near many large towns by operating sewage-farms. In many places, the undiluted sullage has been found to be too strong for healthy plant growth and if it contains readily oxidized organic matter, its use actually reduces nitrates present in the soil. The disadvantages are still greater if sewage is used on land without preliminary treatment. The soil quickly becomes'sewage sick'owing to the mechanical clogging by colloidal matter in the sewage and the development of anaerobic organisms which not only reduce the nitrates already present in the soil but also produce alkalinity. Bacterial contamination makes the eating of raw vegetables on untreated sewage a real danger to health. For these reasons, it is now usual to construct a setting or a septic tank in which sewage is stand allowed to relieve it of the heavier portion of the solid matter in it, or to undergo a preliminary fermentation. The effluent from the settling-tank, however, still carries a large amount of objectionable colloidal matter, and the deposit of sludge that settles in the tank is of small manurial value and is often offensive. These defects can be removed by thoroughly aerating the sewage in the settling-tank by blowing air through it.

The sludge that settles at the bottom in this process is called'activated sludge'It has the remarkable property of bringing about the rapid oxidation of the organic matter present in fresh sewage. It is also in offensive and on dry-weight basis, contains 3 to 6 per cent N, about 2 percent P_2O_5 and 1 per cent K_2O in forms that can become readily available when applied to the soil. Similarly, the effluent is a clear, odourless liquid containing nitrates in solution, from shich most of the pathogenic bacteria originally present have been removed. Both the activated slidge and the effluent can be used with safety for manuring and irrigating crops. However, under no circumstances should any produce grown on a sewage-farm be eaten uncooked.

NIGHT-SOIL OR POUDRETTE

Very few towns India are equipped with complete sewage. The sanitory disposal of night-soil with an effective control of foul smell and fly nuisance is, therefore, a serious problem all over the country. Since human excrements are a potential source4 of soil improvement, public health authorities in many towns make the necessary arrangements for its conservation and conversion into a form in which it can be safely used as a manure. The dehydration of night-soil, as such, or after admixture with absorbing materials,e.g. soil, ash. charcoal and sawdust, produces a poudrette that can be easily used as a manure. The mixing of night-soil with an equal volume of ash and 10 per cent powdered charcoal produces an odourless material,containg 1.32 per cent potash and 24.2 per cent time.

The addition of 40 to 50 per cent of sawdust to the night-soil yields straightway a dry, acidic poudrette which may contain 2 or 3 per cent nitrogen. The annual potential quantity of manurial ingredients in the night-soil from India's present population of 600 millions is estimated at appromimately 8.1 million tonnes of dry matter containing almost 0.4 million tonnes of nitrogen, 0.25 million tonnes of phosphoric acid and 0.17 million tonnes of potash.

GREEN MANURES

Despite special efforts at increasing the supplies of farmyard manure and compost, the supply of farm and other organic manures is scarce and ever becoming more costly. Green-manuring, wherever feasible, is the principal supplementary means of adding organic matter to the soil. It consists in the growing of a quick-growing crop and ploughing it under to incorporate it into the soil.

The green-manure crop supplies organic matter as well as additional nitrogen particularly if it is a legume crop, which has the ability to acquire nitrogen from the air with the help of its root-nodule bacteria. A leguminous crop producing 8 to 25 tonnes of green matter per hectare will add about 60 to 90 kg of nitrogen when ploughed under. Rhis amount would equal an application of three to ten tonnes of farmyard manure on the basis of organic matter and its nitrogen contribution. The green manure crops also exercise a protective action against erosion and leaching.

The crops most commonly used for green-manuring in this country are the following Sunnhemp (Orotalaria juncea), dhaincha(Sesbania aculeata), cluster-bean(Cyamopsis tetragonoloba), senji(Melilotus parviflora), cowpea(Vigna catjang, V. sinensis), horse-gram(Dolichos biflorus), pillipesara(Phaseolus trilobus), berseem or Egyptian clover(Trifolium alexandrinum). Lentil(Lens esculenta) is recommended in Kashmir for green-manuring paddy. Sown in late autumn, it is said to provide a winter cover and make new growth in early spring for ploughing under before the sowing of paddy. Sunnhemp is the most outstanding green-manure crop. It is well suited to almost all parts of the country and fits in well with sugarcane, potatoes, garden crops and the second-season paddy in southern India and with irrigated wheat in the north.

Dhainchais in wide use in Assam, Bengal and Tamil Nadu. It does well on alkaline and water-logged soils. Cluster-bean, berseem and senji do well in Punjab, Uttar Pradesh, Rajas than, Delhi and some parts of Madhya Pradesh. Berseem is well suited for orchards and the irrigated crops of cotton and sugacane sown in spring or early summer, as in Punjab and Uttar Pradesh. Cowpea and horse-gram are used for green locality is naturally the one most suited to its soil and climatic conditions.

Very often, berseem, senji and lucerne (Medicago sativa), and sometimes sunnhemp are grown partly for fodder and partly for fodder and partly for green-manuring. In the case of annual crops of senji and berseem one or more cuttings are taken for use as green-foedder. Lucerne which is allowed to grow for two or three years, is cut seven to eight times for the same purpose.

In the case of sunnhemp, the tops are fed to the cattle. In all these instances, the residues (roots and stumps) are incorporated into the soil. These crop residues contain considerable amounts of nitrogen, phosphorus, potassium and other mineral nutrients, besides organic matter. In the case of orchards, the annual green manure-cum-forage crop should be grown at such a time as to interface the least with tree growth and fruit development.

Pulses form an essential part of the India diet, and are grown commonly as pure crops in rotation or mixed with cereals, oilseeds, and fibre crops. Roots and stubble of these pulse legumes return to the soil small quantities of organic matter rich in nitrogen. The inclusion of groundnut on the cotton-jawar rotation of centra, southern and western India, the growing of a quick-maturing variety of ming(Phaseolus aureus) before wheat in Uttar Pradesh, and wheat and rabi jowar in Marathwada Division of Bombay the sowing of peas in the standing crop of irrigated cotton in Uttar Pradesh, and the sowing of sunnhemp in the standing paddy crop in Andhra Pradesh, Tamil Nadu and Karnataka states, or of val IDolichos lablab) in the coastal paddy areas of Maharashtra, are valuable practices for soil improvement.

All these leguminous crops leave the soil in a better physical conditions and richer in nitrogen. The growing of pulses mixed with cereals all over India, the intercropping of cotton with groundnut of with tur(Cajanus indicus) in central and southern parts of the country, or with cluster-bean, mung and moth(Phaseolus aconitifolus) in Punjab and adjoing states; the growing of wheat mixed with peas and gram in northern and central India; and the growing of fodder sorghum mixed with Dolichos lablab in some parts of Tamil Nadu also enrich the soil.

In localities near forests in Tamil Nadu, Karnataka and Andhra Pradesh, the paddy crop is often manured with green forests leaves. These are incorprated into the soil at the time of puddling. In recent years, extensive efforts have been made in these states to plant Glyricidia maculata and Sesbania speciosa on the borders of paddy fields or in other vacant spaces to provide green leaf for manuring the paddy crop. Grown from seedlings or rooted stumps, planted 2 m apart, each Glyricidia plant is said to give annually tow cuttings each of about 6 to 12 kg of green leaf.

It does well in both red black soils. Similarly Sesbania speciosa seedlings planted 10 cm apart on paddy borders produce 1,000 to 2,500 kg of green leaf for manuring 0.4 ha of paddy. Only 115 g of seed is needed to provide seedlings sufficient for border-planting ot two hectares of paddy. In certain other paddy-growing areas, Pongamia pinnata(karanji), Tephrosia, Terminalia and other trees yielding large quantities of leaves are planted for use as a green manure. In the Malabar districts of Tamil Nadu, Indigofera teysmanni is grown to pfovide green leafy twigs for manuring paddy.

For the proper rotting of the green manure, it is necessary that the green material should be succulent and there should be adequate moisture in the soil. Plant at the flowering stage, contain the greatest bulk of succulent organic matter with a low carbon/nitrogen ratio. The incorporation of the green-manure crop into the soil at this stage allows a quick liberation of nitrogen in the available form. With advancing age, the percentage of carbonaceous matter in the plants increases and that of nitrogen decreases. if the material with a side carbon/nitrogen ratio isploughed under, micro-organisms bring about its decomposition, draw upon the released nitrogen and mineral nutrients and cause a temporary nutirient deficiency.

Sometimes methi or dhaincha is sown in-between the rows of the newly planted sugarcane crop, or cluster-bean is planted between the rows of irrigated American cotton. When the leguminous plants are five to six weeks old, they are incorporated into the soil. Sugarcane and cotton are claimed to benefit from the legumes. The decay of green matter in this case seems to occur when it best serves as a fertilizer for the beneficiary crop.

The increase of yield after green-manuring is usually of the order of 30 to 50 per cent. The fertilizing value of the legume crop can be increased a great deal by manuring it with superphosphate. This practice not only increases the phosphorous content of the green-manure plants, but also encourages their plant growth on the whole, thus converting an inorganic fertilizer into an organic manure. Green manures have a marked residual effect also.

FERTILIZERS

Despite the special steps taken in recent years to increase the supply of farmyard and other bulky organic manures, the available quantities are insufficient to meet the existing and prospective needs. Fertilizers have also the dvantage of smaller bulk, the resultant easy transport, relatively quick availability of their plant-food constituents and the possibility of their application in proportions suited to the actual requirements of different crops and soils. Fertilizers are usually classified according to the three principal elements and may, therefore, be included in more than one group.

Nitrogeneous Fertilizers

According to the manner in which their nitrogen is combined with other elements, the nitrogenous fertilizers are divided into four groups; nitrare, ammonia and ammonium salts, chemical compunds containing nitrogen in the amide form, and plant and animal by-products.

Sodium Nitrates

It is also known as'Chilean'nitrate. It owes its importance as a pioneer nitrogenous fertilizer. It occurs in natural deposits in northern Chile and is refined before shipping. The refined product contains about 16 per cent nitrogen in the nitrate form, which renders it directly available tp plants.

For this reason, it is highly valued as a source of nitrogen when applied as top-and side-dressings, especially to young plants and garden vegetables, which need readily available nitrogen for quick growth leached out from the soil. for wheat, maize, barley, cotton, sugarcane, etc., it is as benefecial as ammonium sulphate. Sodium nitrate is particularly useful for acidic soils. Its continued and abundant use in soils is said to cause deflocculation and develop a bad physical condition in regions of low rainfall. It should be stored in a dry ware-house

AMMONIUM SULPHATE

It is the most widely used fertilizer in the country. It is a white crystalline salt, containing 20 to 21 percent aoomniacal nitrogen. It is easy to handle and it stores well under dry conditions. During the rainy season, it sometimes, forms lumps.

These lumps should be powedered before use. Being soluble in water, it acts quickly, but despite its high solubility, its nitrogen is not readily lost in drainage, because the ammonium ion is retained by the soil particles. it is, therefore, very suitable for wet-land crops, e.g. paddy and jute. It has also been found useful on wheat, cotton, sugarcane, potatoes and many other crops grown on a wide variety of soils. It has, however, an acid effects on the soils. Its long-continued use increases soil acidity and lowers the yield. The application of this fertilizer to acid soils improved the yield of tea plants considerably. It is advisable to use this fertilizer in conjunction with bulky organic manures to safeguard against the ill effects of continued application of ammonium sulphate to field and horticultural crops. Ammonium sulphate can be applied before sowing, at sowing time, or as a top-dressing to the growing crop. it should not be applied along with, or too close to, the seed, because in concentrated form, it affects seed germination very adversely.

AMMONIUM NITRATE

It is a white crystalline salts, containing 33 to 35 per cent nitrogen, half as nitrate nitrogen and half in the ammonium form. in the ammonium form, it cannot be easily leached from the soil.

This fertilizer is quick-acting, but highly hygroscopic and not fit for storage. Granulation of the material and a light coating of the granules with oil reduce hygroscopicity to some extent. It has an acidulating effect on the soil.

Under certian conditions, it is explosive, and, therefore, it should be handled cautiously.'Nitro Chalk'is the trade name of a product formed by mixing ammonium nitrate with about 40 per cent lime-stone or dolomite. It is granulated, non-hazardous and less hygroscopic. it contains 20.5 per cent nitrogen, half in the form form of ammonia and half as nitrate. The presence of lime in it makes it particularly useful for acid soils.

AMMONIUM SULPHATE NITRATE

It is a mixture of ammonium nitrate and ammonium sulphate. it is available in a white crystalline form or as dirty-white granules. This fertilizer contains 26 per cent nitrogen, three-fourths of it in the ammoniacal form and the rest(6.5 per cent) as nitrate nitrogen. It is non-explosive and not as deliquescent as ammonium nitrate. It is readily soluble in water and is very quick-acting. its keeping quality is good and it is useful for all crops. Its keeping quality is good and it is useful for all crops. Its acid effect on the soils is only one-half of that of ammonium sulphate. It can be applied before sowing, at sowing time or as a top-dressing, but it should not be applied along the seed.

AMMONIUM CHLORIDE

It is a white crystalline compound, possessing a good physical condition and containing 26 per cent ammoniacal nitrogen. It is extensively used on paddy in Japan, In India, it is used largely in industries. In general, it is similar to ammonium sulphate in action. It is usually not recommended for tomatoes, tobacco and such other crops as may be injured by chlorine.

UREA

It is a white, crystalline, organic chemical. It is a highly concentrated nitrogenous fertilizer, containing 45 to 46 per cent of non-proteined organic nitrogen. it is fairly hygroscopic and presents considerable difficulty, it is also produced in granular or pellet forms and is coated with a non-hygroscopic inert material. It is highly soluble in water and, therefore, subject to rapid leaching. It is, however, quick-acting. When applied to the soil, its nitrogen is rapidly changed into ammonia. Like ammonium nitrate, urea supplies nothing but nitrogen. It may be applied at sowing time or as a top-dressing, but should not be allowed to come into contact with the seed. It is suitable for most frops and can be applied to all soils.

AMMONIA

It is a gas containing about 80 percent of nitrogen. Under suitable conditions of temperatures and pressure, it becomes liquid(anhydrous ammonia). Another form,'aqueous ammonia', results from the absorption of ammonia gas into water, in which it is soluble. Ammonia is used as a fertilizer in both these forms. Anhydrous ammonia can be applied by introducing it into irrigation water, or directly into the soil from special containers, which makes its use rather expensive. Its possibilities as manure for paddy, sugarcane and cotton are being onvestigated at Bangalore in the Karnataka state. In manurial experiments on cotton in Maharashtra, aqueous ammonia has been found to be as efficient as ammonium sulphate.

Calcium Ammonium Nitrate

Calcium ammonium nitrate is a fine free-flowing, light brown or grey granular fertilzer. It is commercially prepared from ammonium nitrate and ground limestone. It is almost neutral and can be safely applied even to acid soils. Its total nitrogen content may vary from 25 to 28 per cent. Half of this total nitrogenis in the ammoniacal form and half is in nitrate form. According to the prescribed standards, its moisture content should not be more than 0.5per cent by weight. As regards the particle size, 90 per cent of the material should pass through 4 mm IS sieve and be retained on a 1-mm IS sieve but not more than g per cent shall be below 1 mm.

ORGANIC NITROGENOUS FERTILIZERS

These fertilizers include plant and animal by-products, such as oil cakes, fish manure and dried blood from slaughter houses. Before their organic notrogen can be used by the crops, it is converted through bacterial action into readily usable ammonia-nitrogen and nitrate nitrogen. These fertilizers are, therefore, relatively slow-acting, but they supply available nitrogen for a longer perios. Furthermore, they may also small amounts of organic stimulants that they may contain, or of some of the minor elements needed by plant.

Oil-cakes of different kinds are produced in India to the tune of about two million tonnes anually. They contain not only nitrogen but also some phosphoric and potash, besides a large quantity of organic matter. In addition to the three fertilizing constituents (N, P_2O_5 and K_2O), the oil-cakes invariably contain2 to 15 per cent of oil, depending on whether the oil is extracted by using solvent process or with expelers, hydraulic presses or indigeneous ghanis. This residual oil, however, dous not, in practice, affect their manurial value. Owing to the great importance of using edible oil-cakes as a cattle feed, their utilization as fertilizers is undesirable. They should be fed to cattle and the excrements may be used as manure.

Inedible cakes, like castor cake, neem, mahua cake and karanj cake can, however, be recommended for use in conjunction with quicker-acting chemical fertilizers. Dried blood or blood-meal contains 10 to 12 per cent highly available nitrogen and 1 to 2 per cent phosphoric acid. It is a very quick-acting manure and effective on all crops and all types of soils. It should be used in the same way as oil cakes. Fish manure is available either as dried fish or as fish-meal or powder. In regions where fish oil is extracted, the residue can be used as a manure. Depending on the type of fish, its manurial constituents vary from 5 to 8 per cent of organic nitrogen and from 4 to 6 per cent of phosphoric acid. It is quick-acting and suitable for all crops and soils. It should preferably be powdered before use.

PHOSPHATE FERTILIZERS

These are classified as natural phosphates, treated or processed phosphates, and by-product phosphates anc chemical phosphates.

Rock Phosphate

It occurs as natural deposits of rock in Morocco, the United States of America, Poland, Russia, Tunisia, Algeria, Algeria, Brazil, Egypt, Nehru and some islands in the Pacific Ocean and the Indian Ocean. It contains 25 to 35 per cent phosphoric acid, but this phosphorus is insoluble in water. Practically no rock phosphate, as such, is used as a fertilizer in this country, except some quantity in southern India.

In other countries, finely pulverized phosphate rock has been found to give satisfactory results in soils which are very deficient in phosphprus and are acidic. Adequate rainfall and a long growing period of the crop enhance the response. All the same, very little rock phosphate is used is directly as a fertilizer even in these countries. Much more of it is used to manufacture superphosphate, the phosphoric acid of which is water-soluble and is in an available form.

SUPERPHOSPHATE

It is the most widely used fertilizer in India. Prepared formerly by treating bones with sulphuric acid, it is now manufactured largely by treating ground phosphate rock with almost an equalquantity by weight of sulphuric acid.

This treatment produces a brownish-grey mixture containing monocalcium phosphate and calcium sulphate(gypsum) in practically equal quantites. This fertilizer is manufactured in three grades single superphosphate containing 16 to 20 per cent phosphoric acid; dicalcium phosphate, 35 to 38 per cent; and triple superphosphate,44 to 49 per cent. Single superphosphate is the most commonly available grade in the Indian market. Triple superphosphate in the prodiction of which liquid phosphoric acid is used instead of sulphuric acid, contains very little calcium sulphate. Tripple superphosphate is used mostly in the manufacture of concentrated mixed fertilizers.

The phosphoric acid in superphosphate is wholly water-soluble but when applied to the soil, it is immediately converted into insoluble phosphate owing to precipitation as calcium, iron or aluminium phosphate, according as the soil is alkaline or acid. Thus athe fertilizer is not leached, but is slowly dissolved in the soil solution. Fixation losses can be reduced by applying the fertilizer in bands on both sides of the row of seeds at a depth of 10 to 15 cm with a drill. By this means, at least some of the phosphate does not come into direct contact with the soil and thus the available phosphorous is readily released for absorption by the plant roots. The fertilizer is suitable for all crops and can be applied to all soils. In acid soils, it should properly be used in conjunction with organic manure. It should be applied before or at sowing or transplanting.

Basic-slag

This is a by-product of steel factories. Depending on the phosphorous content of the iron ore, it contains from 6 to 20 per cent of phosphoric acid (P_2O_5). Slag from Indian steel-mills is poor in P_2O_5 and is not used as a fertilizer. The high-grade European slag containing 15 to 18 per cent P_2O_5 is a popular phosphatic fertilizer in central Europe. It is not as soluble as superphosphate, but unlike the latter, it is alkaline in reaction and good for acid soils. For effective use, it must be pulverized before application.

Bone-meal

The use of raw bones as amanure for fruit-trees is an age-old practice in this country. Burying the skeleton of an animal under a fruit-tree is known to benefit its growth and bearing. Large quantities of bones used to be exported until a few years ago. Exports have since declined and bone-meal(i.e. ground bone) is now a widely used phosphate fertilizer. It is available in two forms

- Raw bone-mel,
- Steamed bone-meal.

The steaming of bones under pressure removes fats, greases, nitrogen and glue-making substances. Thus, while raw bone-meal contains about 4 per cent slow-acting organic nitrogen and 20 to 25 per cent insoluble phosphoric acid. Steamed bones are more brittle and can be readily ground. This is an advantage, as the rate of availability of phosphoric acid. Steamed bones are more brittle and can be readily ground. This is an advantage, as the rate of availability of phosphoric acid in bones depends largely on their degree of pulverization. Bone-meal contains only 1 to 2 per cent nitrogen but 25 to 30 per cent phosphoric acid.

Steamed bones are more brittle and can be readily ground. This is an advantage, as the rate of availability of phosphoric acid. in bones depends largely on their degree of pulverization. Bone-meal, having particles not larger than3/32 inch, considered suitable for use as a fertilizer, but the more finely powedered it is, the quicker its P_2O_5 becomes available in the soil. Being relatively slow-acting, bone-meal should not be used as a top-dressing; it must be incorporated into the soil in order to become available. It may be applied either at sowing time or a few days before sowing and should be broadcast. It is particularly suitable for acid soils. It is considered a safe manure for all crops. In some parts of the country, charred and powedered bones are used as a manure. Charring destroys about half the nitrogen, but leaves intact practically the whole of P_2O_5 in a quickly available form. In the absence of arrangements for steaming and grinding, charring can be easily carried out even in remote villages.

POTASSIC FERTILIZERS

Most of the Indian fertilizers soils contain a sufficient amount of potash. Potassic fertilizers should, therefore, be applied only to such soils as are definitely known to be defecient in potash or to those which respond to their application, such as sandy soils. They can also be applied to certain crops, such as tobacco, potato, onion, tomato and fruit-trees, to improve the quality and appearance of their produce. Potassic fertilizers in common use are

- Muriate of potash(potassium chloride),
- Sulphate of potash (potassium sulphate).

These salts are important constituents of the waters of oceans and inland seas and of saline deposits derived thereform. The largest known deposits of these salts are at Stassfurt in Germany, in the Caspian Sea region in Russia, in the Dead Sea in Palestine and at some places in California, New Mexico, france and Spain.

Muriate Of Potash

It is a gray crystalline material containing 50 to 63 per cent potash (K_2O), the whole of which is readily available. Though highly soluble in water, it is not lost from the soil, as it is absorbed on the colloidal surfaces. it can be applied at sowing time or before sowing.

Sulphate Of Potash

It is made by treating potassium chloride with magnesium sulphate and is, therefore, more costly. it contains 48 to 52 per cent K_2O. It dissolves readily in water and becomes available to the crop almost immediately. It can be applied at any time up to sowing, but should not be drilled with the seed. It is considered better than muriate of potash for crops, such as tobacco, chillies, potato and fruit-tree, where quality is of prime importance.

OTHER SOURCES

Wood ashes, cattle-dung ash, leaf-mould, tobacco stems and water hyacinth are the available indigenous sources of potash. Unleached wood ash contains 5 to 6 per cent of potash in the form of potassium carbonate (which is alkaline), 1 to 2 per cent phosphoric acid and 25 to 30 per cent lime (Cao). Both potassium carbonate and lime in the wood ashes counteract the acidity in the soil. Groundnut shell, paddy husk and bagasse ashes are available near decorticating factories and rice and sugar-mills.

These also contain a fair amount of potash and some phosphoric acid. Ground tobacco stems contain 2 to 3 per cent of nitrogen and 6 to 10 per cent of potash in quickly available forms. Hyacinth abounds as a weed in fresh-water ponds in Bengal, Assam, Tripura and Malabar in Kerala. When dry, it contains 1 per cent nitrogen, 4 per cent potassium and a small quantity of phosphorous.

Soil Amendments

Lime is generally used for correcting soil acidity, for improving the physical condition of the soil and encouraging bacterial activity. Similarly, gypsum is used for reclaiming'alkali'soils or land from the sea and improving the structure of heavy black clay soils. Hence, these are called soil amendments.

Compund Fertilizers

These fertilizers are multiple nutrient materials, supplying two or three plant nutrients simultaneously. When both nitrogen and phosphorous are deficient in a soil, a compound fertilizer, e.g. ammophos, can be used. It contains 16 per cent N and 20 per cent P2O5. Its use does away with the necessity of purchasing two different fertilizers and mixing them in correct proportion before use. The nutrient contents of some of the other compound fertilizers are shown below

	N	P_2O_5	K_2O
Monoammonium phosphate	11.0	48.0	—
Diammonium phosphate			
Ammoniated superphosphate	2.0	4.0	—
(4 - 7 per cent)	14.0	20.0	
Potassium nitrate	13.0	—	44.0

Mixed Fertilizers

Compound fertilizers contain plant food elements in fixed propertions and are, therefore, not always best adapted ti different kinds of soils. Accordingly, the needs of different soils can generally be met most economically by the use of fertilizer mixtures containing two or more materials in suitable propertions. Mixtures usually meet nutrient deficiences in a more balanced manner and require less labour to apply than straight fertilizers used seperately. Mixtures containing all the three principal used seperately. Mixtures containing all the three principal nutrients (N, P and K) are termed complete fertilizers.

Some manufacturers prepare special mixtures for different crops, such as wheat, sugarcane, pady, potatoes, tabacco, fruit-trees and vegetables. Trade names like Ammophos, Niciphos and Nitrochalk are employed for other proprietary products, In foreign countries, even insecticides, fungicides and weed-killers, such as DDT, BHC and mercury or copper salts and 2, 4-D are sometimes incorporated into fertilizers mixtures.

When a required mixture is not available of the cost of the mixture sold in the market is high in relation to the cost of its individual components, it may be made at home by mixing the constituent fertilizers in correct proportion.

The preparation of mixed fertilizers requires a good knowledge of the properties and mutual reactionof the component straight fertilizers under different climatic and storage conditions. Therefore in preparing such mixtures at the farm or at home, care should be taken to avoid the uneven mixing of incompatible fertilizers, or the mixing which leads to a loss of some of the fertilizing nutrients in the form of gas, converts soluble nutrients into insoluble ones or induces caking. It is unwise to mix the following

- Ammonium sulphate, ammonium chloride, other ammoniacal fertilizers and nitrogenous organic manures with lime.
- Sodium nitrate or potassium nitrate with superphosphate.
- Nitrochalk with superphosphate or lime.
- Ammonium sulphate-nitrate with lime.
- Urea with superphosphate.
- Superphosphate with lime or calcium carbonate or wood ashes.

Ammonium nitrate is an explosive chemical and, therefore, dangerous should be mixing at home. Mixtures containing other nitrates should be made in only such quantities as are to be used immediately, because they absorb water quickly and are not suitable for storage. Bone meal, sulphate of potash and muriate of potash can be mixed with all fertilizers. For further guidance or information on the method of mixing fertilizers, the State Agricultural Chemist or the nearest officer of the Government Agricultural Department should be contacted.

MODE AND TIME OF APPLICATION OF FERTILIZERS AND MANURES

Bulky organic manures should be applied well ahead of sowing, so that the preliminary decomposition takes place before the seeds germinate.

Failing presowing application, they may be applied any time after the seedlings have established themselves. They are best applied in the powedered form.

A sufficient supply of moisture in the soil is essential for their rapid decomposition. In the case of inorganic fertilizers, potassic and phosphatic fertilizers are best applied just before sowing or transplanting. Nitrogenous fertilizers may be applied either at planting and partly later. Split application is particularly desirable for nitrogen when applied to irrigated crops or to crops in heavy-rainfall areas.

Fertilizers applied before sowing should be broadcast uniformly and harrowed in. In the case of fertilizers containing soluble phosphate, the desirability of applying them in 2.5 to 5 cm wide bands on each side of the row of seeds at a depth of 10 to 15 cm with a drill has already been pointed out. This operation reduces the fixation of soluble phosphate in the soil. It is also a good practice to mix superphosphate with farmyard manure at 18 to 22 kg to a tonne before applying the organic manure, particularly of dairy farms land. Sulphate of ammonia used as a top-dressing should not be applied when plant leaves are wet. In the case of irrigated crops, the application of fertilizer should invariably be followed by a watering. In the case of fruit-trees, the fertilizer should be applied to the soil under the crown, a few metres away from the trunk. The area of application should be progressively extended as the trees grow bigger.

In advanced countries, fertilizers are usually applied with the help of machinery of diverse kinds and sometimes with an aeroplane or a helicopter. Combined-planters and fertilizer-distributors are employed when row crops are fertilized at sowing time.

DIAGNOSING THE FERTILIZER NEEDS OF SOILS

There are four methods of determining the fertilizer requirements of soil

- Field experiments,
- Pot tests,
- Biological tests,
- Chemical test.

Field experiments contribute the more relaible method, but being time-consuming and expensive, they are conducted mainly by the research farms and research organisations. Farmers wishing to use field experiments as a Valid means of determining the fertility status of their soils should seek the advice of the state agronomist. Improperly conducted field experiments not only mean economic fertilizer practices.

Pot experiments permit test witha alarge number of manurial treatments within a limited space and in a relatively short time. However, as the conditions of such tests are different from those in the field, the results are not always directly applicable to large-scale farming. Biological tests involve the growth of seedlings or of lower forms of plants, such as fungi and bacteria, under specified conditions and the study of their relative growth or the content of needed nutrients. A peridic testing of plant tissues for nitrates and other nutrients indicates the changing needs of crops for different food elements. But these are slow and costly processes and hence not always practicable.

The chemical analysis of soils or of plants growing on them constituents the modern method of determining the fertility status of a soil. Such analysis give information on the relative abundance or scarcity of the different nutrients required by crops from the soil, but they give no indication regarding the exact quantity, of fertilizer that may be applied to make good the deficiency. The dependability of this, method can, however, be increased a great deal by co-ordinating its results with those obtained from field experiments.

Facilities for rapid soil-testing have been made available in almost all states and can be availed of. At the same time, a very large number of fertilizer experiments with different crops on a variety of soils are conducted annually in all parts of the country under a comprehensive scheme. The results of these field experiments, when calibrated against those of rapid soil tests, will make the latter purely dependable. This will then be a valuable aid in the hands of extension workers in furnishing advice to farmers regarding fertilizer practices.

It is also sometimes possible to obtain a clue to the nutrient deficiences of soil with the help of deficiency symptoms in plants, as described already. However, a correct deagnosis of deficiency symptoms needs extensive experience. Furthermore, such symptoms in plants appear long after the actual occurance of nutrient deficiency in the soil. Therefore, such soil deficiencies must be diagnosed and remedied much earlier by adopting other means.

7

Hydrocarbons in Organic Chemistry

HYDROCARBON

Organic chemistry started as the chemistry of living things. Chemists believed that organic compounds had living things as the only source of organic compounds. Chemists held the belief that there had to be a "vital force" present in living things to make organic compounds. A German chemist, Wohler (1828) produced urea in the lab. Urea is an organic compound found in urine. The urea was produced in the laboratory without a living thing, no vital force present.

Today the chemist views organic chemistry as the study of carbon compounds. There are millions of organic compounds. Why are so many compounds possible? The answer must rest with the carbon atom.

- The carbon atom is a member of group 14 with 4 valence electrons; the ground state electron configuration for carbon is $1s^2 2s^2 2p^2$. The bonding state for carbon is $1s^2 2s^1 2p^3$, one of the 2s electrons is promoted to a 2p orbital, all valence orbitals are half full. The configuration of the valence shell becomes $2s^1 2p_x^{\ 1} 2p_y^{\ 1} 2p_z^{\ 1}$; carbon has 4 bonding electrons and can form 4 single covalent bonds. Carbon is tetravalent and can bond with other elements.
- Carbon can bond with other carbons to form chains of varying lengths.
- The carbons in the chain can be arranged in many different ways.

REACTIONS OF ALKYNES

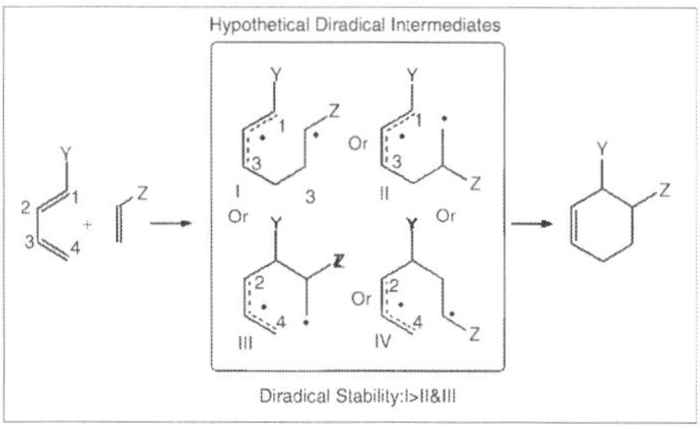

ADDITION REACTIONS OF ALKYNES

A carbon-carbon triple bond may be located at any unbranched site within a carbon chain or at the end of a chain, in which case it is called terminal. Because of its linear configuration (the bond angle of a sp-hybridized carbon is 180°), a ten-membered carbon ring is the smallest that can accommodate this function without excessive strain. Since the most common chemical transformation of a carbon-carbon double bond is an addition reaction, we might expect the same to be true for carbon-carbon triple bonds. Indeed, most of the alkene addition reactions discussed earlier also take place with alkynes, and with similar regio- and stereoselectivity.

Catalytic Hydrogenation

The catalytic addition of hydrogen to 2-butyne not only serves as an example of such an addition reaction, but also provides heat of reaction data that reflect the relative thermodynamic stabilities of these hydrocarbons, as shown in the diagram to the right. From the heats of hydrogenation, shown in blue in units of kcal/mole, it would appear that alkynes are thermodynamically less stable than alkenes to a greater degree than alkenes are less stable than alkanes.

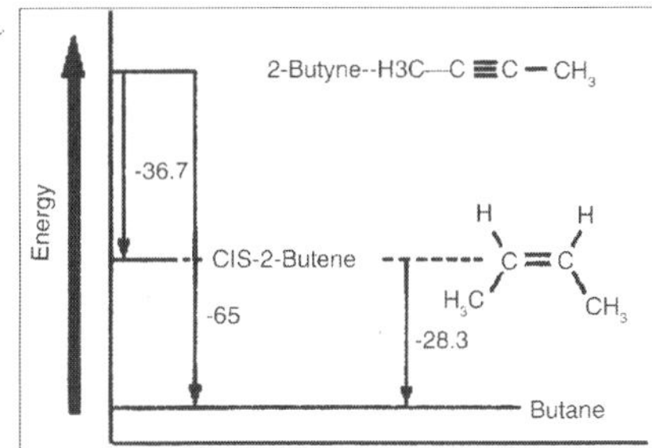

The standard bond energies for carbon-carbon bonds confirm this conclusion. Thus, a double bond is stronger than a single bond, but not twice as strong. The difference (63 kcal/mole) may be regarded as the strength of the π-bond component. Similarly, a triple bond is stronger than a double bond, but not 50per cent stronger. Here the difference (54 kcal/mole) may be taken as the strength of the second π-bond. The 9 kcal/mole weakening of this second π-bond is reflected in the heat of hydrogenation numbers (36.7 - 28.3 = 8.4).

Since alkynes are thermodynamically less stable than alkenes, we might expect addition reactions of the former to be more exothermic and relatively faster than equivalent reactions of the latter. In the case of catalytic hydrogenation, the usual Pt and Pd hydrogenation catalysts are so effective in promoting addition of hydrogen to both double and triple carbon-carbon bonds that the alkene intermediate formed by hydrogen addition to an alkyne cannot be isolated.

A less efficient catalyst, Lindlar's catalyst, prepared by deactivating (or poisoning) a conventional palladium catalyst by treating it with lead acetate and quinoline, permits alkynes to be converted to alkenes without further reduction to an alkane. The addition of hydrogen is stereoselectively syn (e.g. 2-butyne gives cis-2-butene). A complementary stereoselective reduction in the anti mode may be accomplished by a solution of sodium in liquid ammonia. This reaction will be discussed later in this section.

$$R–Ca≡C\text{-}R + H_2 \text{ \& Lindlar catalyst} \rightarrow \textit{cis } R–CH=CH–R$$

$$R–Ca≡C–R + 2\ Na \text{ in } NH_3 \text{ (liq)} \rightarrow \textit{trans } R–CH=CH–R + 2\ NaNH_2$$

Alkenes and alkynes show a curious difference in behaviour towards catalytic hydrogenation. Independent studies of hydrogenation rates for each class indicate that alkenes react more rapidly than alkynes. However, careful hydrogenation of an alkyne proceeds exclusively to the alkene until the former is consumed, at which point the product alkene is very rapidly hydrogenated to an alkane.

This behaviour is nicely explained by differences in the stages of the hydrogenation reaction. Before hydrogen can add to a multiple bond the alkene or alkyne must be adsorbed on the catalyst surface. In this respect, the formation of stable platinum (and palladium) complexes with alkenes has been described earlier. Since alkynes adsorb more strongly to such catalytic surfaces than do alkenes, they preferentially occupy reactive sites on the catalyst. Subsequent transfer of hydrogen to the adsorbed alkyne proceeds slowly, relative to the corresponding hydrogen transfer to an adsorbed alkene molecule.

Consequently, reduction of triple bonds occurs selectively at a moderate rate, followed by rapid addition of hydrogen to the alkene product. The Lindlar catalyst permits adsorption and reduction of alkynes, but does not adsorb alkenes sufficiently to allow their reduction.

Addition by Electrophilic Reagents

When the addition reactions of electrophilic reagents, such as strong Bronsted acids and halogens, to alkynes are studied we find a curious paradox. The reactions are even more exothermic than the additions to alkenes, and yet the rate of addition to alkynes is slower by a factor of 100 to 1000 than addition to equivalently substituted alkenes. The reaction of one equivalent of bromine with 1-penten-4-yne, for example, gave 4,5–dibromo–1–pentyne as the chief product.

$$HC≡C–CH_2–CH=CH_2 + Br_2 \rightarrow HC≡C–CH_2–CHBrCH_2Br$$

Although these electrophilic additions to alkynes are sluggish, they do take place and generally display Markovnikov Rule regioselectivity and anti-stereoselectivity. One problem, of course, is that the products of these additions are themselves substituted alkenes and can therefore undergo further addition. Because of their high electronegativity, halogen substituents on a double bond act to reduce its nucleophilicity, and thereby decrease the rate of electrophilic addition reactions.

Consequently, there is a delicate balance as to whether the product of an initial addition to an alkyne will suffer further addition to a saturated product. Although the initial alkene products can often be isolated and identified, they are commonly present in mixtures of products and may not be obtained in high yield. The following reactions illustrate many of these features. In the last example, 1,2-diodoethene does not suffer further addition inasmuch as vicinal-diiodoalkanes are relatively unstable.

$$C_2H_5-C\equiv C-H \xrightarrow[\text{Slow}]{HCl} (C_2H_5)(Cl)C=CH_2 \xrightarrow[\text{Slow}]{CHI} C_2H_5-CCl_2-CH_2-H$$

$$H_3C-C\equiv C-CH_3 \xrightarrow[\text{Slow}]{HBr} (H_3C)(Br)C=C(H)(CH_3) \xrightarrow[\text{Slow}]{HBr} H_3C-CBr_2-CH_2-CH_3$$

$$C_2H_5-C\equiv C-H \xrightarrow[\text{Slow}]{Br_2} (C_2H_5)(Br)C=C(Br)(H) \xrightarrow[\text{Slow}]{Br_2} C_2H_5-CBr_2-CBr_2-H$$

$$H-C\equiv C-H \xrightarrow[\text{Slow}]{I_2} (H)(I)C=C(I)(H)$$

As a rule, electrophilic addition reactions to alkenes and alkynes proceed by initial formation of a pi-complex, in which the electrophile accepts electrons from and becomes weakly bonded to the multiple bond. Such complexes are formed reversibly and may then reorganize to a reactive intermediate in a slower, rate-determining step. Reactions with alkynes are more sensitive to solvent changes and catalytic influences than are equivalent alkenes.

Why are the reactions of alkynes with electrophilic reagents more sluggish than the corresponding reactions of alkenes? After all, addition reactions to alkynes are generally more exothermic than additions to alkenes, and there would seem to be a higher π-electron density about the triple bond (two π-bonds versus one).

Two factors are significant in explaining this apparent paradox. First, although there are more π-electrons associated with the triple bond, the sp-hybridized carbons exert a strong attraction for these π-electrons, which are consequently bound more tightly to the functional group than are the π-electrons of a double bond. This is seen in the ionization potentials of ethylene and acetylene.

$$\text{Acetylene } HC\equiv CH + \text{Energy} \rightarrow [HC\equiv CH\bullet^{(+)} + e^{(-)} \quad \Delta H = +264 \text{ kcal mole}$$

$$\text{Ethylene } H_2C=CH_2 + \text{Energy} \rightarrow [H_2C=CH_2]\bullet^{(+)} + e^{(-)} \quad \Delta H = +244 \text{ kcal/mole}$$

$$\text{Ethane } H_3C-CH_3 + \text{Energy} \rightarrow [H_3C-CH_3]\bullet^{(+)} + e^{(-)} \quad \Delta H = +296 \text{ kcal/mole}$$

As defined by the preceding equations, an ionization potential is the minimum energy required to remove an electron from a molecule of a compound. Since pi-electrons are less tightly held than sigma-electrons, we expect the ionization potentials of ethylene and acetylene to be lower than that of ethane, as is the case. Gas-phase proton affinities show the same order, with ethylene being more basic than acetylene, and ethane being less basic than either. Since the initial interaction between an electrophile and an alkene or alkyne is the formation of a pi-complex, in which the electrophile accepts electrons from and becomes weakly bonded to the multiple bond, the relatively slower reactions of alkynes becomes understandable.

A second factor is presumed to be the stability of the carbocation intermediate generated by sigma-bonding of a proton or other electrophile to one of the triple bond carbon atoms. This intermediate has its positive charge localized on an unsaturated carbon, and such vinyl cations are less stable than their saturated analogs.

Indeed, we can modify our earlier ordering of carbocation stability to include these vinyl cations in the manner shown below. It is possible that vinyl cations stabilized by conjugation with an aryl substituent are intermediates in HX addition to alkynes of the type Ar–C≡C–R, but such intermediates are not formed in all alkyne addition reactions.

Carbocation Stability	$CH_3^{(+)}$	≈ $RCH{=}CH^{(+)}$	< $RCH_2^{(+)}$	≈ $RCH{=}CR^{(+)}$	< $R2CH^{(+)}$	≈ $CH_2{=}CH\text{-}CH_2^{(+)}$	< $C_6H_5CH_2^{(+)}$	≈ $R_3C^{(+)}$
	Methyl	1°-Vinyl	1°	2°-Vinyl	2°	1°-Allyl	1°-Benzyl	3°

Application of the Hammond postulate indicates that the activation energy for the generation of a vinyl cation intermediate would be higher than that for a lower energy intermediate. This is illustrated for alkenes versus alkynes by the following energy diagrams.

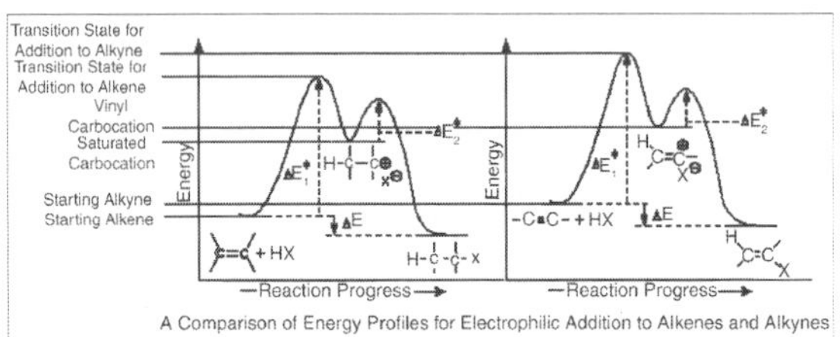

A Comparison of Energy Profiles for Electrophilic Addition to Alkenes and Alkynes

Despite these differences, electrophilic additions to alkynes have emerged as exceptionally useful synthetic transforms. For example, addition of HCl, acetic acid and hydrocyanic acid to acetylene give respectively the useful monomers vinyl chloride, vinyl acetate and acrylonitrile, as shown in the following equations. Note that in these and many other similar reactions transition metals, such as copper and mercury salts, are effective catalysts.

$HC{\equiv}CH + HCl + HgCl_2$ (on carbon) $\rightarrow H_2C{=}CHCl$ vinyl chloride

$HC{\equiv}CCH_2Cl + HCl + HgCl_2 \rightarrow H_2C{=}CClCH_2Cl$ 2,3-dichloropropene

$HC{\equiv}CH + CH_3CO_2H + HgSO_4 \rightarrow H_2C{=}CHOCOCH_3$ vinyl acetate

$HC{\equiv}CH + HCN + Cu_2Cl_2 \rightarrow H_2C{=}CHCN$ acryonitrile

Complexes formed by alkenes and alkynes with transition metals are different from the simple pi-complexes noted above. Here a synergic process involving donation of electrons from a filled π-orbital of the organic ligand into an empty d-orbital of the metal, together with back-donation of electrons from another d-orbital of the metal into the empty π^*-antibonding orbital of the ligand.

Hydration of Alkynes and Tautomerism

As with alkenes, the addition of water to alkynes requires a strong acid, usually sulfuric acid, and is facilitated by mercuric sulfate. However, unlike the additions to double bonds which give alcohol products, addition of water to alkynes gives ketone products (except for acetylene which yields acetaldehyde). The explanation for this deviation lies in enol-keto tautomerization, illustrated by the following equation. The initial product from the addition of water to an alkyne is an enol (a compound having a hydroxyl substituent attached to a double-bond), and this immediately rearranges to the more stable keto tautomer.

$$R-C\equiv C-R \xrightarrow[HgSO_4]{H_2O\ +\ H^+} [R(H)C=C(OH)R] \xrightarrow{\text{Tautomerization}} R-CH_2-C(=O)R$$

Addition — Enol Tautomer — keto tautomer

Tautomers are defined as rapidly interconverted constitutional isomers, usually distinguished by a different bonding location for a labile hydrogen atom (colored red here) and a differently located double bond. The equilibrium between tautomers is not only rapid under normal conditions, but it often strongly favours one of the isomers (acetone, for example, is 99.999per cent keto tautomer).

Even in such one-sided equilibria, evidence for the presence of the minor tautomer comes from the chemical behaviour of the compound. Tautomeric equilibria are catalyzed by traces of acids or bases that are generally present in most chemical samples. The three examples shown below illustrate these reactions for different substitutions of the triple-bond. The tautomerization step is indicated by a red arrow. For terminal alkynes the addition of water follows the Markovnikov rule, as in the second example below, and the final product ia a methyl ketone (except for acetylene, shown in the first example).

For internal alkynes (the triple-bond is within a longer chain) the addition of water is not regioselective. If the triple-bond is not symmetrically located (i.e. if R and R' in the third equation are not the same) two isomeric ketones will be formed.

$$HC\equiv CH + H_2O + HgSO_4 \,\&\, H_2SO_4 \rightarrow [H_2C=CHOH] \rightarrow H_3C-CH=O$$

$$RC\equiv CH + H_2O + HgSO_4 \,\&\, H_2SO_4 \rightarrow [RC(OH)=CH_2] \rightarrow RC(=O)CH_3$$

$$RC\equiv CR' + H_2O + HgSO_4 \,\&\, H_2SO_4 \rightarrow [RHC=(OH)]R' + RC(OH) =$$

$$CHR' \rightarrow RCH_2-C(=O)R' + RC(=0)-CH_2R'$$

Two factors have an important influence on the enol-keto tautomerizations described here. The first is the potential energy difference between the tautomeric isomers. This factor determines the position of the equilibrium state.

The second factor is the activation energy for the interconversion of one tautomer to the other. This factor determines the rate of rearrangement.

Since the potential energy or stability of a compound is in large part a function of its covalent bond energies, we can estimate the relative energy of keto and enol tautomers by considering the bonds that are changed in the rearrangement. From the following diagram, we see that only three significant changes occur, and the standard bond energies for those changes are given to the right of the equation. The keto tautomer has a 17.5 kcal/mole advantage in bond energy, so its predominance at equilibrium is expected.

Enol Tautomer ⟶ Keto Tautomer

Enol bond		Keto bond	Change in Bond Energy
C=C 146	⟶	C–C 83	-63
C–O 85.5	⟶	C=O 178	+92.5
O–H 111	⟶	C–H 99	-12
			+17.5 Kcal/mole

The rapidity with which enol-keto tautomerization occurs suggests that the activation energy for this process is low. We have noted that the rearrangement is acid and base catalyzed, and very careful experiments have shown that interconversion of tautomers is much slower if such catalysts are absent.

A striking example of the influence of activation energy on such transformations may be seen in the following hypothetical rearrangement. Here we have substituted a methyl group (colored maroon) for the proton of a conventional tautomerism, and the methyl shifts from oxygen to carbon just as the proton does in going from an enol to a ketone.

$$H_2C{=}CH\text{-}O\text{-}CH_3 \; \text{–X→} \; CH_3\text{–}CH_2\text{–}CH{=}O$$

The potential energy change for this rearrangement is even more advantageous than for enol-keto tautomerism, being estimated at over 25 kcal/mole from bond energy changes.

Despite this thermodynamic driving force, the enol ether described above is completely stable to base treatment, and undergoes rapid acid-catalyzed hydrolysis with loss of methanol, rather than rearrangement. The controlling difference in this case must be a prohibitively high activation energy for the described rearrangement, combined with lower energy alternative reaction paths.

Hydroboration Reactions

Diborane reacts readily with alkynes, but the formation of substituted alkene products leaves open the possibility of a second addition reaction. A clever technique for avoiding this event takes advantage of the fact that alkynes do not generally suffer from steric hindrance near the triple-bond (the configuration of this functional group is linear). Consequently, large or bulky electrophilic reagents add easily to the triple-bond, but the resulting alkene is necessarily more crowded or sterically hindered and resists further additions. The bulky hydroboration reagent needed for this strategy is prepared by reaction of diborane with 2-methyl-2-butene, a highly branched alkene. Because of the alkyl branching, only two alkenes add to a BH_3 moiety (steric hindrance again), leaving one B-H covalent bond available for reaction with an alkyne, as shown below. The resulting dialkyl borane is called disiamylborane, a contraction of di-secondary-isoamylborane (amyl is an old name for pentyl).

$$2\ (CH_3)_2C{=}CHCH_3 + BH_3 \text{ in ether} \rightarrow [\ (CH_3)_2CH\text{-}CH(CH_3)\ ]_2B\text{-}H$$

disiamylborane

An important application of disiamylborane is its addition reaction to terminal alkynes. As with alkenes, the B-H reagent group adds in an apparently anti-Markovnikov manner, due to the fact that the boron is the electrophile, not the hydrogen. Further addition to the resulting boron-substituted alkene does not occur, and the usual oxidative removal of boron by alkaline hydrogen peroxide gives an enol which rapidly rearranges to the aldehyde tautomer. Thus, by the proper choice of reagents, terminal alkynes may be converted either to methyl ketones (mercuric ion catalyzed hydration) or aldehydes (hydroboration followed by oxidation).

$$RC \equiv CH + (C_5H_{11})_2B - H \rightarrow [RCH = CH - B(C_5H_{11})_2] +$$

$$H_2O_2 \,\&\, NaOH \rightarrow [RCH = CH - OH] \rightarrow RCH_2 - CH = O$$

Hydroboration of internal alkynes is not a particularly useful procedure because a mixture of products will often be obtained, unless the triple-bond is symmetrically substituted. Mercuric ion catalyzed hydration gives similar results.

Oxidations

Reactions of alkynes with oxidizing agents such as potassium permanganate and ozone usually result in cleavage of the triple-bond to give carboxylic acid products. A general equation for this kind of transformation follows. The symbol [O] is often used in a general way to denote an oxidation.

$$RC{\equiv}CR' + [O] \rightarrow RCO_2H + R'CO_2H$$

ALKANE

The simplest organic compounds are composed of carbon and hydrogen. Organic compounds that contain only carbon and hydrogen are called hydrocarbons.

The simplest hydrocarbon would be CH_4 methane, which is a major component of nature gas. Methane is produced by the anaerobic decay of vegetation and is sometimes called "swamp gas".

In methane the carbon atom in the molecule is surrounded by single covalent bonds. As a result, the molecule has a tetrahedral shape and is a non-polar molecule since all of the C-H bond dipoles cancel each other. The structural formula for methane:

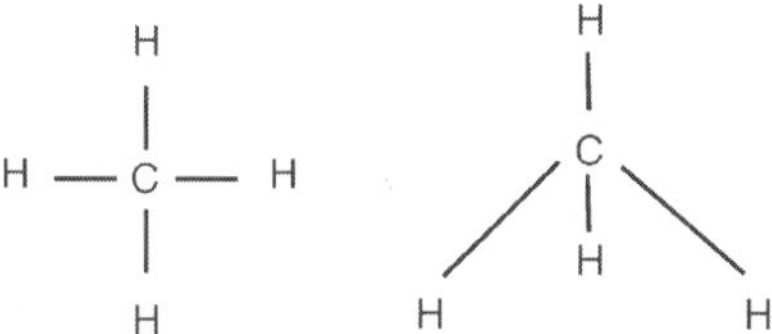

If one of the H is removed, it can be replaced with a C atom, which can hold three additional H atoms. The new molecule would look like:

```
    H   H
    |   |
H — C — C — H
    |   |
    H   H
```

The compound has a molecular formula of C_2H_6 and is called ethane. Note that the C–C is a single covalent bond. If we repeat this process there would be three carbons in the chain and there would be eight hydrogens bonded to these carbons. The molecular formula is C_3H_8 and the structural formula would be:

```
    H   H   H
    |   |   |
H — C — C — C — H
    |   |   |
    H   H   H
```

The molecular formulas for these three compounds are:

CH_4

C_2H_6

C_3H_8

One formula differs from the next formula by CH_2.

The next compound in the series would be $C_3H_8 + CH_2 = C_4H_{10}$ which is butane.

A group of compounds form a homologous series if there is a constant increment from one member of a group to the next member the group. Methane, ethane, propane and butane are members of a homologous series called the alkanes. The following table contains the first ten members of the alkane series (all C–C bonds are single covalent):

Molecular Formula	**Structural Formula**	**IUPAC Name**	**Number of carbons in the chain**
CH_4	CH_4	Meth*ane*	1
C_2H_6	CH_3-CH_3	Eth*ane*	2
C_3H_8	CH_3-CH_2-CH_3	Prop*ane*	3
C_4H_{10}	CH_3-$(CH_2)_2$-CH_3	But*ane*	4
C_5H_{12}	CH_3-$(CH_2)_3$-CH_3	Pent*ane*	5
C_6H_{14}	CH_3-$(CH_2)_4$-CH_3	Hex*ane*	6
C_7H_{16}	CH_3-$(CH_2)_5$-CH_3	Hept*ane*	7

C_8H_{18}	CH_3-$(CH_2)_6$-CH_3	Oct*ane*	8
C_9H_{20}	CH_3-$(CH_2)_7$-CH_3	Non*ane*	9
$C_{10}H_{22}$	CH_3-$(CH_2)_8$-CH_3	Dec*ane*	10

Notice that the alkanes end in "*ane*". This is considered the official IUPAC ending for alkanes. Note that the suffix denotes the number of carbon atoms in the continuous chain. A continuous chain is sometimes called a straight chain. The number of carbons in the straight is counted through carbon – carbon bonds.

If you look at the molecular formula you may notice that there is a mathematical relationship for the alkane series. Let n represent the number of carbon atoms, the number of hydrogen atoms will be given by 2n+2. The general molecular formula for the alkane series is C_nH_{2n+2}.

Branched Chain Alkanes

Not all carbon chains are straight (continuous) chains. Some chains have branches that are created by a carbon substituting for hydrogen. A substituent is an atom or group of atoms that can take the place of a hydrogen atom on a parent chain. The parent chain is the longest continuous carbon chain. Example:

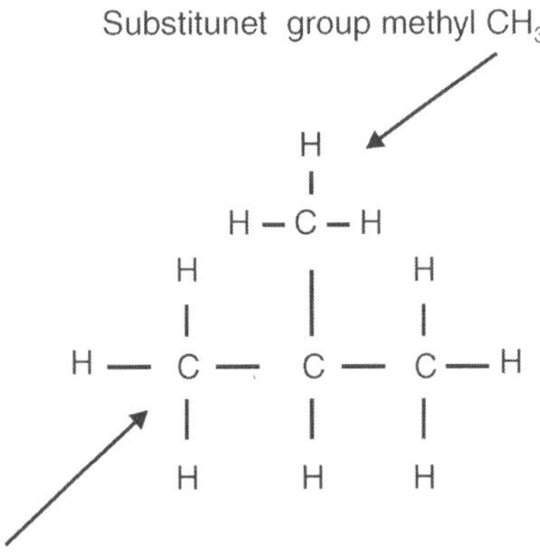

A hydrocarbon substituent is from the alkyl group. The name of the substituent branch is determined from the suffix alkane name and the ending is changed from ane to yl.

CH_3 is methyl (one carbon)

C_2H_5 is ethyl (two carbon)

C_3H_7 is propyl (three carbon)

Because there are so many compounds possible it is very important to identify a particular compound by giving it a unique name. IUPAC has proposed a set of rules to use in naming branched carbon chains.

- Name the parent chain (longest continuous chain).
- Identify each substituent group.

- Number the carbons in the parent chain to give the lowest numbers to indicate the position of substituent groups.
- Use prefixes for groups that appear more than once. The prefix will indicate the number of times the group appears.

The Greek prefixes used are as follows:

di for 2

tri for three

tetra for four

penta for five

hexa for six

hepta for seven

octa for eight

- List the substituent groups in alphabetical order.
- Use proper punctuations, the name is written as one word. Commas separate numbers. Hyphens are used to separate numbers and names.

Use these rules to name the following:

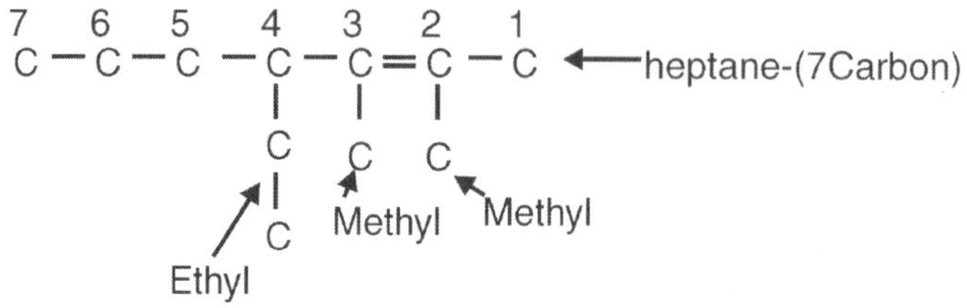

```
7   6   5    4   3   2   1
C — C — C  — C — C = C — C  ←—heptane-(7Carbon)
             |   |   |
             C   C   C
             |  Methyl Methyl
             C
          Ethyl
```

- The longest continuous chain has 7 carbons. (heptane)
- The substituent groups are methyl, methyl and ethyl.
- The position of the substituents. Numbering from the right to left gives 2, 3, 4 and numbering from the left to right gives 4, 5, 6. The lowest numbers are 2, 3, 4, so numbering is done from the right to left.
- There are two methyl substituent groups so it is dimethyl.
- Ethyl comes before methyl.
- Putting it all together: 4-ethyl-2,3-dimethylheptane.

Name the following hydrocarbons:

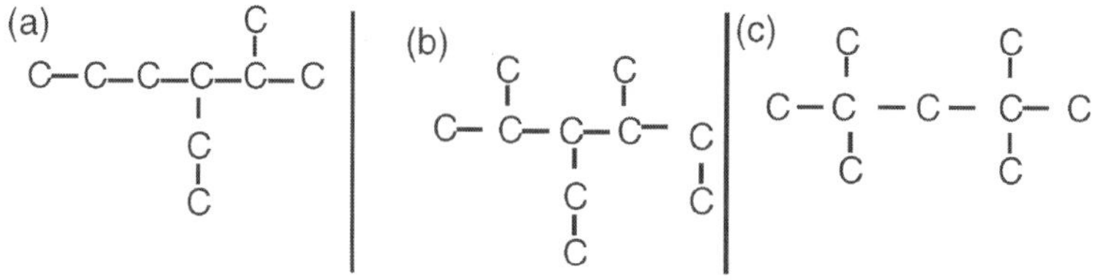

```
(a)             C      (b)    C     C          (c)  C           C
                |             |     |               |           |
C–C–C–C–C–C             C– C– C– C–  C         C–C – C – C– C
      |                       |      |              |           |
      C                       C      C              C           C
      |                       |
      C                       C
```

Structural Formula

When you have the name of the compound it is easy to construct the molecule (structural formula). The following rules may help.

- From the name identify and draw the parent chain (longest continuous chain).
- Number the carbons on the parent chain.
- Attach the substituent groups their numbered positions on the parent chain.

Draw 2,3-dimethylpentane.

The parent chain is pentane (5 carbon)

There are two substituent methyl groups. (C), one on C number 2 and another on C number 3.

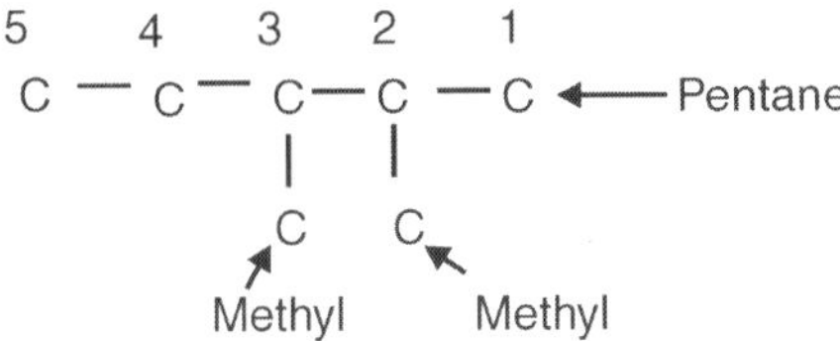

Draw: 3-ethyl-4-methylhexane and 2, 2, 4-trimethylpentane.

Physical Properties of Alkanes

Boiling Points

The boiling point of an organic compound is affected by molecular size and the strength of intermolecular force that exists between the molecules. The boiling points of alkanes are the low because the intermolecular forces holding the alkane molecules together are very weak London Dispersion forces (Van derWaals force).

As a result of these weak intermolecular forces the temperature required to separate the molecules into the vapour state is low. The strength of Van der Waals forces is affected by surface area, the greater the surface area the greater the force. Longer carbon chain leads to a greater surface area, which increases the intermolecular force of attraction and thus a higher boiling temperature. Branching in a compound tends to lower the surface area so branching tends to lower the boiling. Normal (straight chain) butane (C_4H_{10}) boils at 0°C and methylpropane (C_4H_{10}) is a branched chain and has a boiling point of -10°C.

Compounds that have the same molecular formula but different molecular structures are called structural isomers. $C_{10}H_{22}$ has 75 structural isomers and $C_{20}H_{42}$ has 366 319 isomers. A useful pattern for drawing isomers is to start with the normal (straight) chain. Next reduce the length of the longest chain (parent chain) by one and place the methyl group on an interior carbon.

This methyl cannot be placed on an end carbon because it will simply make the chain longer again. Once all isomers of this length are drawn reduce the length of the parent chain by one again and place on an interior carbon. The most common mistake is to draw isomers that on paper look different but are the same. The best way to see if the isomers are different is to use the IUPAC rules to name the isomers. If the isomers have the same name, they are the same. To be different they must have different names.

Example: draw the structural isomers of butane

- Draw a straight chain with 4 C (C–C–C–C name is normal butane)
- Take off one carbon (parent chain now 3C) and place the other C in a position that does not lengthen the chain (this eliminates the C at each end of the chain)

  C–C–C

  |

  C (name is methyl propane)

If we try to reduce the parent chain to two C, the other 2 C can only be only be added to the end, thus making the chain longer (three C which was already done).

UNSUBSTITUTED COMPOUNDS AND RADICALS

- The names of saturated monocyclic hydrocarbons (with no side chains) are formed by attaching the prefix "cyclo" to the name of the acyclic saturated unbranched hydrocarbon with the same number of carbon atoms. The generic name of saturated monocyclic hydrocarbons (with or without side chains) is "cycloalkane".
- *Examples:*

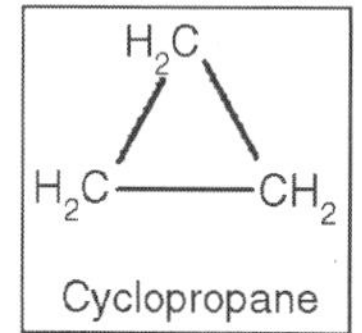

Cyclopropane

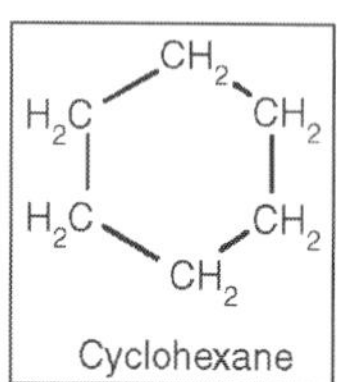

Cyclohexane

- Univalent radicals derived from cycloalkanes (with no side chains) are named by replacing the ending "-ane" of the hydrocarbon name by "-yl", the carbon atom with the free valence being numbered as 1. The generic name of these radicals is "cycloalkyl".
- *Examples:*

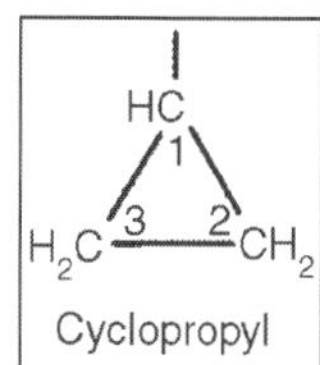

Cyclopropyl

Cyclohexyl

- The names of unsaturated monocyclic hydrocarbons (with no side chains) are formed by substituting "-ene", "-adiene", "-atriene", "-yne", "-adiyne", etc., for "-ane" in the name of the corresponding cycloalkane. The double and triple bonds are given numbers as low as possible.

- *Examples:*

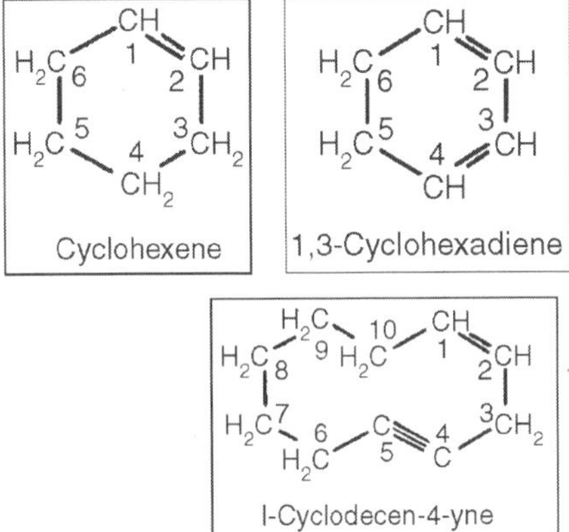

Cyclohexene

1,3-Cyclohexadiene

l-Cyclodecen-4-yne

The name "benzene" is retained.

- The names of univalent radicals derived from unsaturated monocyclic hydrocarbons have the endings "-enyl", "-ynyl", "-dienyl", etc., the positions of the double and triple bonds being indicated according to the principles of Rule.
- The carbon atom with the free valence is numbered as 1, except as stated in the rules for terpenes.
- *Examples:*

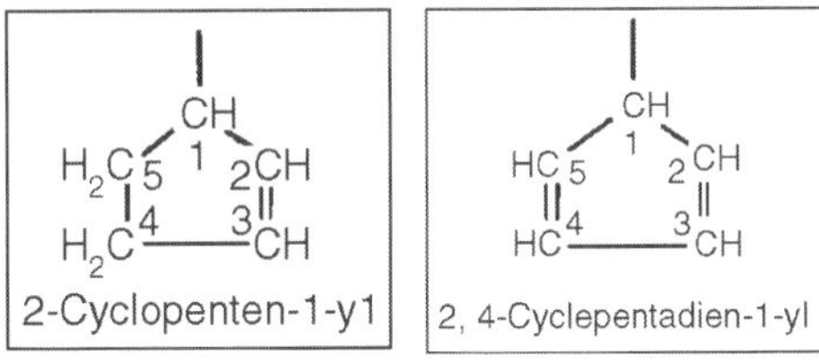

2-Cyclopenten-1-y1

2, 4-Cyclepentadien-1-yl

- The radical name "phenyl" is retained.
- Names of bivalent radicals derived from saturated or unsaturated monocyclic hydrocarbons by removal of two atoms of hydrogen from the same carbon atom of the ring are obtained by replacing the endings "-ane", "-ene", "-yne", by "-ylidene", "-enylidene" and "-ynylidene", respectively.
- The carbon atom with the free valences is numbered as 1 except as stated in the rules for terpenes.
- *Examples:*

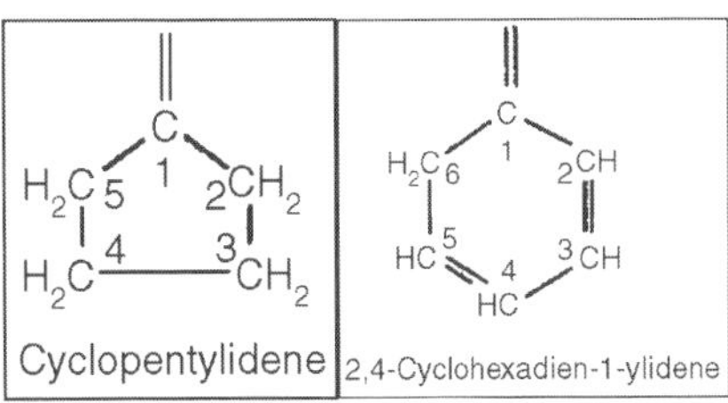

Cyclopentylidene

2,4-Cyclohexadien-1-ylidene

- Bivalent radicals derived from saturated or unsaturated monocyclic hydrocarbons by removing a hydrogen atom from each of two different:carbon atoms of the ring are named by replacing the endings "-ane", "-ene", "-diene", "-yne", etc., of the hydrocarbon name by "-ylene", "-enylene", "-dienylene", "-ynylene", etc.,
- The positions of the double and triple bonds and of the points of attachment being indicated.
- Preference n lowest numbers is given to the carbon atoms having the free valences.

  Examples:

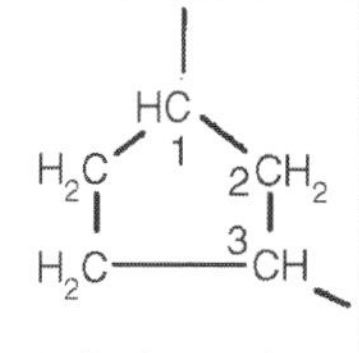

1,3-Cyclopentylene

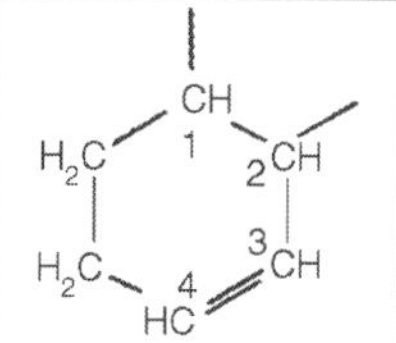

3-Cyclohexen-1,2-ylene

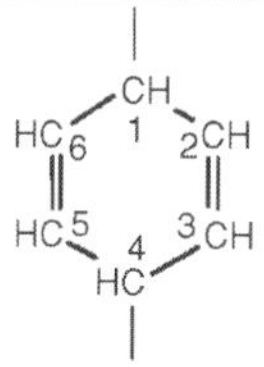

2,3-Cyclohexadien-1,4-ylene

The following name is retained:

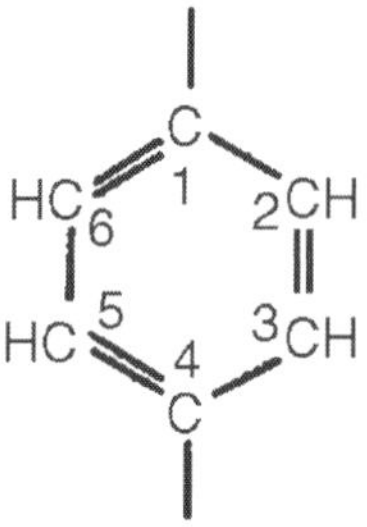

Phenylene (p-shown)

Rule. Substituted Aromatic Compounds

- The following names for monocyclic substituted aromatic hydro carbons are retained:

  Examples:

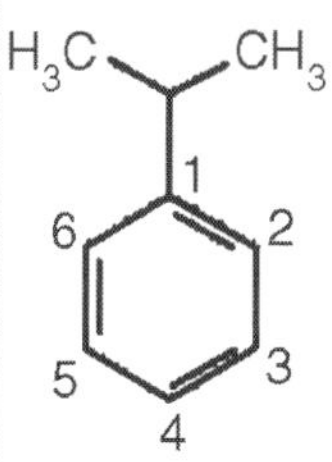

Cumene

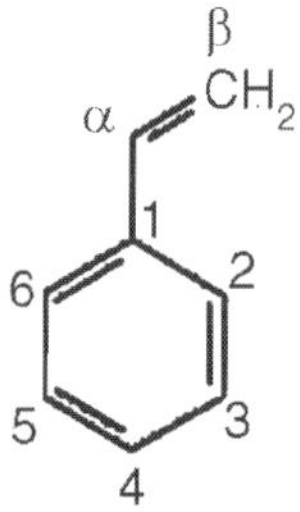

Styrene

Mesitylene | Cymene (p-shown) | Toluene | Xylene (o-shown)

- Other monocyclic substituted aromatic hydrocarbons are named as derivatives of benzene or of one of the compounds listed in Part.1 of this rule. However, if the substituent introduced into such a compound is identical with one already present in that compound, then the substituted compound is named as a derivative of benzene.
- The position of substituents is indicated by numbers except that *o- (ortho), m- (meta)* and *p- (para)* may be used in place of 1,2-, 1,3-, and 1,4-, respectively, when only two substituents are present. The lowest numbers possible are given to substituents, choice between alternatives being governed by Rule so far as applicable except that when names are based on those of compounds listed of this rule the first priority for lowest numbers is given to the substituent(s) already present in those compounds.

  Examples:

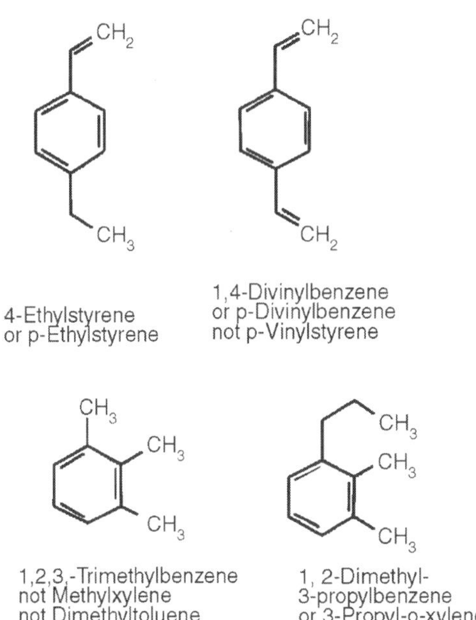

4-Ethylstyrene or p-Ethylstyrene

1,4-Divinylbenzene or p-Divinylbenzene not p-Vinylstyrene

1,2,3,-Trimethylbenzene not Methylxylene not Dimethyltoluene

1, 2-Dimethyl-3-propylbenzene or 3-Propyl-o-xylene

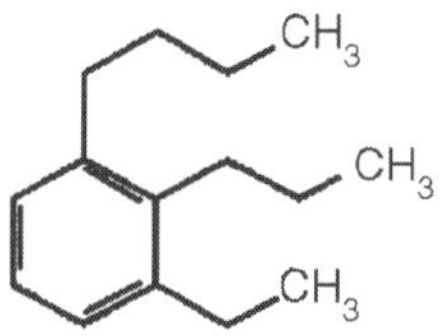

1-Butyl-3-ethyl-2-propylbenzene

- The generic name of monocyclic and polycyclic aromatic hydrocarbons is "arene".

SATURATED UNBRANCHED-CHAIN COMPOUNDS AND UNIVALENT RADICALS

The first four saturated unbranched acyclic hydrocarbons are called methane, ethane, propane and butane. Names of the higher members of this series consist of a numerical term, followed by "-ane" with elision of terminal "a" from the numerical term. Examples of these names are shown in the table below. The generic name of saturated acyclic hydrocarbons (branched or unbranched) is "alkane".

Examples of names:

(n = total number of carbon atoms)

n	n
1 Methane	2 Ethane
3 Propane	4 Butane
5 Pentane	6 Hexane
7 Heptane	8 Octane
9 Nonane	10 Decane
11 Undecane	12 Dodecane
13 Tridecane	14 Tetradecane
15 Pentadecane	16 Hexadecane
17 Heptadecane	18 Octadecane
19 Nonadecane	20 Icosane
21 Henicosane	22 Docosane
23 Tricosane	24 Tetracosane
25 Pentacosane	26 Hexacosane
27 Heptacosane	28 Octacosane
29 Nonacosane	30 Triacontane
31 Hentriacontane	32 Dotriacontane
33 Tritriacontane	40 Tetracontane
50 Pentacontane	60 Hexacontane
70 Heptacontane	80 Octacontane
90 Nonacontane	100 Hectane
132 Dotriacontahectane	

Univalent radicals derived from saturated unbranched acyclic hydrocarbons by removal of hydrogen from a terminal carbon atom are named by replacing the ending "-ane" of the name of the hydrocarbon by "-yl". The carbon atom with the free valence is numbered as 1. As a class, these radicals are called normal, or unbranched chain, alkyls.

Examples:

	5	4	3	2	1
Pentyl	CH_3-	CH_2-	CH_2-	CH_2-	CH_2-

	11	10–2	1
Undecyl	Ch_3-	$[Ch_2]_9-$	CH_2-

Rule. Saturated Branched-chain Compounds and Univalent Radicals

A saturated branched acyclic hydrocarbon is named by prefixing the designations of the side chains to the name of the longest chain present in the formula.

Example:

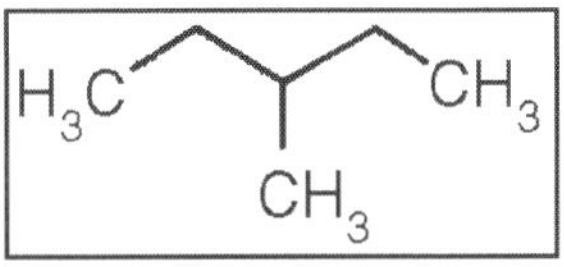

H_3C CH_3
CH_3

The following names are retained for unsubstituted hydrocarbons only:

Name	Structure
Isobutane	H_3C, CH_3, CH_3
Isopentane	H_3C, CH_3, CH_3
Neopentane	$H_3C-C(CH_3)_2-CH_3$
Isohexane	H_3C, CH_3, CH_3

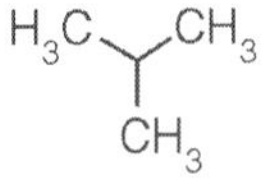

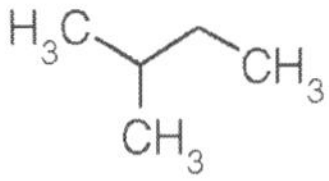

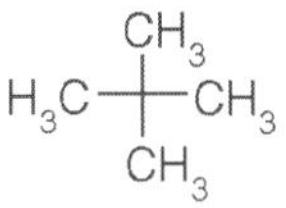

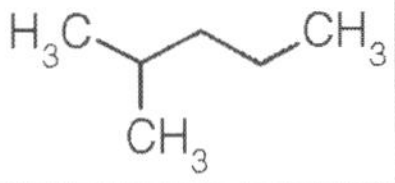

The longest chain is numbered from one end to the other by Arabic numerals, the direction being so chosen as to give the lowest numbers possible to the side chains.

When series of locants containing the same number of terms are compared term by term, that series is "lowest" which contains the lowest number on the occasion of the first difference. This principle is applied irrespective of the nature of the substituents.

Examples:

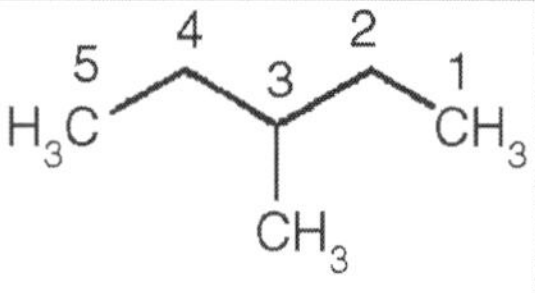

3-Methylpentane

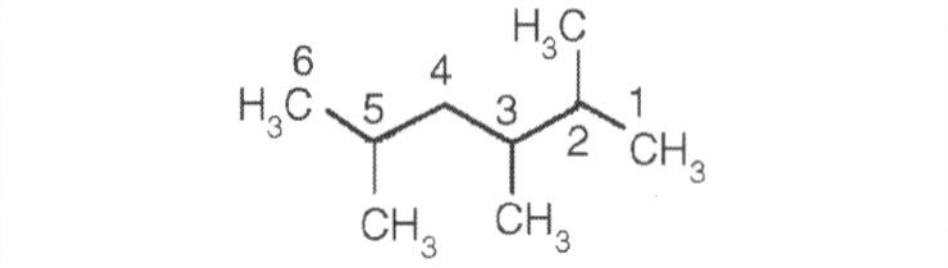

2,3,5-Trimethylhexane (Not 2,4,5-Trimethylhexane)

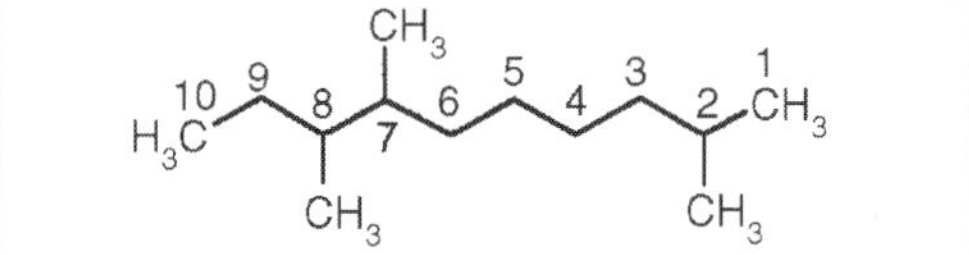

2,7,8-Trimethyldecane (Not 3,4,9-Trimenthyldecane)

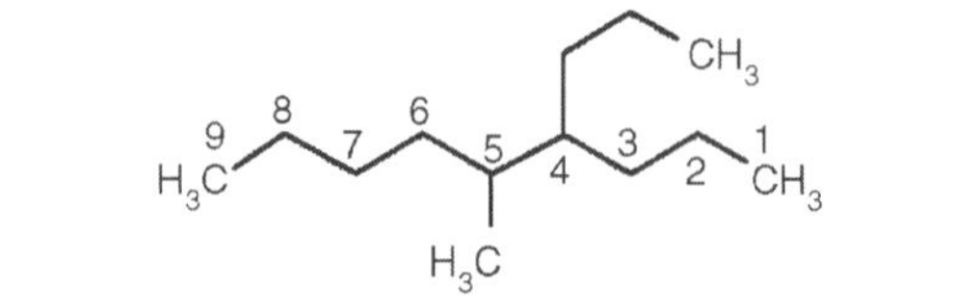

5-Methy1-4-propylnonane
(Not 5-Methyl-6-propylnonane becose 4,5 is Lower then)

Univalent branched radicals derived from alkanes are named by prefixing the designation of the side chains to the name of the unbranched alkyl radical possessing the longest possible chain starting from the carbon atom with the free valence, the said atom being numbered as 1.

Examples:

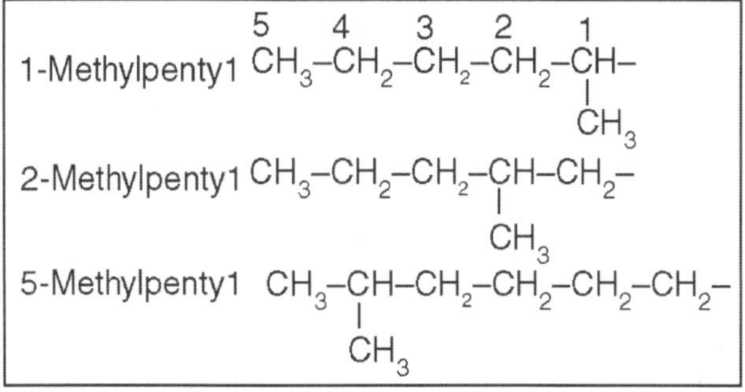

1-Methylpenty1 $\overset{5}{CH_3}-\overset{4}{CH_2}-\overset{3}{CH_2}-\overset{2}{CH_2}-\overset{1}{CH}(CH_3)-$

2-Methylpenty1 $CH_3-CH_2-CH_2-CH(CH_3)-CH_2-$

5-Methylpenty1 $CH_3-CH(CH_3)-CH_2-CH_2-CH_2-CH_2-$

The following names may be used for the unsubstituted radicals only:

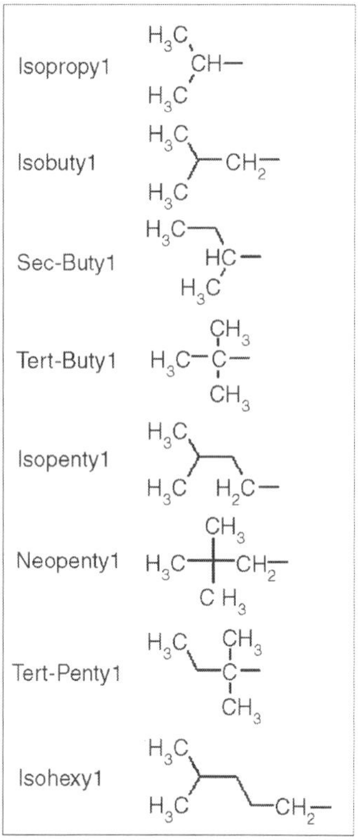

Isopropy1

Isobuty1

Sec-Buty1

Tert-Buty1

Isopenty1

Neopenty1

Tert-Penty1

Isohexy1

- If two or more side chains of different nature are present, they are cited in alphabetical order.
  The alphabetical order is decided as follows:
- The names of simple radicals are first alphabetized and the multiplying prefixes are then inserted.ethyl is cited before methyl, thus 4-Ethyl-3,3-dimethylheptane.
  Examples: 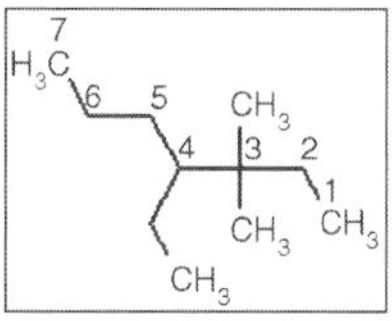
- The name of a complex radical is considered to begin with the first letter of its complete name.
  Examples:

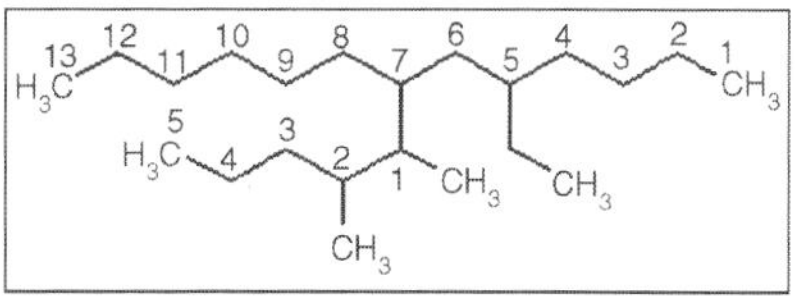

dimethylpentyl (as complete single substituent) is alphabetized under "d", thus 7-(1,2-Dimethylpentyl)-5-ethyltridecane

- In cases where names of complex radicals are composed of identical words, priority for citation is given to that radical which contains the lowest locant at the first cited point of difference in the radical.
  Examples:

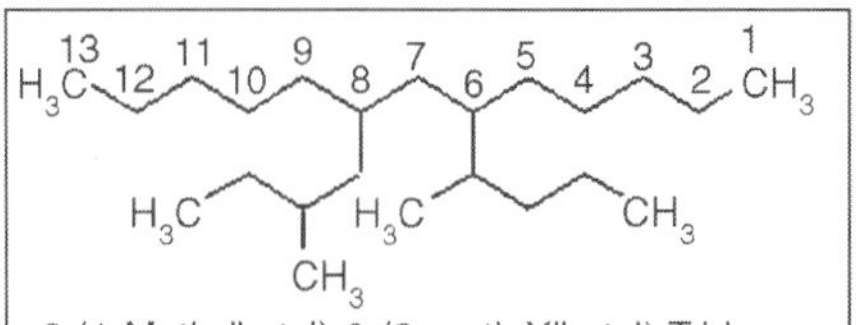

  6-(1-Methylbutyl)-8-(2-meth Ylbutyl) Tridecane

- If two or more side chains are in equivalent positions, the one to be assigned the lower number is that cited first in the name.
  Examples:

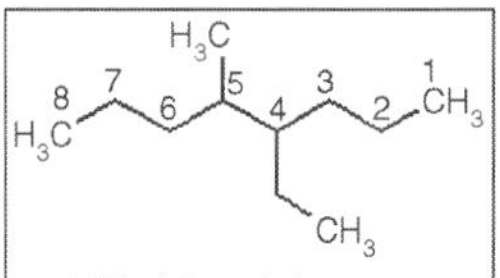

  4-Ethyl-5-methylocatane

- The presence of identical unsubstituted radicals is indicated by the appropriate multiplying prefix di-, tri-, tetra-, penta-, hexa-, hepta-, octa-, nona-, deca-, undeca, *etc.*
  Example:

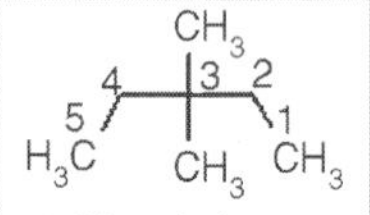

  3,3-Dimethylpentane

- The presence of identical radicals each substituted in the same way may be indicated by the appropriate multiplying prefix bis-, tris-, tetrakis-, pentakis-, etc. The complete expression denoting such a side chain may be enclosed in parentheses or the carbon atoms in side chains may be indicated by primed numbers.
  Examples:

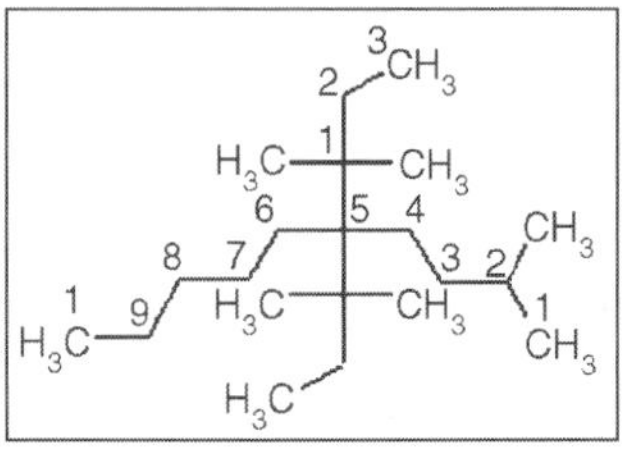

  – *Use of parentheses and unprimed numbers*: 5,5-Bis(1,1-dimethylpropyl)-2-methyldecane
  – *Use of primes*: 5,5-Bis-1',1'-dimethylpropyl-2-methyldecane

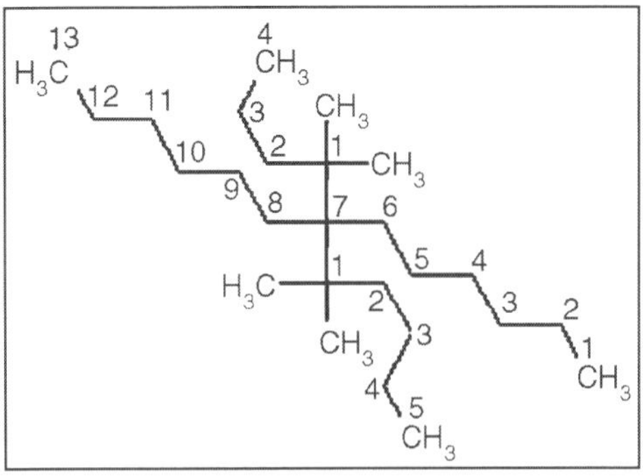

- *Use of parentheses and unprimed numbers*: 7-(1,1-Dimethylbutyl)-7-(1,1-dimethylpentyl)tridecane
- *Use of primes*: 7-1',1'-Dimethylbutyl-7-1",1"-dimethylpentyltridecane

- If chains of equal length are competing for selection as main chain in a saturated branched acyclic hydrocarbon, then the choice goes in series to:

  Example:

  - The chain which has the greatest number of side chains.

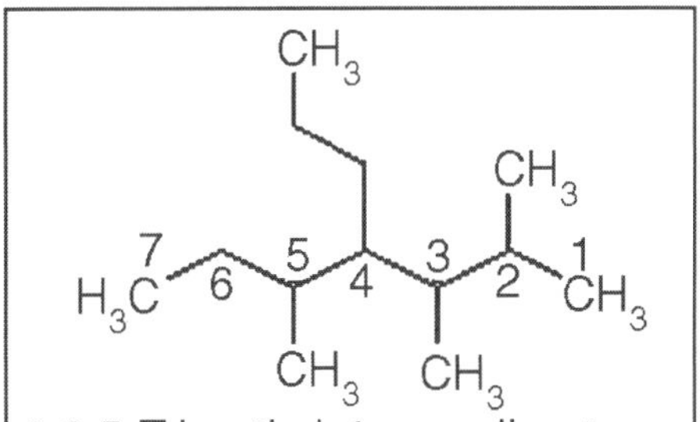

2,3,5-Trimethyl-4-propylheptane

  - The chain whose side chains have the lowest-numbered locants.

    Example:

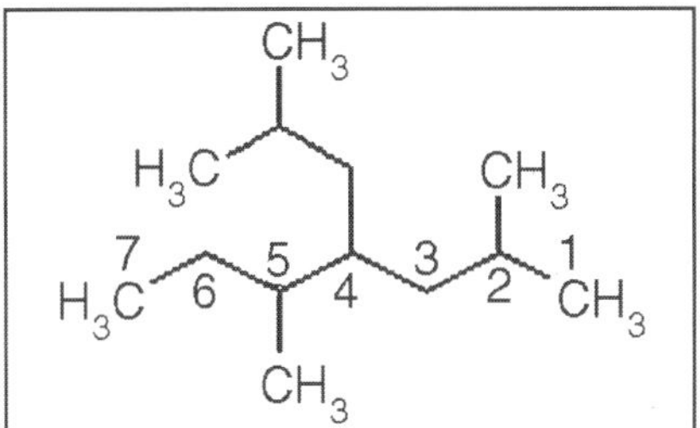

4-Isobutyl-2,5-dimethylheptane

  - The chain having the greatest number of carbon atoms in the smaller side chains.

Example:

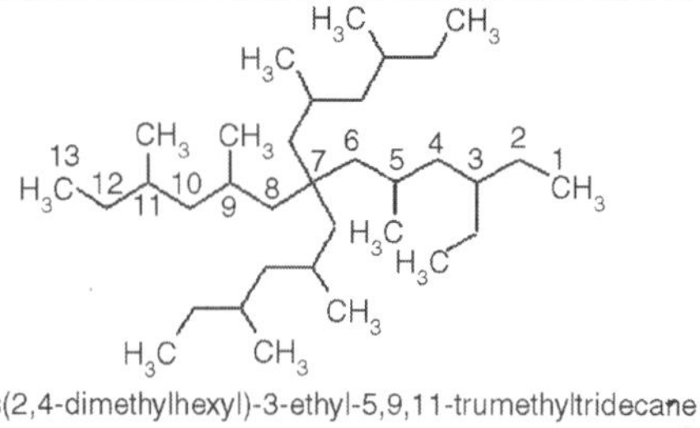

7,7-Bis(2,4-dimethylhexyl)-3-ethyl-5,9,11-trumethyltridecane

– The chain having the least branched side chains.
Example:

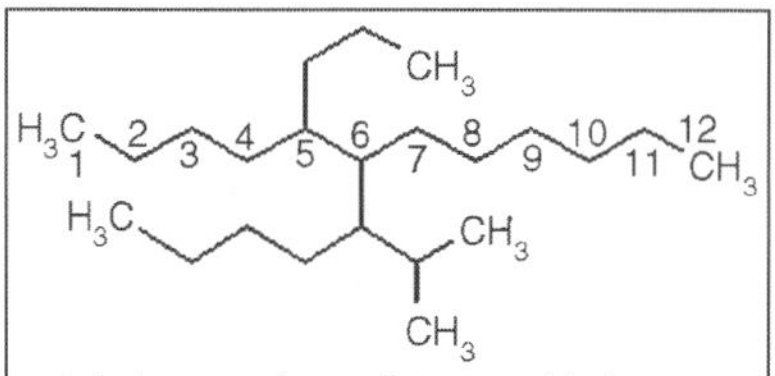

6-(1-Isopropylpentyl)-5-propyldodecane

Rule. Unsaturated Compounds and Univalent Radicals

- Unsaturated unbranched acyclic hydrocarbons having one double bond are named by replacing the ending "-ane" of the name of the corresponding saturated hydrocarbon with the ending "-ene". If there are two or more double bonds, the ending will be "-adiene", "-atriene", *etc.* The generic names of these hydrocarbons (branched or unbranched) are "alkene", "alkadiene", "alkatriene", *etc.* The chain is so numbered as to give the lowest possible numbers to the double bonds. When, in cyclic compounds or their substitution products, the locants of a double bond differ by unity, only the lower locant is cited in the name; when they differ by more than unity, one locant is placed in parentheses after the other.
Examples:

2-Hexene

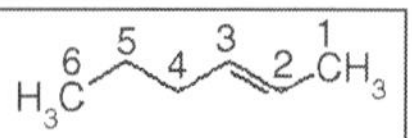

1,4-Hexadiene

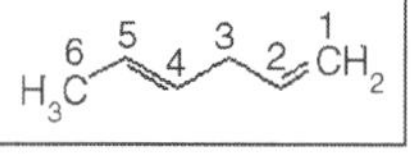

The following non-systematic names are retained:

Ethylene	$H_2C{=}CH_2$
Allene	$H_2C{=}C{=}CH_2$

- Unsaturated unbranched acyclic hydrocarbons having one triple bond are named by replacing the ending "-ane" of the name of the corresponding saturated hydrocarbon with the ending "-yne". If there are two or more triple bonds, the ending will be "-adiyne", "atriyne", *etc.* The generic names of these hydrocarbons (branched or unbranched) are "alkyne", "alkadiyne", "alkatriyne", etcThe chain is so numbered as to give the lowest possible numbers to the triple bonds.

- Only the lower locant for a triple bond is cited in the name of a compound. The name "acetylene" for HC ≡ CH is retained.
- Unsaturated unbranched acyclic hydrocarbons having both double and triple bonds are named by replacing the ending "-ane" of the name of the corresponding saturated hydrocarbon with the ending "-enyne", "-adienyne", "-atrienyne", "-enediyne", etcNumbers as low as possible are given to double and triple bonds even though this may at times give "-yne" a lower number than "-ene". When there is a choice in numbering, the double bonds are given the lowest numbers.

  Examples:

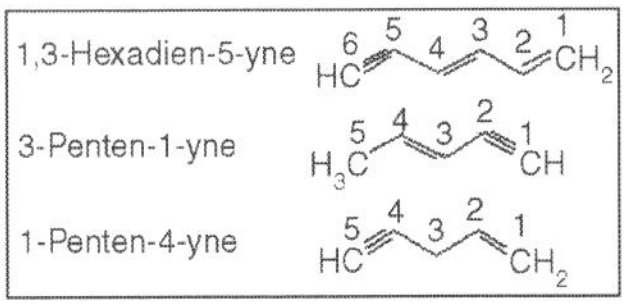

  1,3-Hexadien-5-yne

  3-Penten-1-yne

  1-Penten-4-yne

- Unsaturated branched acyclic hydrocarbons are named as derivatives of the unbranched hydrocarbons which contain the maximum number of double and triple bonds.
- If there are two or more chains competing for selection as the chain with the maximum number of unsaturated bonds, then the choice goes to:
  - That one with the greatest number of carbon atoms;
  - The number of carbon atoms being equal, that one containing the maximum number of double bonds.
- In other respects, the same principles apply as for naming saturated branched acyclic hydrocarbons. The chain is so numbered as to give the lowest possible numbers to double and triple bonds.

  Examples:

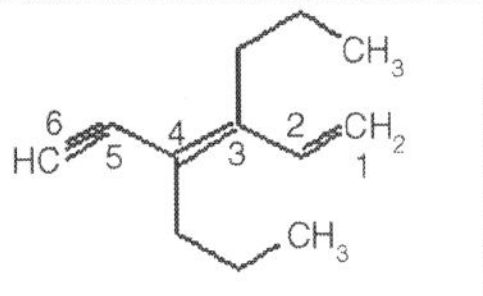

  3,4-Dipropyl-1, 3-hexadien-5-yne

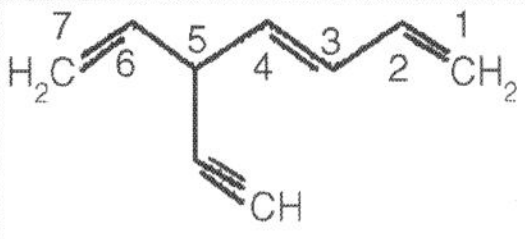

  5-Ethynyl-1,3, 6-heplatriene

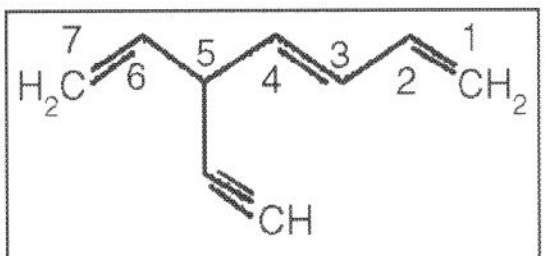

  5-Ethynyl-1,3, 6-heplatriene

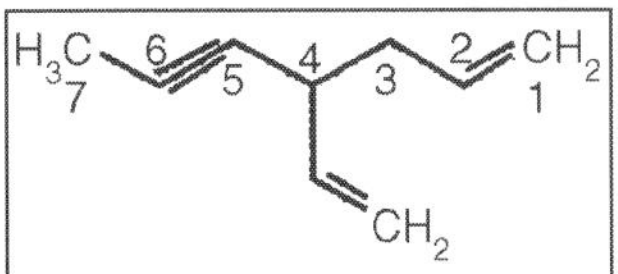

  4-Vinyl-1-hepten-5-yne

The following name is retained for the unsubstituted compound only:

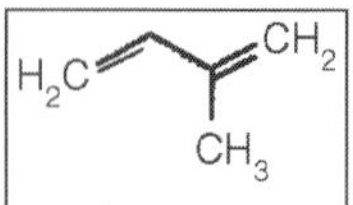

H_2C CH_2
CH_3
Iso prene

- The names of univalent radicals derived from unsaturated acyclic hydrocarbons have the endings "-enyl", "-ynyl", "-dienyl", etc., the positions of the double and triple bonds being indicated where necessary. The carbon atom with the free valence is numbered as 1.
  Examples:

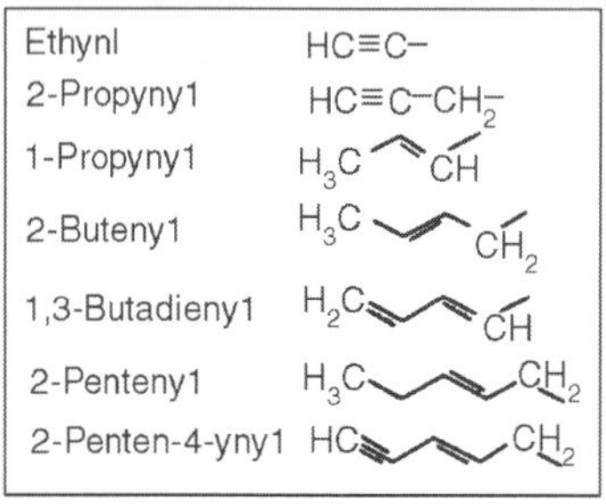

Ethynl	HC≡C–
2-Propyny1	HC≡C–CH_2–
1-Propyny1	H_3C CH
2-Buteny1	H_3C CH_2
1,3-Butadieny1	H_2C CH
2-Penteny1	H_3C CH_2
2-Penten-4-yny1	HC CH_2

  Exceptions:
  The following names are retained (for unsubstituted radical only):

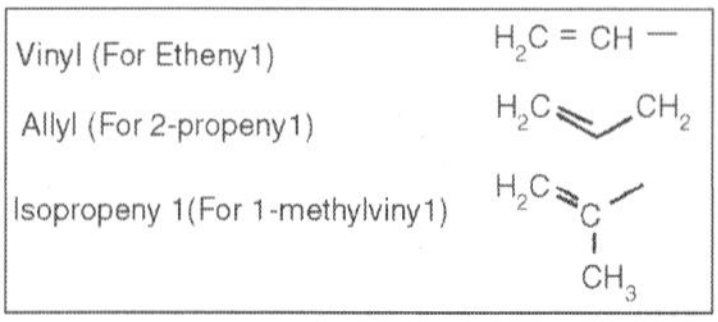

Vinyl (For Etheny1)	H_2C = CH —
Allyl (For 2-propeny1)	H_2C CH_2
Isopropeny 1(For 1-methylviny1)	H_2C C CH_3

- When there is a choice for the fundamental chain of a radical, that chain is selected which contains:
  - The maximum number of double and triple bonds;
  - The largest number of carbon atoms;
  - The largest number of double bonds.

  Examples:

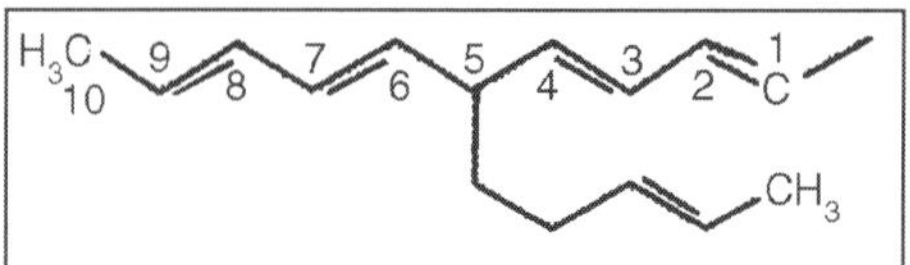

H_3C 10 9 8 7 6 5 4 3 2 C 1
CH_3
5-(3-Pentenyl)-3,6,8-decatrien-1-ynyl

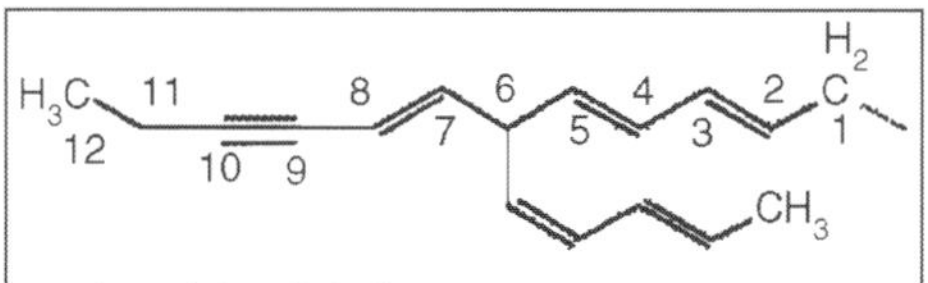

H_2
H_3C 12 11 10 9 8 7 6 5 4 3 2 C 1
CH_3
6-(1,3-Pebtadienyl)-2,4,7-dodecatrien-9-ynyl

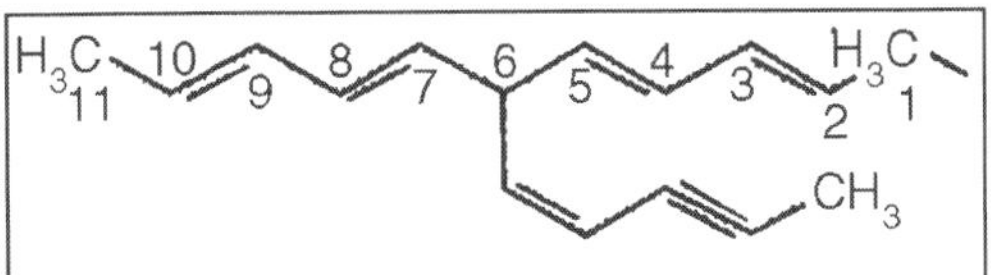

6-(1-Penten-3-ynyl)-2,4,7,9-undecaletraenyl

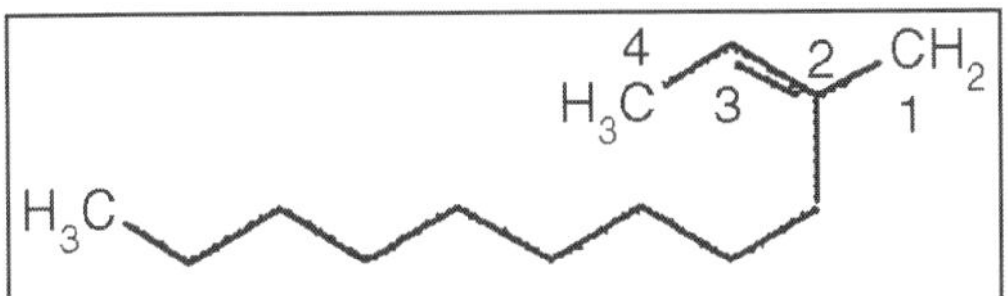

2-Nonyl-2-butenyl

Rule. Bivalent and Multivalent Radicals

- Bivalent and trivalent radicals derived from univalent acyclic hydrocarbon radicals whose authorized names end in "-yl" by removal of one or two hydrogen atoms from the carbon atom with the free valences are named by adding "-idene" or "-idyne", respectively, to the name of the corresponding univalent radical. The carbon atom with the free valence is numbered as 1.
- The name "methylene" is retained for the radical $H_2C =$

  Examples:

Methylidyne	$HC \equiv$
Ethylidene	$H_3C\text{–}HC =$
Ethylidene	$H_3C\text{—}C\equiv$
Vinylidene	$H_2C = C =$
Isopropylidene	$(H_3C)_2C =$

- The names of bivalent radicals derived from normal alkanes by removal of a hydrogen atom from each of the two terminal carbon atoms of the chain are ethylene, trimethylene, tetramethylene, *etc.*

  Examples:

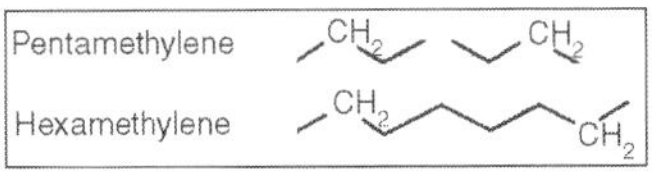

Pentamethylene	$-CH_2-CH_2-CH_2-CH_2-CH_2-$
Hexamethylene	$-CH_2-CH_2-CH_2-CH_2-CH_2-CH_2-$

Example:

Ethylethylene	$-H_2C-CH(-)-H_2C-CH_3$

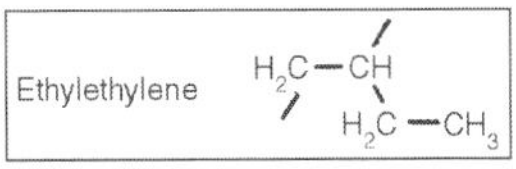

The following name is retained:

Propylene	H_3C–HC–CH_2

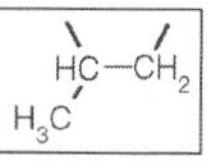

- Bivalent radicals similarly derived from unbranched alkenes, alkadienes, alkynes, *etc.*, by removing a hydrogen atom from each of the terminal carbon atoms are named by replacing the endings "-ene", "-diene", "-yne", *etc.*, of the hydrocarbon name by "-enylene", "-dienylene", "-ynylene", *etc.*, the positions of the double and triple bonds being indicated where necessary.

  Example:
  - Propenylene —H_2C=C=CH—

  The following name is retained:
- Vinylene (for ethen ylene) –HC = CH–

  Example:

4-Propyl-2-pentenylene	H_2C(5)–(4)–(3)=(2)–H_2C(1); CH_3

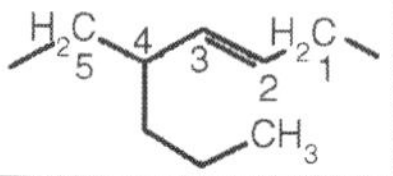

- Trivalent, quadrivalent and higher-valent acyclic hydrocarbon radicals of two or more carbon atoms with the free valences at each end of a chain are named by adding to the hydrocarbon name the terminations "-yl" for a single free valence, "-ylidene" for a double, and "-ylidyne" for a triple free valence on the same atom (the final "e" in the name of the hydrocarbon is elided when followed by a suffix beginning with "-yl"). If different types are present in the same radical, they are cited and numbered in the order "-yl", "-ylidene", "-ylidyne".

  Examples:

Name	Structure
Butanediylidene	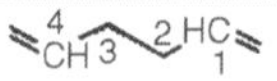 =CH(4)–3–(2)–HC(1)=
Butanediylidyne	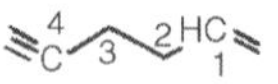 ≡C(4)–3–(2)–HC(1)=
1-Propany1-3-ylidene	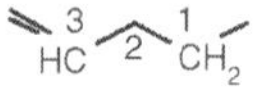 =HC(3)–2–CH_2(1)–
Propadienediylidene	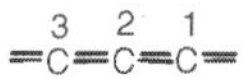 =C(3)=C(2)=C(1)=
2-Pentenediylidyne	 ≡C(5)–4–3=2–C(1)≡
1-Butanyliden-4-ylidyne	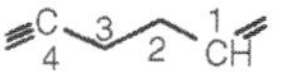 ≡C(4)–3–2–CH(1)=

- Multivalent radicals containing three or more carbon atoms with free valences at each end of a chain and additional free valences at intermediate carbon atoms are named by adding the endings "-triyl", "-tetrayl", "-diylidene", "diyl-ylidene", *etc.*, to the hydrocarbon name.

  Example:

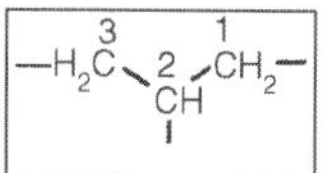

—H_2C(3)–CH(2)–CH_2(1)—
1,2,3-Propanetriyl

SPIRO HYDROCARBONS

A "spiro union" is one formed by a single atom which is the only common member of two rings. A "free spiro union" is one constituting the only union direct or indirect between two rings senior.

The common atom is designated as the "spiro atom". According to the number of spiro atoms present, the compounds are distinguished as monospiro-, dispiro-, trispirocompounds, *etc.* The following rules apply to the naming of compounds containing free spiro unions.

Rule. Compounds: Method 1

- Monospiro compounds consisting of only two alicyclic rings as components are named by placing "spiro" before the name of the normal acyclic hydrocarbon of the same total number of carbon atoms. The number of carbon atoms linked to the spiro atom in each ring is indicated in ascending order in brackets placed between the spiro prefix and the hydrocarbon name.

  Examples:

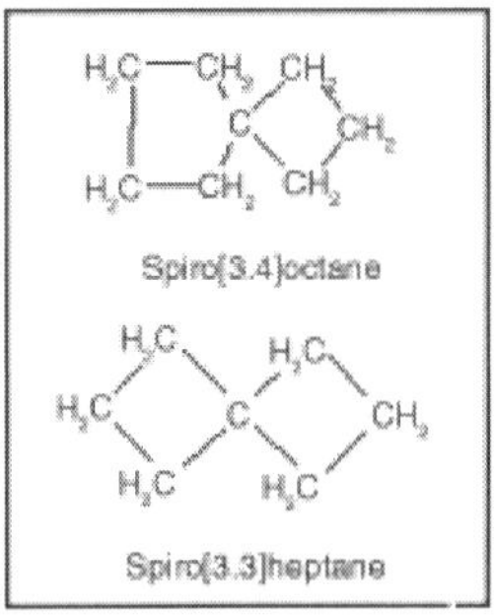

  Spiro[3.4]octane

  Spiro[3.3]heptane

- The carbon atoms in monospiro hydrocarbons are numbered consecutively starting with a ring atom next to the spiro atom, first through the smaller ring (if such be present) and then through the spiro atom and around the second ring.

  Example:

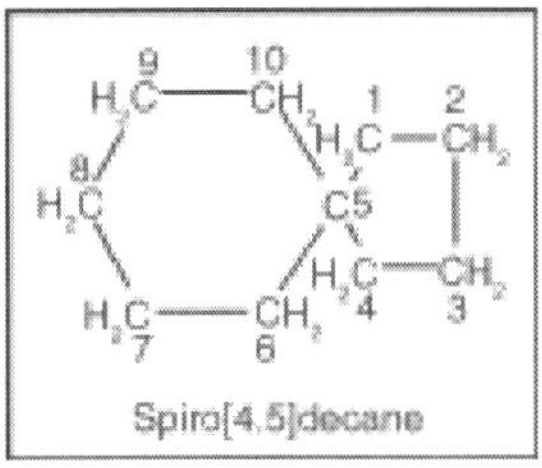

  Spiro[4.5]decane

- When unsaturation is present, the same enumeration pattern is maintained, but in such a direction around the rings that the double and triple bonds receive numbers as low as possible in accordance with Rule.

Example:

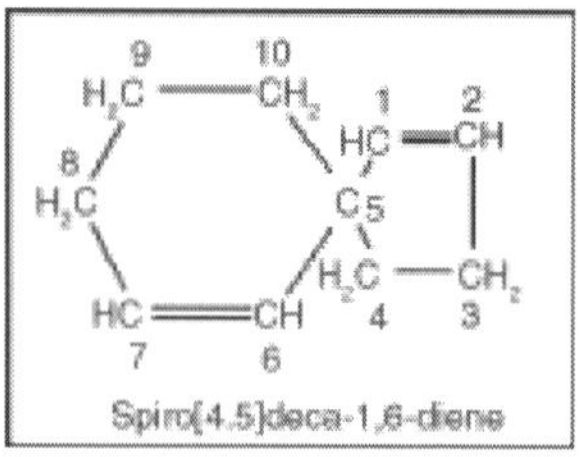

Spiro[4.5]deca-1,6-diene

- If one or both components of the monospiro compound are fused polycyclic systems, "spiro" is placed before the names of the components arranged in alphabetical order and enclosed in brackets. Established numbering of the individual components is retained. The lowest possible number is given to the spiro atom, and the numbers of the second component are marked with primes. The position of the spiro atom is indicated by placing the appropriate numbers between the names of the two components.

  Example:

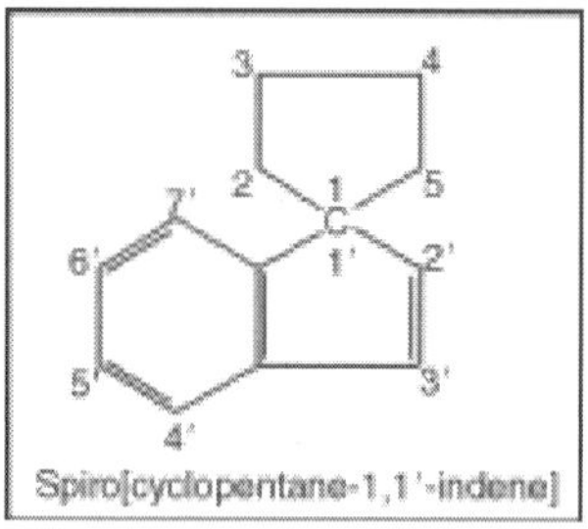

Spiro[cyclopentane-1,1'-indene]

- Monospiro compounds containing two similar polycyclic components are named by placing the prefix "spirobi" before the name of the component ring system. Established enumeration of the polycyclic system is maintained and the numbers of one component are distinguished by primes. The position of the spiro atom is indicated in the name of the spiro compound by placing the appropriate locants before the name.

  Example:

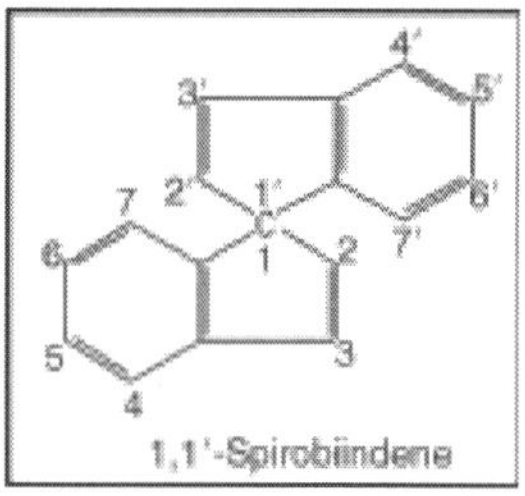

1,1'-Spirobiindene

- Polyspiro compounds consisting of a linear assembly of three or more alicyclic systems are named by placing "dispiro-", "trispiro-", "tetraspiro-", etc., before the name of the unbranched-chain acyclic hydrocarbon of the same total number of carbon atoms. The numbers of carbon atoms linked to the spiro atoms in each ring are indicated in brackets in the same order as the numbering proceeds about the ring. Numbering starts with a ring atom next to a terminal spiro atom and proceeds in such a way as to give the spiro atoms as low numbers as possible after numbering all the carbon atoms of the first ring linked to the terminal spiro atom.

Example:

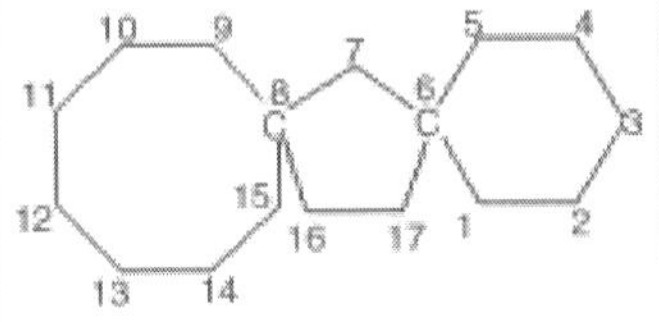

Dispiro[5.1.7.2]heptadecane

- Polycyclic compounds containing more than one spiro atom and at least one fused polycyclic component are named in accordance of this rule by replacing "spiro" with "dispiro", "trispiro", etc., and choosing the end components by alphabetical order.

  Example:

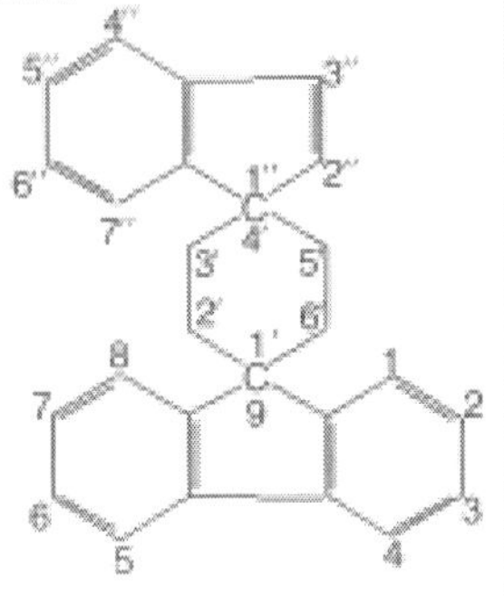

  Dispiro[fluorene-9,1'-cyclohexane-4',1''-indene]

Rule. Compounds: Method 2

- When two dissimilar cyclic components are united by a spiro union, the name of the larger component is followed by the affix "spiro" which, in turn, is followed by the name of the smaller component. Between the affix "spiro" and the name of each component system is inserted the number denoting the spiro position in the appropriate ring system, these numbers being as low as permitted by any fixed enumeration of the component. The components retain their respective enumerations but numerals for the component mentioned second are primed. Numerals I may be omitted when a free choice is available for a component.

Examples:

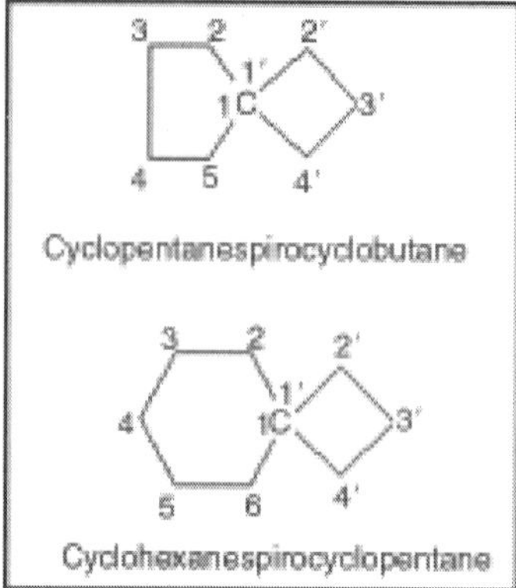

Cyclopentanespirocyclobutane

Cyclohexanespirocyclopentane

- Rule applies also with appropriate different enumeration, where nomenclature is according to Rule, but the spiro junction has priority for lowest numbers over unsaturation.
  Example:

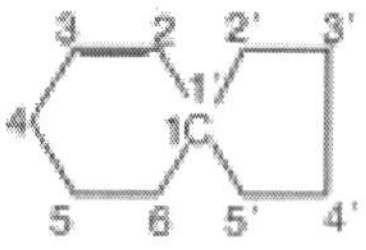

2-Cyclohexanespiro-(2′-cyclopentene)

- The nomenclature of Rule is applied also to monocyclic components with identical saturation, the spiro union being numbered 1.
  Example:

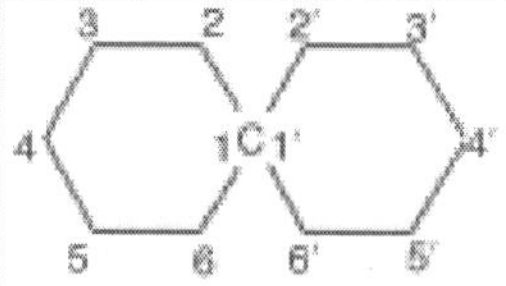

Spirobicyclohexane

but

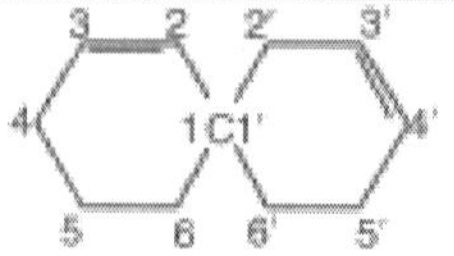

2-Cyclohexenespiro-(3′-cyclohexene)

- Polycyclic compounds containing more than one spiro atom are named in accordance with Rule starting from the senior end-component irrespective of whether the components are simple or fused rings.

Examples:

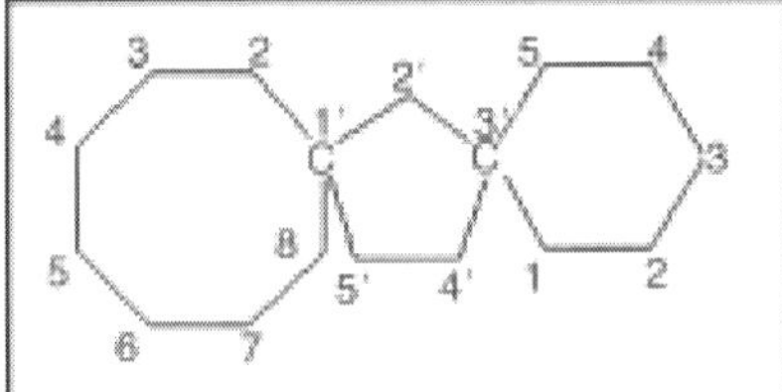

Dispiro[5.1.7.2]heptadecane

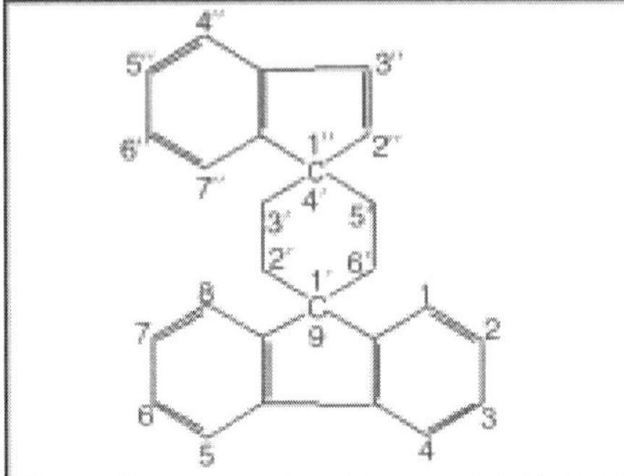

Dispiro[fluorene-9,1'-cyclohexane-4',1''-indne]

Rule. Radicals

- Radicals derived from spiro hydrocarbons are named according to the principles set forth in Rules.
  Examples:

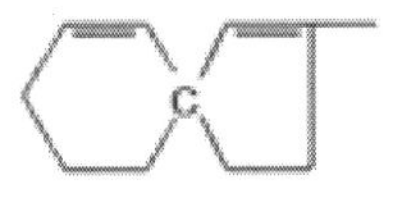

Spiro[4,5]deca-1,6-diene-2-yl

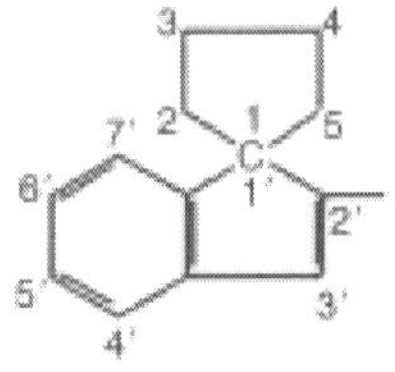

Spiro[cyclopentane-1,1'-inden]-2'-yl

BRIDGED HYDROCARBONS

RULE. BICYCLIC SYSTEMS

- Saturated alicyclic hydrocarbon systems consisting of two rings only, having two or more atoms in common, take the name of an open chain hydrocarbon containing the same total number of carbon atoms preceded by the prefix "bicyclo-". The number of carbon atoms in each of the three bridges connecting the two tertiary carbon atoms is indicated in brackets in descending order.

Examples:

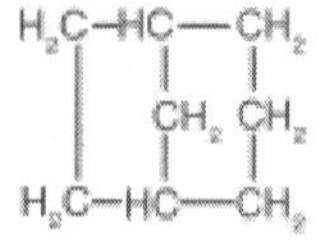

Bicyclo[3.2.1]octane

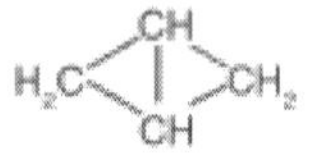

Bicyclo[1.1.0]butane

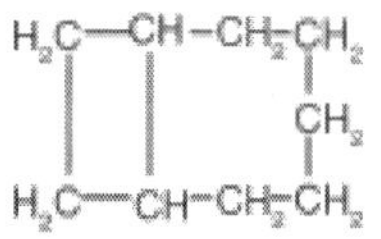

Bicyclo[5.2.0]nonane

- The system is numbered commencing with one of the bridgeheads, numbering proceeding by the longest possible path to the second bridgehead; numbering is then continued from this atom by the longer unnumbered path back to the first bridgehead and is completed by the shortest path from the atom next to the first bridgehead.

  Examples:

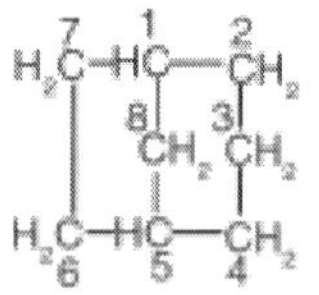

Bicyclo[3.2.1]octane

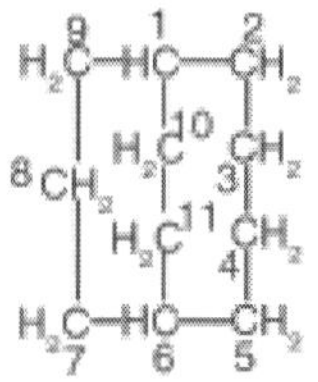

Bicyclo[4.3.2.]undecane

- Unsaturated hydrocarbons are named in accordance with the principles set forth in Rule. When after applying Rule a choice in numbering remains unsaturation is given the lowest numbers.

Examples:

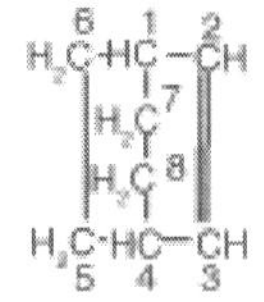

Bicyclo[2.2.2]oct-2-ene

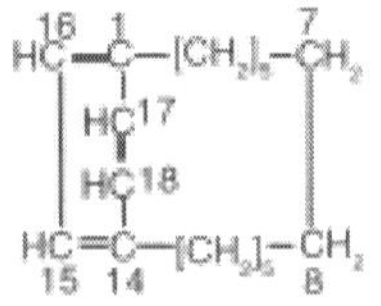

Bicyclo[2.2.2]octadeca-1(16),14,17-triene
or Bicyclo[12.2.2]octadeca-14,16(1),17-triene

- Radicals derived from bridged hydrocarbons are named in accordance with the principles set forth in Rule The numbering of the hydrocarbon is retained and the point or points of attachment are given numbers as low as is consistent with the fixed numbering of the saturated hydrocarbon.
  Examples:

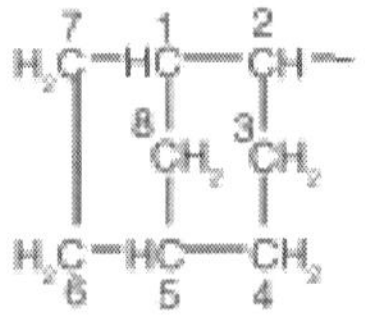

Bicyclo[3.2.1]oct-2-yl

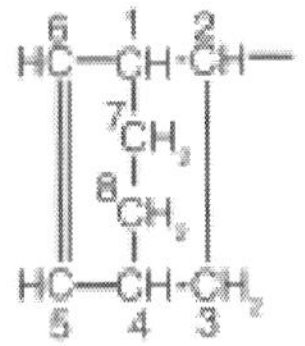

Bicyclo[3.2.1]oct-5-en-2-yl

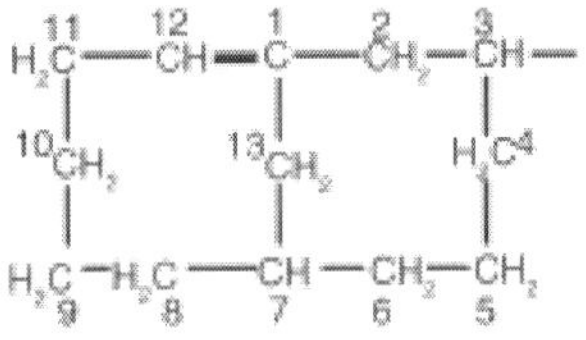

Bicyclo[5.5.1]tridec-1(12)-en-3-yl
or Bicyclo[5.5.1]tridec-12(1)-en-3-yl

Rule. Polycyclic Systems

- Cyclic hydrocarbon systems consisting of three or more rings may be named in accordance with the principles stated in Rule.
  The appropriate prefix "tricyclo-", "tetracyclo-", *etc., is* substituted for "bicyclo-" before the name of the open-chain hydrocarbon containing the same total number of carbon atoms.
  Radicals derived from these hydrocarbons are named according to the principles set forth in Rule.
- A polycyclic system is regarded as containing a number of rings equal to the number of scissions required to convert the system into an open-chain compound.
- The word "cyclo" is followed by brackets containing, in decreasing order, numbers indicating the number of carbon atoms in:
  - The two branches of the main ring,
  - The main bridge,
  - The secondary bridges.

  Examples:

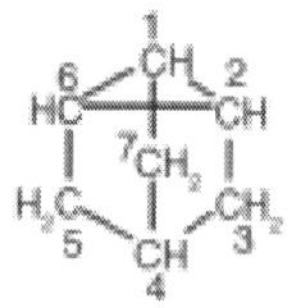

  Tricyclo[2.2.1.0]heptane

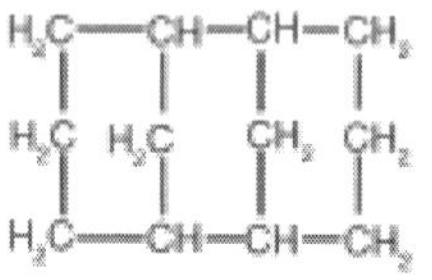

  Tricyclo[5.3.1.1]dodecane

- The main ring and the main bridge form a bicyclic system whose numbering is made in compliance with Rule.
- The location of the other or so-called secondary bridges is shown by superscripts following the number indicating the number of carbon atoms in the said bridges.
- For the purpose of numbering, the secondary bridges are considered in decreasing order. The numbering of any bridge follows from the part already numbered, proceeding from the highest-numbered bridge head. If equal bridges are present, the numbering begins at the highest. numbered bridgehead.
- When there is a choice, the following criteria are considered in turn until a decision is made:

- The main ring shall contain as many carbon atoms as possible, two of which must serve as bridgeheads for the main bridge.
  Examples:

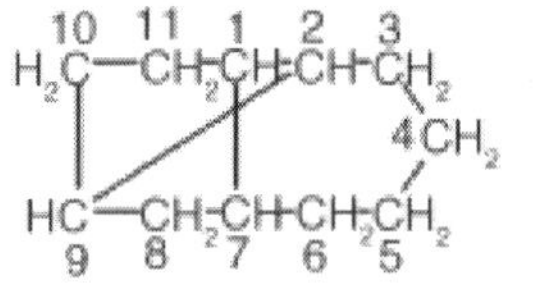

Tricyclo[5.4.0.$0^{2,9}$]undecane
Correct Numbering

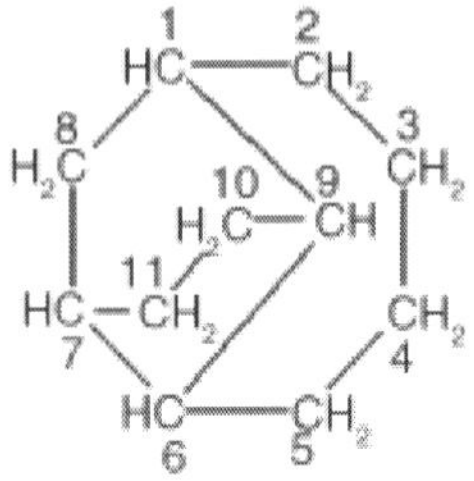

Tricyclo[4.2.1.$2^{7,9}$]undecane
Incorrect Numbering

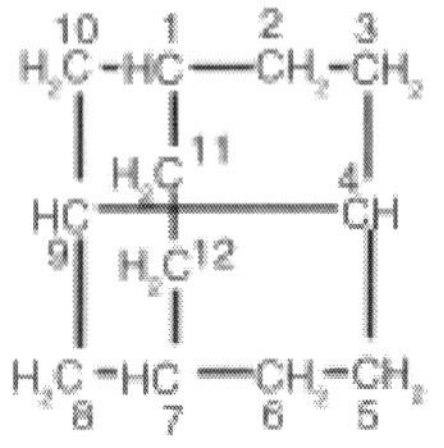

Tricyclo[5.3.2.$0^{4,9}$]dodecane
Correct Numbering

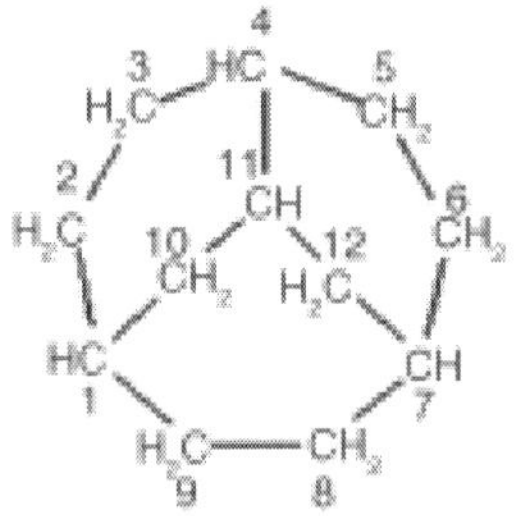

Tricyclo[5.3.2.$0^{4,11}$]dodecane
Incorrect Numbering

- The main bridge shall be as large as possible.

Examples:

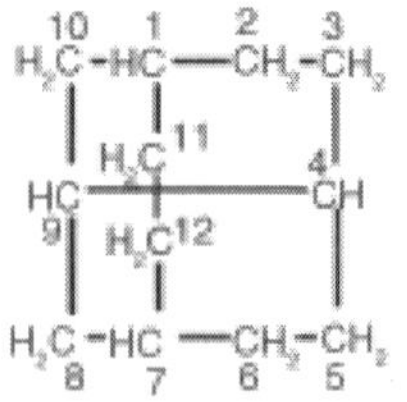

Tricyclo[5.3.2.$0^{4,9}$]dodecane
Correct Numbering

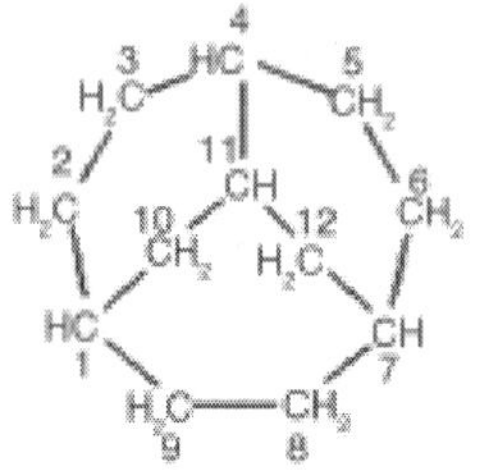

Tricyclo[5.3.2.$0^{4,11}$]dodecane
Incorrect Numbering

– The main ring shall be divided as symmetrically as possible by the main bridge.

Examples:

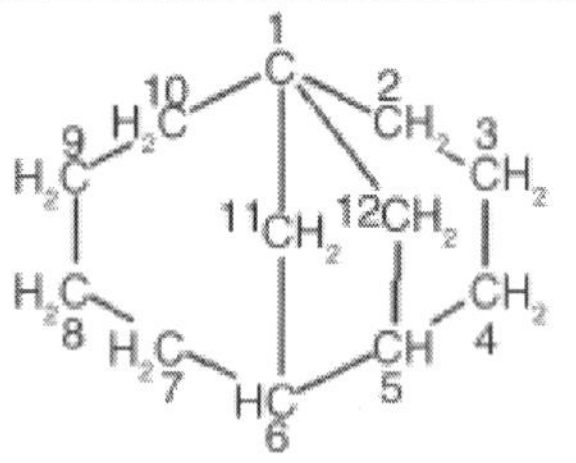

Tricyclo [4.4.1.$1^{1,5}$]dodecane
Correct Numbering

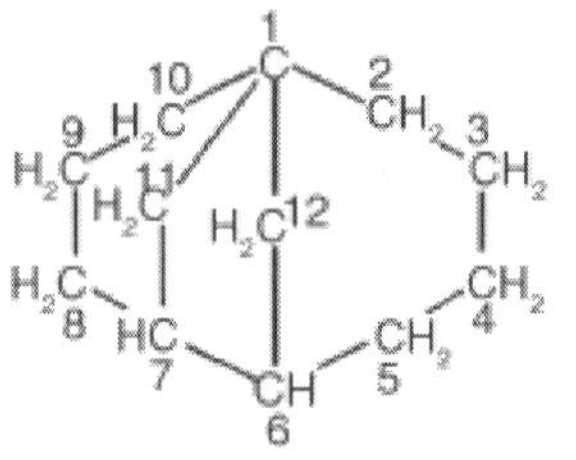

Tricyclo [5.3.1.$1^{1,6}$]dodecane
Incorrect Numbering

- The superscripts locating the other bridges shall be as small as possible (in the sense indicated in Rule).

  Example:

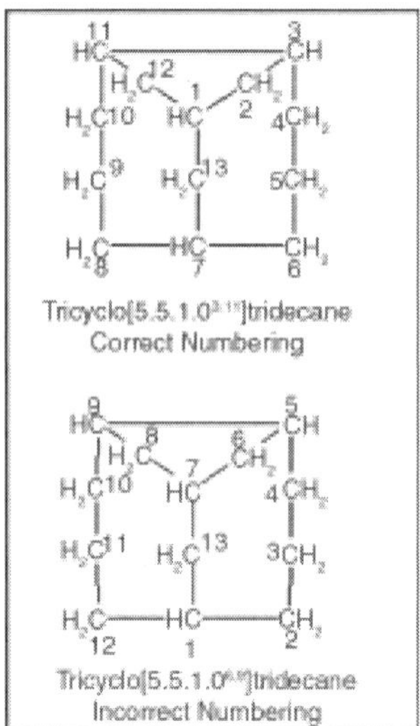

Tricyclo[5.5.1.$0^{3,11}$]tridecane
Correct Numbering

Tricyclo[5.5.1.0[illegible]]tridecane
Incorrect Numbering

Rule. Hydrocarbon Bridges

- Polycyclic hydrocarbon systems which can be regarded as "*ortho*-fused" or "*ortho*- and *peri*-fused" systems according to Rule and which, at the same time, have other bridges, are first named as "*ortho*-fused" or "*ortho*- and *peri*-fused" systems. The other bridges are then indicated by prefixes derived from the name of the corresponding hydrocarbon by replacing the final "-ane", "-ene", etc., by "-ano", "-eno", *etc.,* and their positions are indicated by the points of attachment in the parent compound. If bridges of different types are present, they are cited in alphabetical order.

  Examples:

  - Butano
    $—CH_2—CH_2—CH_2—CH_2—$
  - Benzeno (*o*-, *m*-, *p*-)
    $—C_6H_4—$
  - Ethano
    $—CH_2—CH_2—$
  - Etheno
    $—CH=CH—$
  - Methano
    $—CH_2—$
  - Propano
    $—CH_2—CH_2—CH_2—$

Examples:

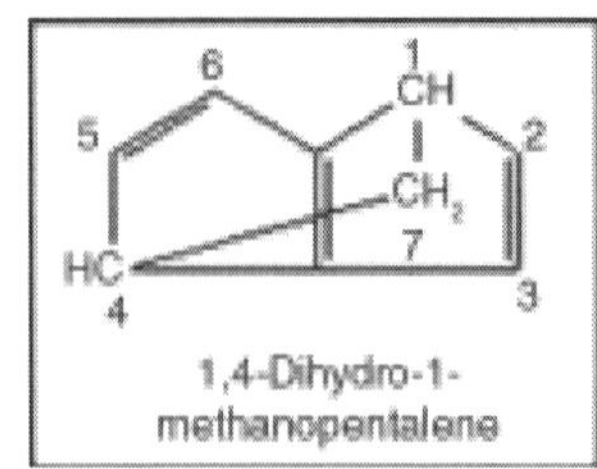

1,4-Dihydro-1-methanopentalene

9,10-Dihydro-9,10-[2]butenoanthracene

7,14-Dihydro-7,14-ethano-dibenz[a,h]anthracene

- The parent "*ortho*-fused" or "*ortho*- and *peri*-fused" system is numbered as prescribed in Rule. Where there is a choice, the position numbers of the bridgeheads should be as low as possible. The remaining bridges are then numbered in turn starting each time with the bridge atom next to the bridgehead possessing the highest number.

Example:

Perhydro-1,4-ethano-anthracene

Not

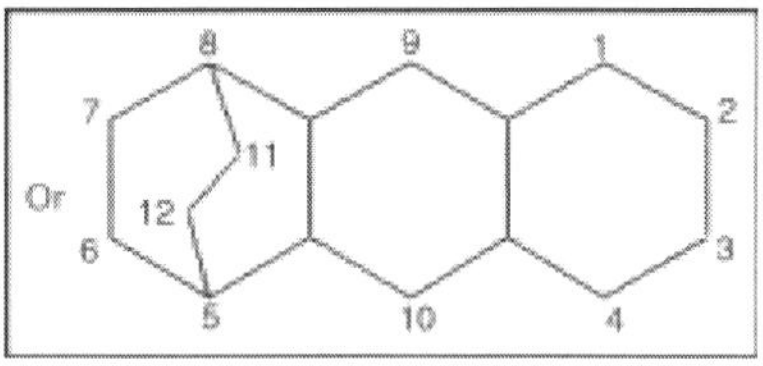

- When there is a choice of position numbers for the points of attachment for several individual bridges, the lowest numbers are assigned to the bridgeheads in the order of citation of the bridges and the bridge atoms are numbered according to the preceding rule.
  Example:

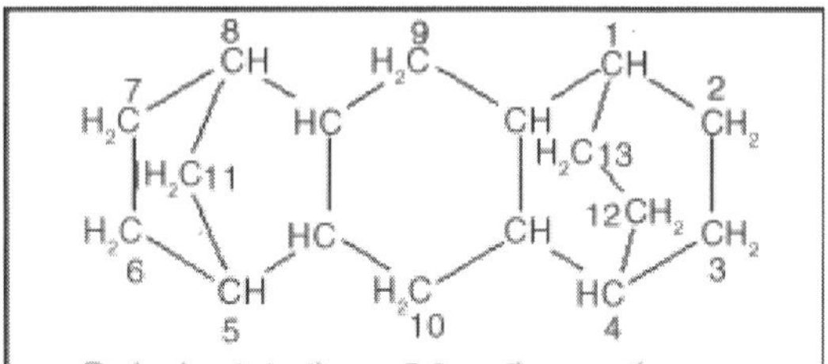

Perhydro-1,4-ethano-5,8-methanoanthracene

- When the bridge is formed from a bivalent cyclic hydrocarbon radical, low numbers are given to the carbon atoms constituting the shorter bridge and numbering proceeds around the ring.
  Examples:

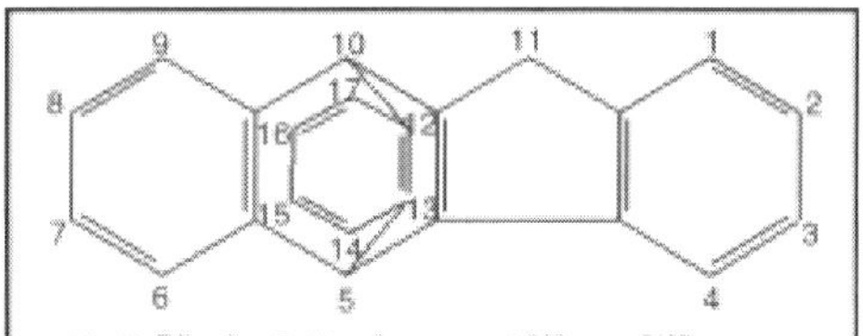

10,11-Dihydro-5,10-o-benzeno-5*H*-benzo[*b*]fluorene

- Names for radicals derived from the bridged hydrocarbons considered in Rule are constructed in accordance with the principles set forth in Rule. The abbreviated radical names naphthyl, anthryl, phenanthryl, naphthylene, etc., permitted as exceptions to Rules, are replaced in such cases by the regularly formed names naphthalenyl, anthracenyl, phenanthrenyl, naphthalenediyl, *etc.*
  Examples:

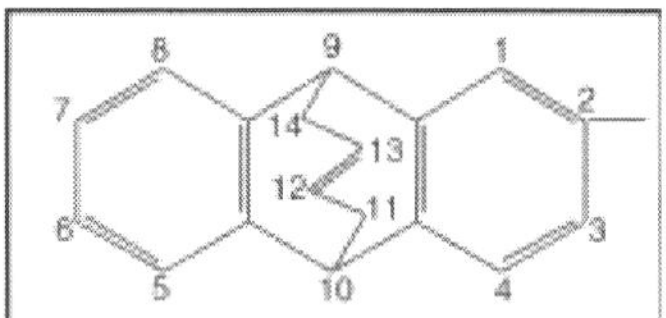

9,10-Dihydro-9-10-[2]Butenoanthracen-2-

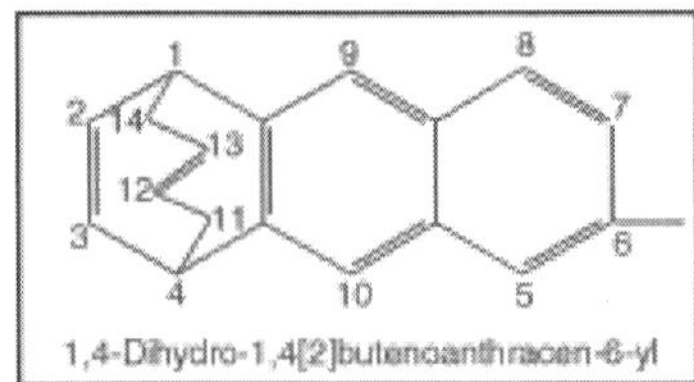

1,4-Dihydro-1,4[2]butenoanthracen-6-yl

HYDROCARBON RING ASSEMBLIES

Rule. Definition

- Two or more cyclic systems (single rings or fused systems) which are directly joined to each other by double or single bonds are named "ring assemblies" when the number of such direct ring junctions is one less than the number of cyclic systems involved.
  Examples:

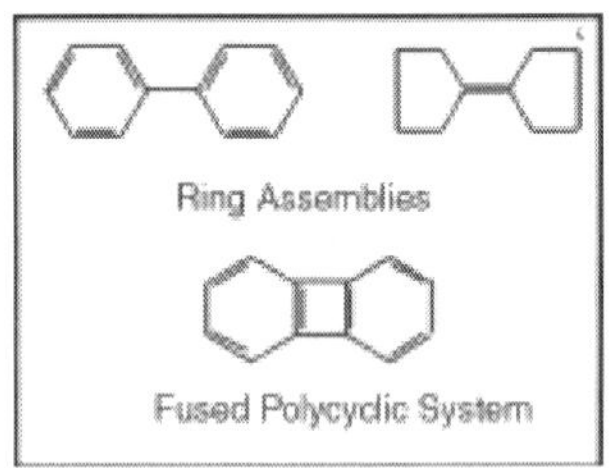

Rule. Two Identical Ring Systems

- Assemblies of two identical cyclic hydrocarbon systems are named in either of two ways:
  - By placing the prefix "bi-" before the name of the corresponding radical,
  - For systems joined by a single bond by placing the prefix "bi-" before the name of the corresponding hydrocarbon.
- In each case, the numbering of the assembly is that of the corresponding radical or hydrocarbon, one system being assigned unprimed numbers and the other primed numbers. The points of attachment are indicated by placing the appropriate locants before the name.
  Examples:

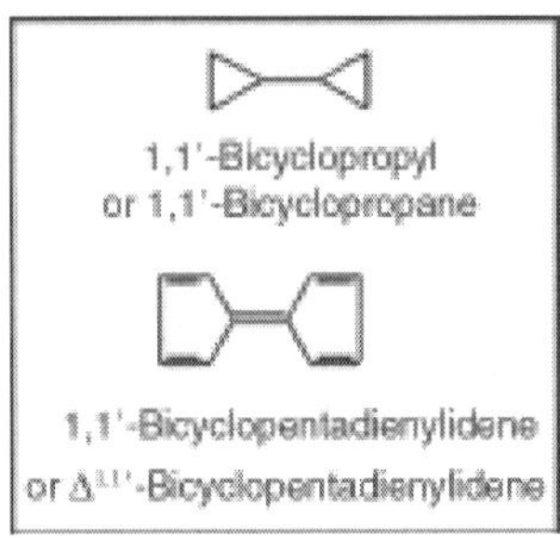

- If there is a choice in numbering, unprimed numbers are assigned to the system which has the lower-numbered point of attachment.
  Example:

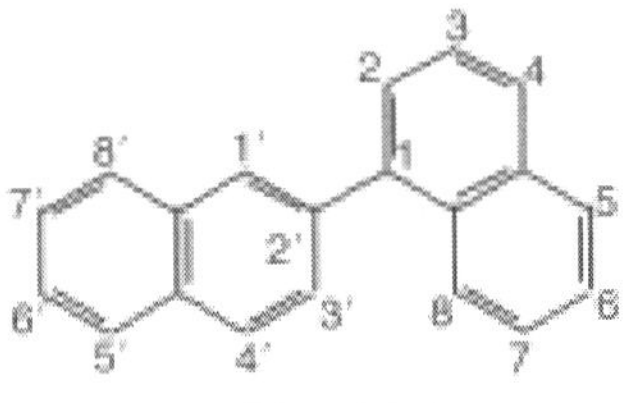

  1,2'-Binaphthyl
  or 1,2'-Binaphthalene

- If two identical hydrocarbon systems have the same point of attachment and contain substituents at different positions, the locants of these substituents are assigned according to Rule for this purpose an unprimed number is considered lower than the same number when primed. Assemblies of primed and unprimed numbers are arranged in ascending numerical order.
  Example:

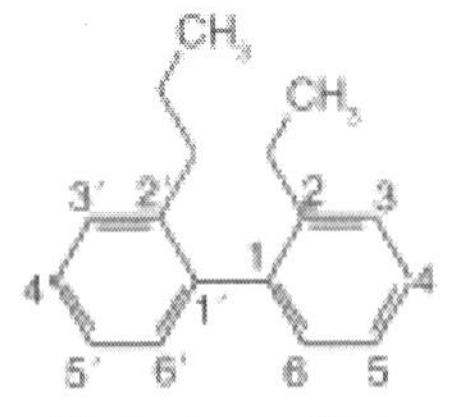

  2-Ethyl-2'-propylbiphenyl

- The name "biphenyl" is used for the assembly consisting of two benzene rings.
  Example:

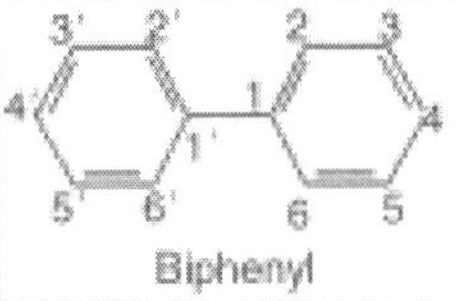

  Biphenyl

Rule. Non-Identical Ring Systems

- Other hydrocarbon ring assemblies are named by selecting one ring system as the base component and considering the other systems as substituents of the base component. Such substituents are arranged in alphabetical order. The base component is assigned unprimed numbers and the substituents are assigned numbers with primes.

- The base component is chosen by considering the following characteristics in turn until a decision is reached:
  - The system containing the larger number of rings.
    Example:

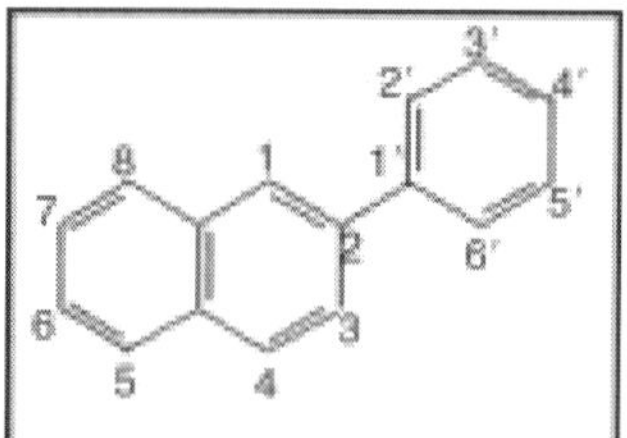

    2-Phenylnaphthalene

  - The system containing the larger ring.
    Example:

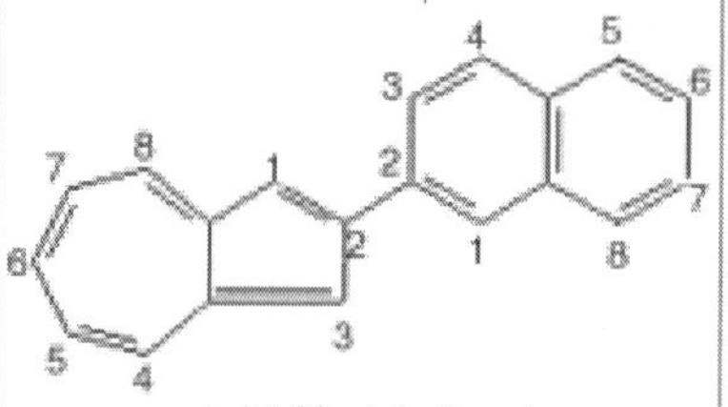

    2-(2'-Naphthyl)azulene

  - The system in the lowest state of hydrogenation.
    Example:

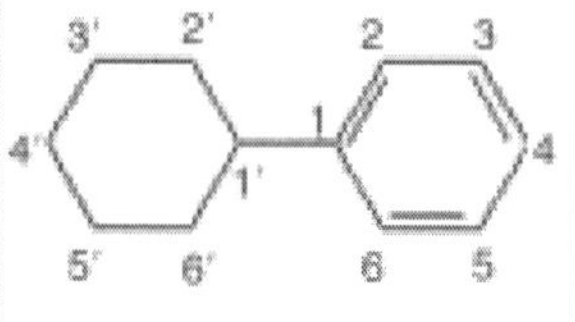

    Cyclohexylbenzene

  - The order of ring systems as set forth in the list of Rule.
- Compounds covered by this rule may also be named as hydrogenation products according to Rule
  Example:

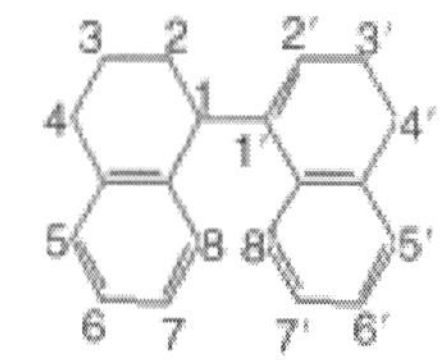

  1,2,3,3',4,4'-Hexaxydro
  -1,1'-binaphthyl
  or1,2,3,3',4,4',-Hexaxydro
  -1,1'-binaphthalene

Rule. Three or More Identical Ring Systems

- Unbranched assemblies consisting of three or more identical hydrocarbon ring systems are named by placing an appropriate numerical prefix before the name of the hydrocarbon corresponding to the repetitive unit.

  The following numerical prefixes are used:

  - Ter-
  - Quater-
  - Quinque-
  - Sexi-
  - Septi-
  - Octi-
  - Novi-
  - Deci-

  Example:

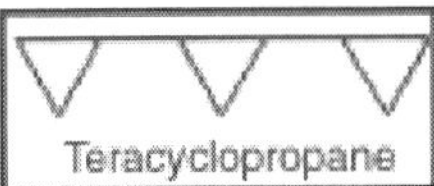
Teracyclopropane

- Unprimed numbers are assigned to one of the terminal systems, the other systems being primed serially. Points of attachment are assigned the lowest numbers possible.

  Example:

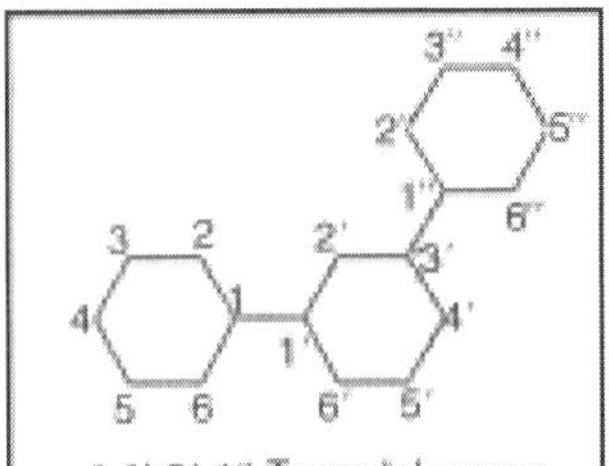

1,1':3',1''-Tercyclohexane

- As exceptions, unbranched assemblies consisting of benzene rings are named by using the appropriate prefix with the radical name "phenyl".

  Example:

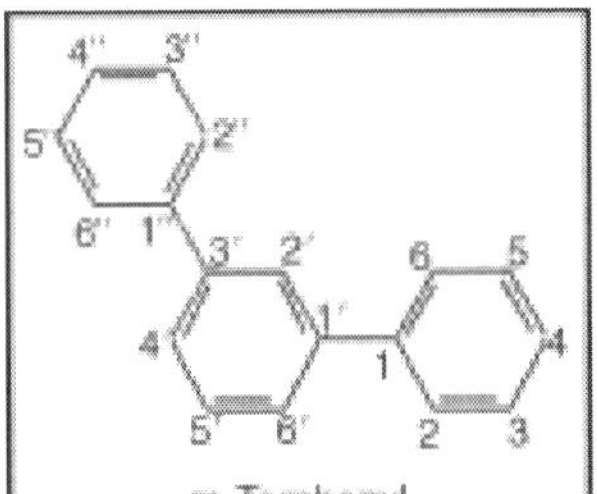

m-Terphenyl
or1,1'3'-1''-Terphenyl

Rule. Radicals for Identical Ring Systems

- Univalent and multivalent radicals derived from assemblies of identical hydrocarbon ring systems are named by adding "-yl", "-ylene" or "-diyl", "-triyl", *etc.,* to the name of the ring assembly.
  Example:

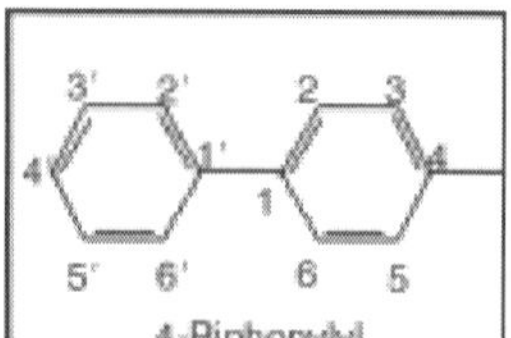

4-Biphenylyl

Rule. Radicals for Non-benzenoid Ring Systems

- Radicals derived from hydrocarbon ring assemblies other that benzene ring assemblies by removal of one or more hydrogen atoms from only one ring are named with that ring as the parent radical, the remaining rings being named as substituents.
  Example:

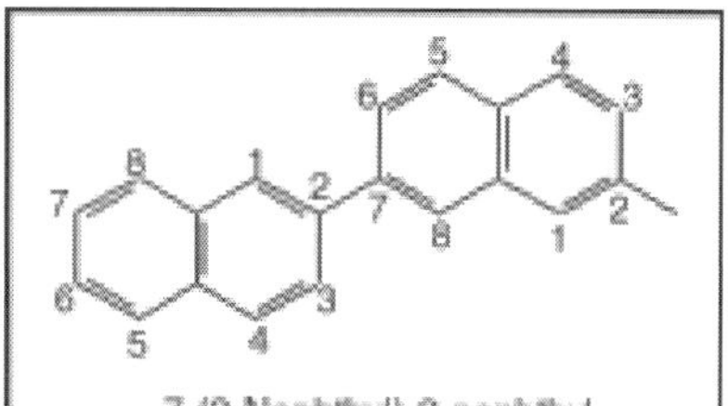

7-(2-Naphthyl)-2-naphthyl

Note: This method is used for assemblies of non-identical systems; also it is sometimes preferable to that of Rule.for assemblies of identical systems when a group to be specified as a suffix or as a separate word is present in a chain attached to the ring assembly.

CYCLIC AND AROMATIC HYDROCARBONS

In some hydrocarbons the carbons at the ends of the chain join together to form a ring and are called cyclic hydrocarbons.

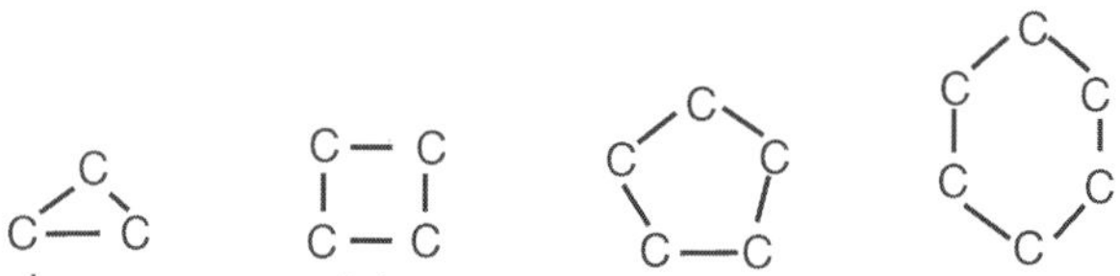

cyclopropane cyclobutane cyclopentane cyclohexane

Bonding angles for carbon should be 109°; if the bonding angle is less than this value there is a strain on the bond that makes the molecule unstable. The most stable (most abundant) rings have 5 or 6 carbons (smallest ring strain). Cycloalkanes are named using the alkane parent with the suffix *cyclo*. The general formula for the cycloalkanes is C_nH_{2n}. Cyclic hydrocarbons with multiple carbon – carbon double bonds form a special group of compounds called aromatic hydrocarbons. The simplest member of this group is benzene (C_6H_6). The structure of benzene is given by:

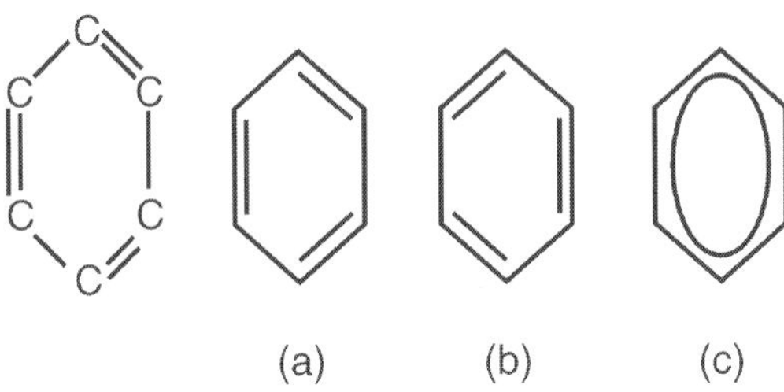

The carbon – carbon double bonds can alternate position but neither isomer can be isolated. The chemist will use a combination of both structures (resonance structure) to represent the structure of benzene (c).

Substituent groups replace the hydrogen and are named as derivatives of benzene (benzene is the parent chain). When there are two substituent groups there are three possible isomers as shown below:

CH_3 CH_3 CH_3 CH_3 CH_3 CH_3 CH_3

methyl-benzene   1, 2-dimethyl benzene   1, 3-dimethyl benzene   1, 4-dimethyl benzene

IUPAC Name	Common Name	Suffix
1,2-dimethylbenzene	o-dimethylbenzene	o – ortho
1,3-dimethylbenzene	m-dimethylbenzene	m – meta
1,4-dimethylbenzene	p-dimethylbenzene	p – para

Para-dichlorobenzene is often used as a bathroom deodorizer; draw the structure for this compound.

The petroleum industry is an important source of hydrocarbons.

Crude oil is a mixture of hydrocarbon chains that are continuous chains and branched chains. The crude oil hydrocarbon chains vary in length and can therefore be separated by their different boiling points. Fractional distillation uses the different boiling points to separate the crude oil mixture. The boiling point is determined by the strength of van der Waals force; which is determined by length and shape of the molecule. Each fraction contains several hydrocarbon chains similar in length. These fractions serve as the starting materials for the manufacture of many plastics, fabrics and construction materials (paints, glues chalking asphalt to mention a few). Larger (longer) chains can be broken down into smaller more useful chains by a process called cracking.

The general molecular formulas for the Alkane, Alkene, and Alkyne and Cycloalkane families:

Family	Molecular Formula
Alkane	C_nH_{2n+2}
Alkene	C_nH_{2n}
Cycloalkane	C_nH_{2n}
Alkyne	C_nH_{2n-2}

Functional Groups and Organic Reactions

In a previous section on organic chemistry you studied hydrocarbons that are essential components of every organic compound. Hydrocarbons molecules are chemically inert (fairly unreactive). Most organic reactions involve substituents on the hydrocarbon chain. These substituents determine the major chemical properties of the individual organic compounds.

The substituent, which is the focus of the present section of organic chemistry, is called a functional group. The term functional group is used because it is the substituent that determines how the compound functions chemically (how the compound reacts).

The type of chemical reactions will be determined by the functional group because it is the chemically active part (site) on the molecule. Organic chemists classify organic compounds according to their functional group. The chemist uses the symbol R- to represent a hydrocarbon chain (alkyl) or ring (cyclic). The following table shows the functional group and structure that will be studied in detail.

Halocarbon

Halocarbons are a class of organic compounds containing a hydrocarbon parent covalently bonded with fluorine, chlorine, bromine or iodine.

The IUPAC rules for naming halocarbons uses the same rules as is used for Alkanes if the parent chain is saturated, or for Alkenes or Alkynes if the parent chain is unsaturated. The halogen(s) is treated as a branched group and is located on the continuous chain of carbons as you would locate and name any alkyl branch.

For example, the structure below would be called

```
    Br
    |
C - C - C - C - C - C - C  2,3-dibromoheptane
        |
        Br
```

The longest continuous chain of carbons with the halogen atoms attached is seven. The chain is numbered from the end that will give the lowest carbon number designators for the halogen atoms and any existing alkyl branch.

Physical Properties

Boiling Points

Halocarbons are more polar than the hydrocarbon families. Therefore the boiling points will be larger for the same sized molecules. The London Dispersion and the dipole-dipole interactions are the intermolecular forces holding the halocarbon molecules together.

Water Solubility

The halocarbons are insoluble in water because there is very little attraction that the halide molecules have for the solvent water. Hydrogen bonding is not present and for that reason most halocarbons are insoluble in water but tend to be soluble in hydrocarbon solvents and other alkyl halide solvents.

Density of Alkyl Halides

The densities of halocarbons tend to be greater than that of water because of the relatively massive halogen atoms.

Production of Halocarbons

Halocarbons can be produced using two methods:

- Substitution Reaction (inorganic chemistry calls this single replacement)

A halogen is mixed with a saturated hydrocarbon; the halogen atom substitutes hydrogen from the hydrocarbon chain. This reaction can be shown by the general equation:

$X_2 + R - H \rightarrow R - X + HX$

Example:

$Cl_2(g) + CH_3 - CH_3(g) \rightarrow CH_3\text{-}CH_2Cl(g) + HCl(g)$

chloroethane (only one possible isomer)

- Addition Reaction (inorganic chemistry calls this combination or composition)

A halogen is mixed with an unsaturated hydrocarbon; a halogen atom adds on to each of the two carbons from the carbon-carbon double bond in the hydrocarbon chain. This reaction can be shown by the general equation:

$$X_2 + R - C = C - R' \rightarrow R - \underset{}{\overset{X}{\overset{|}{C}}} - \overset{X}{\overset{|}{C}} - R'$$ R and R' represent alkyl groups

Example:

$Cl_2(g) + CH_3 - CH_3(g) \rightarrow CH_2Cl - CH_2Cl(g)$

1,2-dichloroethane (more than one isomer possible but only one can be produced by this reaction)

Alcohol

Alcohols are a family of organic compounds that share a common chemically active group called the hydroxyl group (– O-H). Do not confuse the hydroxyl group with the hydroxide ion (OH^-) (a negatively charged ion). The hydroxyl group is electrically neutral and forms a covalent bond. The hydroxyl group (–OH) represents the position in an alcohol molecule where the chemical change takes place.

The general symbol to represent any alcohol with only one hydroxyl group attached to one of its carbons is R –OH. Alcohols can have two or more hydroxyl groups attached to the molecule and are called "diols", "triols".

Naming

Simple alcohols can be named using the common name approach of identifying and naming the alkyl portion and follow it with the word alcohol. For example, CH_3 – OH could be called methyl alcohol since the CH_3 group is the methyl group.

This method works as long as you have simple alcohols. When the alcohols become more complex and more branched, then IUPAC has come up with a set of rules that are used to name any alcohol regardless of its complexity. These rules are as follows:

- Identify the longest continuous (parent) chain of carbons with the hydroxyl group attached to one of the carbons in the chain.
- Number the chain so that the hydroxyl group is attached to the lowest numbered carbons.
- Identify and locate the other branches on the chain so that they are named alphabetically and their carbon number is hyphenated onto the front of the name. If more than one of the same group is present use the Greek prefix attached to the branch name. (di=2, tri=3, etc)
- After all the branches have been named and located then attach the carbon number that is attached to the hydroxyl group onto the alkane name associated with the number of carbons found in the parent chain in step 1. Drop the "e" on the alkane name and attach the IUPAC ending ol to the rest of the alkane name.
- For alcohols with more than one hydroxyl group you would have to locate two or more carbons which the hydroxyl groups were attached and hyphenate those carbon numbers to the front of the alkane name and attach "diol" if two are involved, "triol" if three hydroxyl groups were involved.

Examples:

CH$_3$
OH

| |

CH$_3$–CH–CH$_2$–CH$_2$–CH$_2$–CH$_2$–OH   C–C–C–C–OH
5-methyl-1-hexanol   1,2-butandiol

Physical Properties Boiling Points

The boiling points of alcohols are in general much higher than comparably sized hydrocarbons. There are extra intermolecular forces between the alcohol molecules due to the hydrogen bonding.

There is no such hydrogen bonding between the hydrocarbon molecules because all the hydrogen atoms are bonded to carbon and not oxygen or nitrogen. Because there is no hydrogen bonding, the forces between the hydrocarbon molecules are much weaker and the molecules can be much more easily vapourized. As an example, the boiling point of ethanol, $CH_3–CH_2–OH$, is 78°C; ethane has a boiling point of –88.5°C.

Water Solubility

The alcohols are soluble in water because there is very strong attraction between alcohol molecules and the solvent water molecules. Hydrogen bonding is present in both and for that reason most alcohols are soluble in water. Alcohols also have a nonpolar part of the molecule that tends to make them soluble in hydrocarbon solvents.

Aldehydes and Ketones

Aldehydes are organic compounds that have a carbonyl group (–C=O) attached to the end carbon in a chain and has the general formula, R–C=O. There is hydrogen bonded to the carbonyl carbon (double bonded carbon).

Ketones are a part of another carbonyl family with the carbonyl group (–C=O), attached to two carbons so that the carbonyl group is in the central part of the carbon chain. The general formula for the ketone would be:

R'
|
R–C=O

There are two alkyl groups, R and R', connected to the carbonyl carbon. To name Aldehydes, use the name of the longest carbon (parent) chain containing the carbonyl group and change the –e ending to –al. C–C–C=O is propanal. To name Ketones, use the name of the longest carbon (parent) chain containing the carbonyl group and change the –e ending to –one.

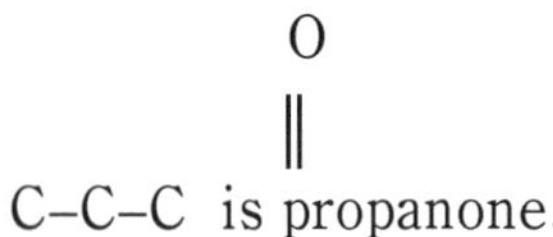

Aldehydes and Ketones have similar chemistry activity because of the fact that they both have a carbonyl carbon.

Carboxylic Acid

The Carboxylic Acid Family is a family of organic compounds with the functional group being the carboxyl group, –COOH. This group is attached to one of the carbons in the rest of the molecule. The carboxyl group is actually a carbonyl group, C=O, bonded to a hydroxyl group, –OH. Taking the first four letters of the word carbonyl and the last four letters of the word hydroxyl you get the word carboxyl. This family is a weak acid.

To name carboxylic acids

- Determine the longest continuous chain of carbons that have the Carbonyl carbon as part of the continuous chain.
- Number the carbons in the chain beginning with the carbonyl carbon as the C-1 carbon.
- Identify the various branching groups attached to this continuous chain of carbons by name and name the branched groups as we did before.
- If it is a saturated acid use the normal alkane name corresponding to the number of carbons in the continuous chain. If it has an unsaturated centre that is a double bond then use the alkene name and prefix the name with the lowest carbon involving the double bond.

  If a triple bond is involved then use the alkyne name and prefix the name with the lowest carbon number involving the triple bond
- Drop the last "e" in "ane", "ene", or "yne" ending and add the IUPAC ending for this family which is "oic acid".

Identify the IUPAC name for the following:

CH_3-CH_2-CH_2-CH_2-CH_2-COOH Hexanoic Acid

CH_3-CH_2-CH_2-COOH 2-methylbutanoic acid

|

C H_3

Physical Properties

Boiling Points

The boiling points of carboxylic acids are the high because of the hydrogen bonding involved in the hydroxyl group. Carboxylic acids have higher boiling points than alcohols of the same number of carbons. The solubilities of carboxylic acids are very similar to the alcohols. Both families have molecules that are polar.

Carboxylic acids are weak acids: R-COOH → $RCOO^-$ + H^+.

ESTERS

The reaction of a carboxylic acid with an alcohol is called an esterification (results in an ester being formed). The reaction is relatively slow sometimes requiring a catalyst and also a period of refluxing (vapourizing and condensing repeatedly). Net Reaction: RCOOH + R'OH → RCOOR' + H_2O

```
O                          O
‖                          ‖
R-C –OH + H-O-R' →  R-C–O-R' + H2O
```

The mechanism has the alcohol molecule releasing a H^+ and the carboxyl group releasing the hydroxide (OH^-) group. The carboxyl group hydroxide ion removes the hydrogen ion the alcohol to produce the ester.

To name an ester you name it as the salt of the carboxylic acid used to produce the ester, find the longest chain that contains the carbonyl group (this is the part from the acid). The acid ending –ic changes to –ate. The prefix from the alcohol, the alkyl branch name, is used as the first part of the ester name.

```
  O                        O
  ‖                        ‖
C-C-OH + C-OH  →    C-C-O-C       +        H2O
Ethanoic Acid + methanol → methylethanoate + water
```

8

Industrial Applications of Enzyme in Organic Chemistry

ENZYMES AND INDUSTRIAL APPLICATIONS

Maps produces industrial enzymes originating from microorganisms in the soil. Microorganisms are usually bacteria, fungi or yeast. One microorganism contains over 1, 000 different enzymes. A long period of trial and error in the laboratory is needed to isolate the best microorganism for producing a particular type of enzyme. When the right microorganism has been found, it has to be modified so that it is capable of producing the desired enzyme at high yields. Then the microorganism is 'grown' in trays or huge fermentation tanks where it produces the desired enzyme. With the latest technological advancements of fermenting microorganisms, it possible to produce enzymes economically and in virtually unlimited quantities.

The end product of fermentation is a broth from which the enzymes are extracted. After this, the remaining fermentation broth is centrifuged or filtered to remove all solid particles. The resulting biomass, or sludge in everyday language, contains the residues of microorganisms and raw materials, which can be a very good natural fertilizer. The enzymes are then, used for various industrial applications.

ENZYMES IN TEXTILES

Biotechnology is the application of living organisms and their components to industrial products and processes. In 1981 the European federation of biotechnology defined Biotechnology as. Integrated use of Biochemistry, Microbiology and Chemical Engineering in order to achieve the technological application of the capacities of microbes and cultured tissue cells.. The rapid developments in the field of genetic engineering have given a new impetus to biotechnology. This introduces the possibility of. tailoring. organisms in order to optimise the production of established or novel metabolites of commercial importance and of transferring genetic material (genes) from one organism to another.

Biotechnology also offers the potential for new industrial processes that require less energy and are based on renewable raw materials. It is important to note that biotechnology is not just concerned with biology, but it is a truly interdisciplinary subject involving the integration of natural and engineering sciences.

Defining the scope of biotechnology is not easy because it overlaps with so many industries, such as, the chemical industry or food industry being the majors, but biotechnology has found many applications in textile industry also, especially in genetic engineering, textile processing and effluent management.

FARMING WITH BUGS

Simple cellular organisms, such as, yeast have been used for millennia, knowingly or otherwise, to make bread, beer and wine. Most industrial applications of biotechnology are still based upon fermentation processes using bacteria and enzymes to digest, transform and synthesise natural materials from one form into another. It is not surprising perhaps that biotechnology is often described as. farming with bugs. For example, the active agent in many transformation processes is an enzyme rather than the cellular living organism itself. Enzymes are not alive themselves but are complex chemical catalysts, which can, in principle, be produced by a number of different methods, including non-biological synthetic routes.

THE SYSTEMATIC APPLICATION OF BIOLOGICAL SCIENCE

The contribution of science has been to understand to a much greater extent what exactly are the active components and mechanisms of the. Bugs. and their derivatives and therefore to begin to control, manipulate and reproduce their capabilities in a more systematic and intelligent fashion. Modern biotechnology has also brought forward a number of techniques, which do not rely on microbes and enzymes at all, but which directly modify the power of DNA molecule first and foremost among these is genetic engineering.

GENETIC EENGINEERING

With an improved understanding of how different genes are responsible for the various characteristics and properties of a living organism, techniques have been developed for isolating these active components (in particular, the DNA which carries the genetic code) and manipulating them outside of the cell. The next step has been to introduce fragments of DNA obtained from one organism into another, thereby transferring some of the properties and capabilities of the first to the second. For example, scientists working for the leading enzyme producer, Novo of Denmark, discovered that an enzyme produced in minute quantities by one particular fungus had very desirable properties for dissolving fats. The relevant genes where. spliced. into another micro-organism which was capable of producing the desired enzyme at much higher yield. Genetic engineering methods are being investigated for their potential to produce new kinds of textile fibres. The systems fall into two main groups. There are those systems that can produce monomeric protein molecules in solution from appropriately engineered genes and include expression in bacteria, cell cultures or in the milk of transgenic animals such as goats or sheep. The protein monomers are then isolated from the chosen system and spun and drawn into fibres.

The other approach is to modify keratin fibres such as wool by expressing other proteins in the internal components by trangenesis. Biopesticides based on a strain of soil bacteria known as Bt are already being used for control of caterpillar and beetle pests in a wide variety of fruits, vegetables and crops. More stable, longer lasting and more active Bts are now being developed for the suppression of loopers, bollworms and budworms in cotton.

The next stage will be to introduce greater insect and herbicide resistance by direct genetic engineering of the cotton plant itself. Practical results achieved so far include development of a cotton fibre with 50 per cent greater strength than its. parent.. Coloured cottons are also being developed, not only by conventional genetic selection but also by direct DNA engineering to produce, for example, deep blue cotton for denim production. The prospect is even being held out of encouraging natural polyesters such as polyhydroxy butyrate (PHB) to grow within the central hollow channel of the cotton fibre, thereby creating. natural. polyester-cotton.

MONOCLONAL ANTIBODIES

Monoclonal antibodies are protein molecules with an amasing ability to. recognise. specific substances, even at extremely low concentrations. They were first developed for use in medicine to detect and target cancer cells so called. magic bullet approach.; they have also been used for pregnancy testing. Recently, Biocode Company has developed monoclonal antibodies as very sensitive marking tool for the prevention of counterfeiting. The markers themselves are cheap and safe substances, which can be applied to foodstuffs, drinks and textiles in concentrations of a few parts per million or less. The. codes. embodied in these markers are completely secure but can readily be detected by customs or trading standards inspectors using simple equipment in the field. The technology has already been evaluated for the marking of branded denims. Methods have been perfected for use in nylon and acrylic resins and markers can also be incorporated into dyestuffs or applied to surfaces using ink jet printers.

DNA PROBES

DNA probes are another technology, which has grown out of genetic engineering. Short pieces of DNA can be designed to stick very specifically to other pieces of DNA and thereby, to help identify target species. The technique can be applied, for example, to distinguish Cashmere from Wool and other goat fibres. The initial impetus for application of DNA probes in the textile industry has come from importers and processors of specialty animal hairs who have seen a surge in trading and labelling fraud, especially in the wake of recent high fibre prices. Now, similar probes are being identified to distinguish between cotton, ramie, kapok, coir, flax, jute and hemp.

BIOSENSORS

Another way in which biological systems can be used as extremely sensitive analytical and control tools is biosensors. These employ some change produced by very small quantities of biologically active agents to measure and therefore, in principle, to control chemical and physical reactions. BTTG has been working on the use of certain fungi, which are capable of absorbing and concentrating heavy metal ions such as lead, copper and cadmium. Resultant changes in the conductivity and dielectric properties of the fungi can be used to measure these species in a process or effluent stream relatively cheaply and easily. Application can be envisaged which incorporate biosensitive materials into textiles, for example, to produce. intelligent. filter media or protective clothing which detects as well as protects against chemicals, gases and biological agents.

APPLICATIONS TO PROCESSING

The use of enzymes in textile processing and after care is already the best established example of the application of biotechnology to textiles and is likely to continue to provide some of the most immediate and possibly dramatic illustrations of its potential in the near-to medium-term future.

Fibre Preparation

Linen is a cellulosic fibre obtained from the flax plant. These fibres are formed in the cortex between the lignified core and the outer layers of the stem, they are separated from the stems by retting, in which matrix components, mainly pectin and lignin are removed and the fibres are separated. Recently, considerable efforts have been put to use enzymes in the retting process to control the process to produce linen fibres of consistent quality. Pre-treatment of the flax with sulphur dioxide gas brings about sufficient breakdown of the woody straw material to speed up enzyme retting whilst preventing excessive bacterial or fungal deterioration of the fibre.

The carbonisation process in which vegetable matter in wool is degraded by treatment with strong acid and then subjected to mechanical crushing can, in principle, be replaced by selective enzyme degradation of the impurities.

Fabric Preparation

Desising using amylase enzymes has been well established for many years. However, there is still considerable scope for improving the speed, economics and consistency of the process, including the development of more temperature stable enzymes as well as a better understanding of how to characterise their activity and performance with respect to different fabrics, sizes, and processing conditions, eg, for pad batch as opposed to jigger desising. The current application in the textile industry involves mainly hydrolases and now to some extent is Oxidoreductase.

The Tables exemplify such textile applications. Another desirable development would be enzymes capable of destroying honeydew sugars, insect secretions that cause stickiness and severe processing problems for cotton spinners. An already established application is the use of catalase enzymes to breakdown residual hydrogen peroxide after, for example, pre-bleach of cotton that is to be dyed a pale or medium shade.

Table. Application of Hydrolase Enzyme in Fabric Preparation

SI No.	Enzyme Name	Substrate Attacked	TextileApplication
1	Amylase	Starch	Starch desising
2	Cellulase	Cellulose	1. Stone wash-Biop-polishing (Bio singeing) 2. Bio finishing for handle modification 3. Carbonisation of wool
3	Pectinase	Pectin	Bio scour replacing caustic
4	Catalase	Peroxides	In situ peroxide decomposition without any rinse in bleach bath
5	Lipase	Fats and oils	Improve hydrophilicity of PET in place of alkaline hydrolysis

Table. Application of oxidoreductase in fabric preparation.

SI No.	Enzyme Name	Substrate Attacked	Textile Application
1	Laccase	Colour coloured effluent and pigments lignin containing jute indigo of in denim	1. Discolouration of chromophore 2. Bio-bleaching of fibres like kenaf and 3. Bio-bleaching or for various effects
2	Peroxidase	colour pulp	Bio-bleaching of wood chromophore and pigments
3	Glucose	Pigments	In situ generation of H_2O and bio-bleaching of cotton

Reactive dyes are especially sensitive to peroxide and currently require extended rinsing and/or use of chemical scavengers.

The enzyme catalase is added after oxidative bleaching and allowed to react for 15 minutes at 30° C-40° C. It degrades the residual peroxide in water and oxygen. The results obtained were compared with the conventional process and it was found that the outcome of the enzymatic process was excellent. The best suitable conditions are the temperature range of 20° C-60° C, pH 5-10 and the application time is 10 min to 15 min.

FINISHING

Biostoning and the closely related process of bio-polishing are perhaps attracting most current attention in the area of enzyme processing. They are also an excellent illustration of how different industry structural and market considerations can affect the uptake of enzyme technology.

Conventional stone washing uses abrasive pumice stones in a tumbling machine to abrade and remove particles of indigo dyestuff from the surfaces of denim yarns and fabric. Cellulase enzymes can also cut through cotton fibres and achieve much the same effect without the damaging abrasion of the stones on both garment and machine. Disadvantages can include degradation of the fabric and loss of strength as well as. back staining.

A slight reddening of the original indigo shade can also occur. Now processors are learning to play more sophisticated tunes such as achieving a peach skin finish by use of a combination of stones and natural cellulase.

Bio-polishing employs basically the same cellulose action to remove fine surface fuzz and fibrils from cotton and viscose fabrics. The polishing action thus achieved helps to eliminate pilling and provides better print definition, colour brightness, surface texture, drapeability, and softness without any loss of absorbency. Bio-polishing can be used to clean up the fabric surface after the primary fibrillation of a peach skin treatment and prior to a secondary fibrillation process which imparts interesting fabric aesthetics.

A weight loss in the base fabric of some 3 per cent-5 per cent is typical but reduction in fabric strength can be controlled to within 2 per cent-7 per cent by terminating the treatment after about 30 min-40 min using a high temperature or low pH. enzyme stop.. One area that still poses problems is that of tubular cotton finishing. Here, the fibre residues tend to be trapped inside the fabric rather than washed away.

WOOL PROCESSING APPLICATIONS

The international wool secretariat (IWS) together with, Novo, been developing the use of protease enzymes for a range of wool finishing treatments aimed at increased comfort (reduced prickle, greater softness) as well as improved surface appearance and pilling performance. The basic mechanisms closely parallel those of bio-polishing. The improved enzyme treatments will allow more selective removal of parts of the wool cuticle, there by modifying the luster, handle and felting characteristics without degradation or weakening of the wool fibre as a whole and without the need for environmentally damaging pre-chlorination treatment.

OTHER PROTEASE APPLICATIONS

Protease enzymes similar to those being developed for wool processing are already being used for the degumming of silk and for producing sand washed effects on silk garments.

Treatment of Silk-Cellulosic blend is claimed to produce some unique effects. Proteases are also being used to wash down printing screens after use in order to remove the proteinaceous gums, which are used for thickening of printing pastes.

TEXTILE AFTER-CARE

Enzymes have been widely used in domestic laundering detergents since the 1960s. Some of the major classes of enzymes and their effectiveness against common stains are summarised in Table.

Early problems of allergic reactions to some of these enzymes have now largely been overcome by the use of advanced granulation technology. Modern enzyme systems have reduced the use of sodium perborate in detergents by 25 per cent along with the release of harmful salts into the environment.

However, enzymes still have to make a corresponding impact upon the commercial laundering market. One of the problems here has been the level of investment in. continuous-batch. or tunnel washers.

These typically afford a residence time of 6 min-12 min which is not long enough for present enzyme systems to perform adequately. More efficient methods of. enzyme kill. are also required because of the extent of water recycling in modern washers.

Table. Types of enzymes and their Effectiveness Against various stains

Enzyme	Effective for
Proteases	Grass, Blood, Egg, Sweat stains
Lipases	Lipstick, Butter, Salad oil, Sauces
Amylases	Spaghetti, Custard, chocloate
Celluloses	Colour brighterning, Softening, Soil removal

Further developments in the field of textile after-care may include treatments to reverse wool shrinkage as well as alternatives to dry cleaning.

CARING FOR THE ENVIRONMENT

Natural and enhanced microbial process have been used to treat waste materials and effluent streams from the textile industry. Conventional activated sludge and other systems are generally well able to meet BOD and related discharge limits on most cases.

The industry faces some specific problems like colour removal from dyestuff effluent and handling of toxic wastes including PCPs and heavy metals. The synthetic dyes are designed in such a way that they become resistant to microbial degradation under the aerobic conditions. Also, the water solubility and the high molecular weight inhibit the permeation through biological cell membranes.

Anaerobic processes convert the organic contaminants principally into methane and carbon dioxide, usually occupy less space, treat wastes containing up to 30 000 mg/l of COD, have lower running costs and produce less sludge. A novel approach to promoting aerobic degradation in contaminated lagoons and preventing the development of malodorous and unpleasant anaerobic processes. The development based on a 3-D. biomat. of knitted polyester monofilament is used as a support for the micro-organisms. The mat is stable and resistant to compression; its open supporting structure counteracts the build-up of anaerobic sludges on the bottom of the lagoon.

NEW FIBRE SOURCES

Several possibilities exist for producing entirely new fibre materials, so called biopolymers, using biotechnological process routes, Naturally occurring polyester, PHB is produced by bacterial fermentation of a sugar feed stock and commercially available as. Biopol..

The polymer is stable under normal conditions but biodegrades completely in any microbially active environment. Other biopolymers with textile potential include polylactates and polycaprolactones, which are investigated for medical applications.

Bacterial Cellulose The speciality papers and non-wovens are produced based on bacterially grown cellulose fibres these are extremely fine and resilient and are used as specialised filters, odour absorbers and reinforcing blends with aramids. Genetically Modified Micro-Organisms Attempts have been made to transfer certain advantageous textile properties into micro-organisms where they can be more readily reproduced by bulk fermentation processes. The spider DNA is transferred into bacteria with the air of manufacturing proteins with the strength and resilience of spider silk for use in bulletproof vests.

DYESTUFFS AND INTER MEDIATES

Attempts have been made to synthesise bacterial forms of indigo as well as fungal pigments for use in the textile industry. Certain micro fungi are capable of yielding up to 30 per cent of their biomass as pigment. Potential non-textile applications include food industry colourants This note of caution needs to be echoed across the whole spectrum of biotechnology developments. Although biological systems after many attractive possibilities and new approaches to all sorts of problems and needs, considerable advances are still being made in. conventional. technologies, such as, catalysis, chemical synthesis and physical fibre modification which need to be kept in perspective. There is also still great concern in society about the unbridled advance of biotechnology, especially with regard to the modification of natural species with possible unknown long-term consequences.

CLASSIFICATION AND NOMENCLATURE OF ENZYMES IN CHEMISTRY

Because of their close interdependence, it is convenient to deal with the classification and nomenclature together.

The *first general principle* of these 'Recommendations' is that names purporting to be names of enzymes, especially those ending in-*ase*, should be used only for single enzymes, *i.e.,* single catalytic entities. They should not be applied to systems containing more than one enzyme. When it is desired to name such a system on the basis of the overall reaction catalysed by it, the word *system* should be included in the name. For example, the system catalysing the oxidation of succinate by molecular oxygen, consisting of succinate dehydrogenase, cytochrome oxidase, and several intermediate carriers, should not be named *succinate oxidase*, but it may be called the *succinate oxidase system*. Other examples of systems consisting of several structurally and functionally linked enzymes (and cofactors) are the *pyruvate dehydrogenase system*, the similar *2-oxoglutarate dehydrogenase system*, and the *fatty acid synthase system*. In this context it is appropriate to express disapproval of a loose and misleading practice that is found in the biological literature.

It consists in designation of a natural substance (or even of an hypothetical active principle), responsible for a physiological or biophysical phenomenon that cannot be described in terms of a definite chemical reaction, by the name of the phenomenon in conjugation with the suffix-*ase*, which implies an individual enzyme. Some examples of such *phenomenase* nomenclature, which should be discouraged even if there are reasons to suppose that the particular agent may have enzymic properties, are: *permease, translocase, reparase, joinase, replicase, codase, etc.*.

The *second general principle* is that enzymes are principally classified and named according to the reaction they catalyse. The chemical reaction catalysed is the specific property that distinguishes one enzyme from another, and it is logical to use it as the basis for the classification and naming of enzymes.

Several alternative bases for classification and naming had been considered, *e.g.,* chemical nature of the enzymes (whether it is a flavoprotein, a hemoprotein, a pyridoxal-phosphate protein, a copper protein, and so on), or chemical nature of the substrate (nucleotides, carbohydrates, proteins, *etc.*). The first cannot serve as a general basis, for only a minority of enzymes have such identifiable prosthetic groups. The chemical nature of the enzyme has, however, been used exceptionally in certain cases where classification based on specificity is difficult, for example, with the peptidases. The second basis for classification is hardly practicable, owing to the great variety of substances acted upon and because it is not sufficiently informative unless the type of reaction is also given. It is the overall reaction, as expressed by the formal equation, that should be taken as the basis.

Thus, the intimate mechanism of the reaction, and the formation of intermediate complexes of the reactants with the enzyme is not taken into account, but only the observed chemical change produced by the complete enzyme reaction. For example, in those cases in which the enzyme contains a prosthetic group that serves to catalyse transfer from a donor to an acceptor (*e.g.*, flavin, biotin, or pyridoxal-phosphate enzymes) the name of the prosthetic group is not normally included in the name of the enzyme. Nevertheless, where alternative names are possible, the mechanism may be taken into account in choosing between them. A consequence of the adoption of the chemical reaction as the basis for naming enzymes is that a systematic name cannot be given to an enzyme until it is known what chemical reaction it catalyses. This applies, for example, to a few enzymes that have so far not been shown to catalyse any chemical reaction, but only isotopic exchanges; the isotopic exchange gives some idea of one step in the overall chemical reaction, but the reaction as a whole remains unknown.

A second consequence of this concept is that a certain name designates not a single enzyme protein but a group of proteins with the same catalytic property. Enzymes from different sources (various bacterial, plant or animal species) are classified as one entry. The same applies to isoenzymes. However, there are exceptions to this general rule. Some are justified because the mechanism of the reaction or the substrate specificity is so different as to warrant different entries in the enzyme list. This applies, for example, to the two cholinesterases, the two citrate hydro-lyases, and the two amine oxidases. Others are mainly historical, *e.g.*, acid and alkaline phosphatases.

A *third general principle* adopted is that the enzymes are divided into groups on the basis of the type of reaction catalysed, and this, together with the name(s) of the substrate(s) provides a basis for naming individual enzymes. It is also the basis for classification and code numbers.

Special problems attend the classification and naming of enzymes catalysing complicated transformations that can be resolved into several sequential or coupled intermediary reactions of different types, all catalysed by a single enzyme (not an enzyme system). Some of the steps may be spontaneous non-catalytic reactions, while one or more intermediate steps depend on catalysis by the enzyme. Wherever the nature and sequence of intermediary reactions is known or can be presumed with confidence, classification and naming of the enzyme should be based on the first enzyme-catalysed step that is essential to the subsequent transformations, which can be indicated by a supplementary term in parentheses, *e.g.*, *acetyl-CoA:glyoxylate C-acetyltransferase (thioester-hydrolysing, carboxymethyl-forming)*. To classify an enzyme according to the type of reaction catalysed, it is occasionally necessary to choose between alternative ways of regarding a given reaction. Some considerations of this type are outlined in section of this chapter. In general, that alternative should be selected which fits in best with the general system of classification and reduces the number of exceptions.

One important extension of this principle is the question of the direction in which the reaction is written for the purposes of classification. To simplify the classification, the direction chosen should be the same for all enzymes in a given class, even if this direction has not been demonstrated for all. Thus the *systematic* names, on which the classification and code numbers are based, may be derived from a written reaction, even though only the reverse of this has been actually demonstrated experimentally. In the list in this volume, the reaction is written to illustrate the classification, *i.e.,* in the direction described by the systematic name. However, the *common* name may be based on either direction of reaction, and is often based on the presumed physiological direction.

Many examples of this usage are found in section of the list. The reaction is written as an oxidation of xylitol by NAD^+, in parallel with all other oxidoreductases in subgroup, and the systematic name is accordingly, *xylitol:NAD*$^+$ *2-oxidoreductase (*D*-xylulose-forming)*. However, the common name, based on the reverse direction of reaction, is D-*xylulose reductase*.

COMMON AND SYSTEMATIC NAMES

The first Enzyme Commission gave much thought to the question of a systematic and logical nomenclature for enzymes, and finally recommended that there should be two nomenclatures for enzymes, one systematic, and one working or trivial. The systematic name of an enzyme, formed in accordance with definite rules, showed the action of an enzyme as exactly as possible, thus identifying the enzyme precisely. The trivial name was sufficiently short for general use, but not necessarily very systematic; in a great many cases it was a name already in current use. The introduction of (often cumbersome) systematic names was strongly criticised. In many cases the reaction catalysed is not much longer than the systematic name and can serve just as well for identification, especially in conjunction with the code number.

The Commission for Revision of Enzyme Nomenclature discussed this problem at length, and a change in emphasis was made. It was decided to give the trivial names more prominence in the Enzyme List; they now follow immediately after the code number, and are described as Common Name. Also, in the index the common names are indicated by an asterisk. Nevertheless, it was decided to retain the systematic names as the basis for classification for the following reasons:

- The code number alone is only useful for identification of an enzyme when a copy of the Enzyme List is at hand, whereas the systematic name is self-explanatory;
- The systematic name stresses the type of reaction, the reaction equation does not;
- Systematic names can be formed for new enzymes by the discoverer, by application of the rules, but code numbers should not be assigned by individuals;

- Common names for new enzymes are frequently formed as a condensed version of the systematic name; therefore, the systematic names are helpful in finding common names that are in accordance with the general pattern.

It is recommended that for enzymes that are not the main subject of a paper or abstract, the common names should be used, but they should be identified at their first mention by their code numbers and source. Where an enzyme is the main subject of a paper or abstract, its code number, systematic name, or, alternatively, the reaction equation and source should be given at its first mention; thereafter the common name should be used. In the light of the fact that enzyme names and code numbers refer to reactions catalysed rather than to discrete proteins, it is of special importance to give also the source of the enzyme for full identification; in cases where multiple forms are known to exist, knowledge of this should be included where available.

When a paper deals with an enzyme that is not yet in the Enzyme List, the author may introduce a new name and, if desired, a new systematic name, both formed according to the recommended rules. A number should be assigned only by the Nomenclature Committee of IUBMB. The Enzyme List contains one or more references for each enzyme.

It should be stressed that no attempt has been made to provide a complete bibliography, or to refer to the first description of an enzyme. The references are intended to provide sufficient evidence for the existence of an enzyme catalysing the reaction as set out.

Where there is a major paper describing the purification and specificity of an enzyme, or a major review article, this has been quoted to the exclusion of earlier and later papers. In some cases separate references are given for animal, plant and bacterial enzymes.

SCHEME FOR THE CLASSIFICATION OF ENZYMES AND THE GENERATION OF EC NUMBERS

The first Enzyme Commission, in its report in 1961, devised a system for classification of enzymes that also serves as a basis for assigning code numbers to them. These code numbers, prefixed by EC, which are now widely in use, contain four elements separated by points, with the following meaning:

- The first number shows to which of the six main divisions (classes) the enzyme belongs,
- The second figure indicates the subclass,
- The third figure gives the sub-subclass,
- The fourth figure is the serial number of the enzyme in its sub-subclass.

The subclasses and sub-subclasses are formed according to principles indicated below. The main divisions and subclasses are:

CLASS 1. OXIDOREDUCTASES

To this class belong all enzymes catalysing oxidoreduction reactions. The substrate that is oxidised is regarded as hydrogen donor. The systematic name is based on *donor:acceptor oxidoreductase*. The common name will be *dehydrogenase*, wherever this is possible; as an alternative, *reductase* can be used. *Oxidase* is only used in cases where O_2 is the acceptor. The second figure in the code number of the oxidoreductases, indicates the group in the hydrogen (or electron) donor that undergoes oxidation: 1 denotes a-CHOH-group, 2 a-CHO or-CO-COOH group or carbon monoxide, and so on, as listed in the key.

The third figure, except in subclasses, indicates the type of acceptor involved: 1 denotes $NAD(P)^+$, 2 a cytochrome, 3 molecular oxygen, 4 a disulfide, 5 a quinone or similar compound, 6 a nitrogenous group, 7 an iron-sulfur protein and 8 a flavin. In subclasses a different classification scheme is used and sub-subclasses are numbered from 11 onwards. It should be noted that in reactions with a nicotinamide coenzyme this is always regarded as acceptor, even if this direction of the reaction is not readily demonstrated. The only exception is the subclass, in which NAD(P)H is the donor; some other redox catalyst is the acceptor. Although not used as a criterion for classification, the two hydrogen atoms at carbon-4 of the dihydropyridine ring of nicotinamide nucleotides are not equivalent in that the hydrogen is transferred stereospecifically.

CLASS 2. TRANSFERASES

Transferases are enzymes transferring a group, *e.g.,* a methyl group or a glycosyl group, from one compound (generally regarded as donor) to another compound (generally regarded as acceptor). The systematic names are formed according to the scheme *donor:acceptor grouptransferase*. The common names are normally formed according to *acceptor grouptransferase* or *donor grouptransferase*. In many cases, the donor is a cofactor (coenzyme) charged with the group to be transferred. A special case is that of the transaminases.

Some transferase reactions can be viewed in different ways. For example, the enzyme-catalysed reaction X-Y + Z = X + Z-Y may be regarded either as a transfer of the group Y from X to Z, or as a breaking of the X-Y bond by the introduction of Z.

Where Z represents phosphate or arsenate, the process is often spoken of as 'phosphorolysis' or 'arsenolysis', respectively, and a number of enzyme names based on the pattern of *phosphorylase* have come into use. These names are not suitable for a systematic nomenclature, because there is no reason to single out these particular enzymes from the other transferases, and it is better to regard them simply as *Y-transferases*.

In the above reaction, the group transferred is usually exchanged, at least formally, for hydrogen, so that the equation could more strictly be written as:

$$X\text{-}Y + Z\text{-}H = X\text{-}H + Z\text{-}Y.$$

Another problem is posed in enzyme-catalysed transaminations, where the-NH_2 group and-H are transferred to a compound containing a carbonyl group in exchange for the =O of that group, according to the general equation:

R^1-CH(-NH_2)-R^2 + R^3-CO-R^4 $\rightarrow$ R^1-CO-R^2 + R^3-CH(-NH_2)-R^4. The reaction can be considered formally as oxidative deamination of the donor (*e.g.,* amino acid) linked with reductive amination of the acceptor (*e.g.,* oxo acid), and the transaminating enzymes (pyridoxal-phosphate proteins) might be classified as oxidoreductases.

However, the unique distinctive feature of the reaction is the transfer of the amino group (by a well-established mechanism involving covalent substrate-coenzyme intermediates), which justified allocation of these enzymes among the transferases as a special subclass (*transaminases*).

The second figure in the code number of transferases indicates the group transferred; a one-carbon group, an aldehydic or ketonic group, an acyl group and so on.

The third figure gives further information on the group transferred; *e.g.,* subclass is subdivided into *methyltransferases*, *hydroxymethyl*-and *formyltransferases* and so on; only in subclass does the third figure indicate the nature of the acceptor group.

CLASS 3. HYDROLASES

These enzymes catalyse the hydrolytic cleavage of C-O, C-N, C-C and some other bonds, including phosphoric anhydride bonds. Although the systematic name always includes *hydrolase*, the common name is, in many cases, formed by the name of the substrate with the suffix-*ase*. It is understood that the name of the substrate with this suffix means a hydrolytic enzyme.

A number of hydrolases acting on ester, glycosyl, peptide, amide or other bonds are known to catalyse not only hydrolytic removal of a particular group from their substrates, but likewise the transfer of this group to suitable acceptor molecules. In principle, all hydrolytic enzymes might be classified as transferases, since hydrolysis itself can be regarded as transfer of a specific group to water as the acceptor. Yet, in most cases, the reaction with water as the acceptor was discovered earlier and is considered as the main physiological function of the enzyme. This is why such enzymes are classified as hydrolases rather than as transferases.

Some hydrolases (especially some of the esterases and glycosidases) pose problems because they have a very wide specificity and it is not easy to decide if two preparations described by different authors (perhaps from different sources) have the same catalytic properties, or if they should be listed under separate entries. An example is *vitamin A esterase*. To some extent the choice must be arbitrary; however, separate entries should be given only when the specificities are sufficiently different.

Another problem is that proteinases have 'esterolytic' action; they usually hydrolyse ester bonds in appropriate substrates even more rapidly than natural peptide bonds. In this case, classification among the peptide hydrolases is based on historical priority and presumed physiological function.

The second figure in the code number of the hydrolases indicates the nature of the bond hydrolysed; the *esterases*; the *glycosylases*, and so on.

The third figure normally specifies the nature of the substrate, *e.g.,* in the esterases the *carboxylic ester hydrolases*, *thiolester hydrolases*, *phosphoric monoester hydrolases*; in the glycosylases the *O-glycosidases*, *N-glycosylases*, *etc*. Exceptionally, in the case of the peptidyl-peptide hydrolases the third figure is based on the catalytic mechanism as shown by active centre studies or the effect of pH.

CLASS 4. LYASES

Lyases are enzymes cleaving C-C, C-O, C-N, and other bonds by elimination, leaving double bonds or rings, or conversely adding groups to double bonds. The systematic name is formed according to the pattern *substrate group-lyase*. The hyphen is an important part of the name, and to avoid confusion should not be omitted, *e.g., hydro-lyase* not 'hydrolyase'. In the common names, expressions like *decarboxylase, aldolase, dehydratase* (in case of elimination of CO_2, aldehyde, or water) are used. In cases where the reverse reaction is much more important, or the only one demonstrated, *synthase* (not synthetase) may be used in the name. Various subclasses of the lyases include pyridoxal-phosphate enzymes that catalyse the elimination of a b-or g-substituent from an a-amino acid followed by a replacement of this substituent by some other group. In the overall replacement reaction, no unsaturated end-product is formed; therefore, these enzymes might formally be classified as *alkyl-transferases*. However, there is ample evidence that the replacement is a two-step reaction involving the transient formation of enzyme-bound a,b(or b,g)-unsaturated amino acids. According to the rule that the first reaction is indicative for classification, these enzymes are correctly classified as *lyases*. Examples are *tryptophan synthase* and *cystathionine b-synthase*.

The second figure in the code number indicates the bond broken: carbon-carbon lyases, carbon-oxygen lyases and so on. The third figure gives further information on the group eliminated (*e.g.,* CO_2, H_2O).

CLASS 5. ISOMERASES

These enzymes catalyse geometric or structural changes within one molecule. According to the type of isomerism, they may be called *racemases, epimerases, cis-trans-isomerases, isomerases, tautomerases, mutases* or *cycloisomerases*. In some cases, the interconversion in the substrate is brought about by an intramolecular oxidoreduction; since hydrogen donor and acceptor are the same molecule, and no oxidised product appears, they are not classified as oxidoreductases, even though they may contain firmly bound $NAD(P)^+$. The subclasses are formed according to the type of isomerism, the sub-subclasses to the type of substrates.

CLASS 6. LIGASES

Ligases are enzymes catalysing the joining together of two molecules coupled with the hydrolysis of a diphosphate bond in ATP or a similar triphosphate. The systematic names are formed on the system *X:Y ligase (ADP-forming)*. In earlier editions of the list the term *synthetase* has been used for the common names. Many authors have been confused by the use of the terms *synthetase* (used only for Group 6) and *synthase* (used throughout the list when it is desired to emphasis the synthetic nature of the reaction). Consequently NC-IUB decided in 1983 to abandon the use of synthetase for common names, and to replace them with names of the type *X-Y ligase*. In a few cases in Group 6, where the reaction is more complex or there is a common name for the product, a synthase name is used.

It is recommended that if the term *synthetase* is used by authors, it should continue to be restricted to the ligase group. The second figure in the code number indicates the bond formed: EC 6.1 for C-O bonds (enzymes acylating tRNA), EC 6.2 for C-S bonds (acyl-CoA derivatives), *etc.* Sub-subclasses are only in use in the C-N ligases. In a few cases it is necessary to use the word *other* in the description of subclasses and sub-subclasses. They have been provisionally given the figure in order to leave space for new subdivisions.

From time to time, some enzymes have been deleted from the List, while some others have been renumbered. However, the old numbers have *not* been allotted to new enzymes; rather the place has been left vacant and cross-reference is made according to the following scheme: Entries for reclassified enzymes transferred from one position in the List to another are followed, for reference, by a comment indicating the former number. It is regarded as important that the same policy be followed in future revisions and extensions of the Enzyme List, which may become necessary from time to time.

ENZYMES CATALYSE

Enzymes are biomolecules that catalyse (*i.e.,* increase the rates of) chemical reactions. Almost all enzymes are proteins. In enzymatic reactions, the molecules at the beginning of the process are called substrates, and the enzyme converts them into different molecules, the products. Almost all processes in a biological cell need enzymes in order to occur at significant rates. Since enzymes are extremely selective for their substrates and speed up only a few reactions from among many possibilities, the set of enzymes made in a cell determines which metabolic pathways occur in that cell.

Like all catalysts, enzymes work by lowering the activation energy (E_a or Ä$G^‡$) for a reaction, thus dramatically increasing the rate of the reaction. Most enzyme reaction rates are millions of times faster than those of comparable uncatalysed reactions.

As with all catalysts, enzymes are not consumed by the reactions they catalyse, nor do they alter the equilibrium of these reactions. However, enzymes do differ from most other catalysts by being much more specific. Enzymes are known to catalyse about 4,000 biochemical reactions. A few RNA molecules called ribozymes catalyse reactions, with an important example being some parts of the ribosome. Synthetic molecules called artificial enzymes also display enzyme-like catalysis. Enzyme activity can be affected by other molecules. Inhibitors are molecules that decrease enzyme activity; activators are molecules that increase activity.

Many drugs and poisons are enzyme inhibitors. Activity is also affected by temperature, chemical environment (*e.g.*, pH), and the concentration of substrate. Some enzymes are used commercially, for example, in the synthesis of antibiotics. In addition, some household products use enzymes to speed up biochemical reactions (*e.g.*, enzymes in biological washing powders break down protein or fat stains on clothes; enzymes in meat tenderisers break down proteins, making the meat easier to chew). The use of enzymes in the diagnosis of disease is one of the important benefits derived from the intensive research in biochemistry since the 1940's. Enzymes have provided the basis for the field of clinical chemistry.

It is, however, only within the recent past few decades that interest in diagnostic enzymology has multiplied. Many methods currently on record in the literature are not in wide use, and there are still large areas of medical research in which the diagnostic potential of enzyme reactions has not been explored at all.

This section has been prepared by Worthington Biochemical Corporation as a practical introduction to enzymology. Because of its close involvement over the years in the theoretical as well as the practical aspects of enzymology, Worthington's knowledge covers a broad spectrum of the subject. Some of this information has been assembled here for the benefit of laboratory personnel.

This section summarises in simple terms the basic theories of enzymology.

HISTORY OF ENZYMOLOGY

As early as the late 1700s and early 1800s, the digestion of meat by stomach secretions and the conversion of starch to sugars by plant extracts and saliva were known. However, the mechanism by which this occurred had not been identified. In the 19th century, when studying the fermentation of sugar to alcohol by yeast, Louis Pasteur came to the conclusion that this fermentation was catalysed by a vital force contained within the yeast cells called "ferments", which were thought to function only within living organisms. He wrote that "alcoholic fermentation is an act correlated with the life and organisation of the yeast cells, not with the death or putrefaction of the cells." In 1878 German physiologist Wilhelm Kühne first used the term *enzyme*, which comes from Greek *åíæõìïí* "in leaven", to describe this process.

The word *enzyme* was used later to refer to non-living substances such as pepsin, and the word *ferment* used to refer to chemical activity produced by living organisms. In 1897 Eduard Buchner began to study the ability of yeast extracts that lacked any living yeast cells to ferment sugar. In a series of experiments at the University of Berlin, he found that the sugar was fermented even when there were no living yeast cells in the mixture. He named the enzyme that brought about the fermentation of sucrose "zymase". In 1907 he received the Nobel Prize in Chemistry "for his biochemical research and his discovery of cell-free fermentation". Following Buchner's example; enzymes are usually named according to the reaction they carry out. Typically the suffix-*ase* is added to the name of the substrate (*e.g.*, lactase is the enzyme that cleaves lactose) or the type of reaction (*e.g.*, DNA polymerase forms DNA polymers).

Having shown that enzymes could function outside a living cell, the next step was to determine their biochemical nature. Many early workers noted that enzymatic activity was associated with proteins, but several scientists (such as Nobel laureate Richard Willstätter) argued that proteins were merely carriers for the true enzymes and that proteins *per se* were incapable of catalysis. However, in 1926, James B. Sumner showed that the enzyme urease was a pure protein and crystallised it; Sumner did likewise for the enzyme catalase in 1937. The conclusion that pure proteins can be enzymes was definitively proved by Northrop and Stanley, who worked on the digestive enzymes pepsin, trypsin and chymotrypsin. These three scientists were awarded the 1946 Nobel Prize in Chemistry. This discovery that enzymes could be crystallised eventually allowed their structures to be solved by x-ray crystallography. This was first done for lysozyme, an enzyme found in tears, saliva and egg whites that digests the coating of some bacteria; the structure was solved by a group led by David Chilton Phillips and published in 1965. This high-resolution structure of lysozyme marked the beginning of the field of structural biology and the effort to understand how enzymes work at an atomic level of detail.

ENZYMES AND LIFE PROCESSES

The living cell is the site of tremendous biochemical activity called metabolism. This is the process of chemical and physical change which goes on continually in the living organism. Build-up of new tissue, replacement of old tissue, conversion of food to energy, disposal of waste materials, reproduction-all the activities that we characterise as "life."

This building up and tearing down takes place in the face of an apparent paradox. The greatest majority of these biochemical reactions do not take place spontaneously. The phenomenon of catalysis makes possible biochemical reactions necessary for all life processes. Catalysis is defined as the acceleration of a chemical reaction by some substance which itself undergoes no permanent chemical change. The catalysts of biochemical reactions are enzymes and are responsible for bringing about almost all of the chemical reactions in living organisms. Without enzymes, these reactions take place at a rate far too slow for the pace of metabolism.

The oxidation of a fatty acid to carbon dioxide and water is not a gentle process in a test tube-extremes of pH, high temperatures and corrosive chemicals are required. Yet in the body, such a reaction takes place smoothly and rapidly within a narrow range of pH and temperature. In the laboratory, the average protein must be boiled for about 24 hours in a 20 per cent HCl solution to achieve a complete breakdown. In the body, the breakdown takes place in four hours or less under conditions of mild physiological temperature and pH. It is through attempts at understanding more about enzyme catalysts-what they are, what they do, and how they do it-that many advances in medicine and the life sciences have been brought about.

EARLY ENZYME DISCOVERIES

The existence of enzymes has been known for well over a century. Some of the earliest studies were performed in 1835 by the Swedish chemist Jon Jakob Berzelius who termed their chemical action catalytic. It was not until 1926, however, that the first enzyme was obtained in pure form, a feat accomplished by James B. Sumner of Cornell University. Sumner was able to isolate and crystallise the enzyme urease from the jack bean. His work was to earn him the 1947 Nobel Prize. John H. Northrop and Wendell M. Stanley of the Rockefeller Institute for Medical Research shared the 1947 Nobel Prize with Sumner. They discovered a complex procedure for isolating pepsin. This precipitation technique devised by Northrop and Stanley has been used to crystallize several enzymes.

CHEMICAL NATURE OF ENZYMES

All known enzymes are proteins. They are high molecular weight compounds made up principally of chains of amino acids linked together by peptide bonds.

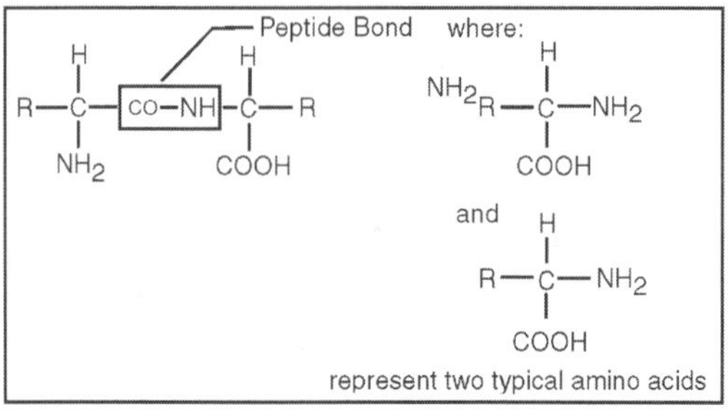

Peptide Bond

R—CH(NH_2)—CO—NH—CH(COOH)—R

where:

NH_2R—CH(COOH)—NH_2

and

R—CH(COOH)—NH_2

represent two typical amino acids

Enzymes can be denatured and precipitated with salts, solvents and other reagents. They have molecular weights ranging from 10,000 to 2,000,000. Many enzymes require the presence of other compounds-cofactors-before their catalytic activity can be exerted. This entire active complex is referred to as the holoenzyme; *i.e.*, apoenzyme (protein portion) plus the cofactor (coenzyme, prosthetic group or metal-ion-activator) is called the holoenzyme.

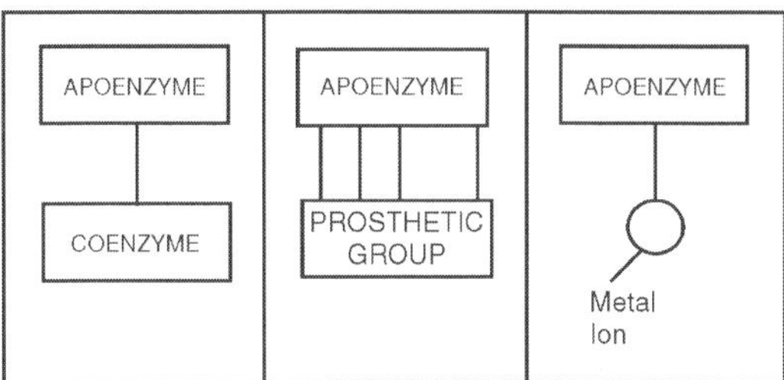

Fig. Holoenzymes-Apoenzymes plus Various types of Cofactors

Apoenzyme + Cofactor = Holoenzyme

According to Holum, the cofactor may be:

- *A coenzyme*-a non-protein organic substance which is dialysable, thermostable and loosely attached to the protein part.
- *A prosthetic group*-an organic substance which is dialysable and thermostable which is firmly attached to the protein or apoenzyme portion.
- *A metal-ion-activator*-these include K^{+}, Fe^{++}, Fe^{+++}, Cu^{++}, Co^{++}, Zn^{++}, Mn^{++}, Mg^{++}, Ca^{++}, and Mo^{+++}.

SPECIFICITY OF ENZYMES

One of the properties of enzymes that makes them so important as diagnostic and research tools is the specificity they exhibit relative to the reactions they catalyse. A few enzymes exhibit absolute specificity; that is, they will catalyse only one particular reaction. Other enzymes will be specific for a particular type of chemical bond or functional group. In general, there are four distinct types of specificity:

- *Absolute specificity*-the enzyme will catalyse only one reaction.
- *Group specificity*-the enzyme will act only on molecules that have specific functional groups, such as amino, phosphate and methyl groups.
- *Linkage specificity*-the enzyme will act on a particular type of chemical bond regardless of the rest of the molecular structure.
- *Stereochemical specificity*-the enzyme will act on a particular steric or optical isomer.

Though enzymes exhibit great degrees of specificity, cofactors may serve many apoenzymes. For example, nicotinamide adenine dinucleotide (NAD) is a coenzyme for a great number of dehydrogenase reactions in which it acts as a hydrogen acceptor. Among them are the alcohol dehydrogenase, malate dehydrogenase and lactate dehydrogenase reactions.

RULES FOR CLASSIFICATION AND NOMENCLATURE

GENERAL RULES FOR SYSTEMATIC NAMES AND GUIDELINES FOR COMMON NAMES

Rule 1 (Common Names)

Generally accepted trivial names of substrates may be used in enzyme names. The prefix D-should be omitted for all D-sugars and L-for individual amino acids, unless ambiguity would be caused. In general, it is not necessary to indicate positions of substituents in common names, unless it is necessary to prevent two different enzymes having the same name. The prefix *keto* is no longer used for derivatives of sugars in which-CHOH-has been replaced by-CO-; they are named throughout as dehydro-sugars.

(Systematic Names)

To produce usable systematic names, accepted trivial names of substrates forming part of the enzyme names should be used. Where no accepted and convenient trivial names exist, the official IUPAC rules of nomenclature should be applied to the substrate name. The 1,2,3 system of locating substituents should be used instead of the a,b,g system, although group names such as b-aspartyl-, g-glutamyl-, and also b-alanine and g-lactone are permissible; a,b should normally be used for indicating configuration, as in a-D-glucose. For nucleotide groups, *adenylyl* (not adenyl), *etc*. should be the form used. The name oxo acids (not keto acids) may be used as a class name, and for individual compounds in which-CH_2-has been replaced by-CO-, oxo should be used.

Rule 2

Where the substrate is normally in the form of an anion, its name should end in-*ate* rather than-*ic; e.g., lactate dehydrogenase,* not 'lactic dehydrogenase' or 'lactic acid dehydrogenase'.

Rule 3

Commonly used abbreviations for substrates, *e.g.* ATP, may be used in names of enzymes, but the use of new abbreviations (not listed in recommendations of the IUPAC-IUB Commission on Biochemical Nomenclature) should be discouraged. Chemical formulae should not normally be used instead of names of substrates. Abbreviations for names of enzymes, *e.g.* GDH, should not be used.

Rule 4

Names of substrates composed of two nouns, such as glucose phosphate, which are normally written with a space, should be hyphenated when they form part of the enzyme names, and thus become adjectives, *e.g., glucose-6-phosphate 1-dehydrogenas*.

Rule 5

The use as enzyme names of descriptions such as *condensing enzyme, acetate-activating enzyme, pH 5 enzyme* should be discontinued as soon as the catalysed reaction is known. The word *activating* should not be used in the sense of converting the substrate into a substance that reacts further; all enzymes act by activating their substrates, and the use of the word in this sense may lead to confusion.

Rule 6 (Common Names)

If it can be avoided, a common name should not be based on a substance that is not a true substrate, *e.g.,* enzyme should not be called 'crotonase', since it does not act on crotonate.

Rule 7 (Common Names)

Where a name in common use gives some indication of the reaction and is not incorrect or ambiguous, its continued use is recommended. In other cases a common name is based on the same general principles as the systematic name but with a minimum of detail, to produce a name short enough for convenient use. A few names of proteolytic enzymes ending in-*in* are retained; all other enzyme names should end in-*ase*.

(Systematic Names)

Systematic names consist of two parts. The first contains the name of the substrate or, in the case of a bimolecular reaction, of the two substrates separated by a colon. The second part, ending in-*ase*, indicates the nature of the reaction.

Rule 8

A number of generic words indicating a type of reaction may be used in either common or systematic names: *oxidoreductase, oxygenase, transferase* (with a prefix indicating the nature of the group transferred), *hydrolase, lyase, racemase, epimerase, isomerase, mutase, ligase.*

Rule 9 (Common Names)

A number of additional generic words indicating reaction types are used in common names, but not in the systematic nomenclature, *e.g., dehydrogenase, reductase, oxidase, peroxidase, kinase, tautomerase, deaminase, dehydratase, etc..*

Rule 10

Where additional information is needed to make the reaction clear, a phrase indicating the reaction or a product should be added in parentheses after the second part of the name, *e.g. (ADP-forming), (dimerising), (CoA-acylating).*

Rule 11 (Common Names)

The direct attachment of-*ase* to the name of the substrate will indicate that the enzyme brings about hydrolysis.

(Systematic Names)

The suffix-*ase* should never be attached directly to the name of the substrate.

Rule 12 (Common Names)

The name 'dehydrase' which was at one time used for both dehydrogenating and dehydrating enzymes, should not be used. *Dehydrogenase* will be used for the former and *dehydratase* for the latter.

Rule 13 (Common Names)

Where possible, common names should normally be based on a reaction direction that has been demonstrated, *e.g., dehydrogenase* or *reductase, decarboxylase* or *carboxylase*.

(Systematic Names)

In the case of reversible reactions, the direction chosen for naming should be the same for all the enzymes in a given class, even if this direction has not been demonstrated for all. Thus, systematic names may be based on a written reaction, even though only the reverse of this has been actually demonstrated experimentally.

Rule 14 (Systematic Names)

When the overall reaction includes two different changes, *e.g.,* an oxidative demethylation, the classification and systematic name should be based, whenever possible, on the one (or the first one) catalysed by the enzyme; the other function(s) should be indicated by adding a suitable participle in parentheses, as in the case of *sarcosine:oxygen oxidoreductase (demethylating)*; D-*aspartate:oxygen oxidoreductase (deaminating)*; L-*serine hydro-lyase (adding indoleglycerol-phosphate)*.

Other examples of such additions are *(decarboxylating), (cyclising), (acceptor-acylating), (isomerising).*

Rule 15

When an enzyme catalyses more than one type of reaction, the name should normally refer to one reaction only.

Each case must be considered on its merits, and the choice must be, to some extent, arbitrary. Other important activities of the enzyme may be indicated in the List under 'Reaction' or 'Comments'.

Similarly, when any enzyme acts on more than one substrate (or pair of substrates), the name should normally refer only to one substrate (or pair of substrates), although in certain cases it may be possible to use a term that covers a whole group of substrates, or an alternative substrate may be given in parentheses.

Rule 16

A group of enzymes with closely similar specificities should normally be described by a single entry. However, when the specificity of two enzymes catalysing the same reactions is sufficiently different (the degree of difference being a matter of arbitrary choice) two separate entries may be made. Separate entries are also appropriate for enzymes having similar catalytic functions, but known to differ basically with regard to reaction mechanism or to the nature of the catalytic groups, *e.g., amine oxidase (flavin-containing)* and *amine oxidase (copper-containing).*

Rules and Guidelines for Particular Classes of Enzymes

Class 1 Rule 17 (Common Names)

The terms *dehydrogenase* or *reductase* will be used much as hitherto. The latter term is appropriate when hydrogen transfer from the substance mentioned as donor in the systematic name is not readily demonstrated. *Transhydrogenase* may be retained for a few well-established cases. *Oxidase* is used only for cases there O_2 acts as an acceptor, and *oxygenase* only for those cases where the O_2 molecule (or part of it) is directly incorporated into the substrate. *Peroxidase* is used for enzymes using H_2O_2 as acceptor. *Catalase* must be regarded as exceptional. Where no ambiguity is caused, the second reactant is not usually named; but where required to prevent ambiguity, it may be given in parentheses, *e.g alcohol dehydrogenase* and *alcohol dehydrogenase (NADP$^+$).*

(Systematic Names)

All enzymes catalysing oxidoreductions should be named *oxidoreductases* in the systematic nomenclature, and the names formed on the pattern *donor:acceptor oxidoreductase.*

Rule 18 (Systematic Names)

For oxidoreductases using NAD$^+$ or NADP$^+$, the coenzyme should always be named as the acceptor except for the special case of Section (enzymes whose normal physiological function is regarded as reoxidation of the reduced coenzyme). Where the enzyme can use either coenzyme, this should be indicated by writing NAD(P)$^+$.

Rule 19

Where the true acceptor is unknown and the oxidoreductase has only been shown to react with artificial acceptors, the word *acceptor* should be written in parentheses, *succinate:(acceptor) oxidoreductase.*

Rule 20 (Common Names)

Oxidoreductases that bring about the incorporation of molecular oxygen into one donor or into either or both of a pair of donors are named *oxygenase*. If only one atom of oxygen is incorporated the term *monooxygenase* is used; if both atoms of O_2 are incorporated, the term *dioxygenase* is used.

(Systematic Names)

Oxidoreductases bringing about the incorporation of oxygen into one of paired donors should be named on the pattern *donor,donor:oxygen oxidoreductase (hydroxylating)*.

Class 2. Rule 21 (Common Names)

Only one specific substrate or reaction product is generally indicated in the common names, together with the group donated or accepted. The forms *transaminase, etc.,* may be replaced if desired by the corresponding forms *aminotr ansferase, etc.*. A number of special words are used to indicate reaction types, *e.g., kinase* to indicate a phosphate transfer from ATP to the named substrate (not 'phosphokinase'), *diphosphokinase* for a similar transfer of diphosphate.

(Systematic Names)

Enzymes catalysing group-transfer reactions should be named *transferase* and the names formed on the pattern *donor:acceptor group-transferred-transferase, e.g. ATP:acetate phosphotransferas*. A figure may be prefixed to show the position to which the group is transferred, *e.g. ATP:*D*-fructose 1-phosphotransferase*.

The spelling 'transphorase' should not be used. In the case of the phosphotransferases, ATP should always be named as the donor. In the case of the transaminases involving 2-oxoglutarate, the latter should always be named as the acceptor.

Rule 22 (Systematic Names)

The prefix denoting the group transferred should, as far as possible, be non-committal with respect to the mechanism of the transfer, *e.g., phospho-*, rather than *phosphate-*.

Class 3. Rule 23 (Common Names)

The direct addition of-*ase* to the name of the substrate generally denotes a hydrolase. Where this is difficult, *e.g.,* for EC 3.1.2.1, the word *hydrolase* may be used. Enzymes should not normally be given separate names merely on the basis of optimal conditions for activity. The acid and alkaline phosphatases should be regarded as special cases and not as examples to be followed. The common name *lysozyme* is also exceptional.

(Systematic Names)

Hydrolysing enzymes should be systematically named on the pattern *substrate hydrolase*. Where the enzyme is specific for the removal of a particular group, the group may be named as a prefix, *e.g., adenosine aminohydrolase* (EC 3.5.4.4). In a number of cases this group can also be transferred by the enzyme to other molecules, and the hydrolysis itself might be regarded as a transfer of the group to water.

Class 4. Rule 24 (Common Names)

The old names *decarboxylase, aldolase, etc*., are retained; and *dehydratase* (not 'dehydrase') is used for the hydro-lyases. 'Synthetase' should not be used for any enzymes in this class. The term *synthase* may be used instead for any enzyme in this class (or any other class) when it is desired to emphasise the synthetic aspect of the reaction.

(Systematic Names)

Enzymes removing groups from substrates non-hydrolytically, leaving double bonds (or adding groups to double bonds) should be called *lyases* in the systematic nomenclature. Prefixes such as *hydro-, ammonia-*should be used to denote the type of reaction, *e.g.* (S)*-malate hydro-lyase*. Decarboxylases should be regarded as *carboxy-lyases*. A hyphen should always be written before *lyase* to avoid confusion with hydrolases, carboxylases, *etc*.

Rule 25 (Common Names)

Where the equilibrium warrants it, or where the enzyme has long been named after a particular substrate, the reverse reaction may be taken as the basis of the name, using *hydratase, carboxylase, etc., e.g., fumarate hydratase* (in preference to 'fumarase', which suggests an enzyme hydrolysing fumarate).

(Systematic Names)

The complete molecule, not either of the parts into which it is separated, should be named as the substrate.

The part indicated as a prefix to-*lyase* is the more characteristic and usually, but not always, the smaller of the two reaction products. This may either be the removed (saturated) fragment of the substrate molecule, as in *ammonia-, hydro-, thiol-lyases, etc*. or the remaining unsaturated fragment, *e.g.,* in the case of *carboxy-, aldehyde-*or *oxo-acid-lyases*.

Rule 26

Various subclasses of the lyases include a number of strictly specific or group-specific pyridoxal-5-phosphate enzymes that catalyse *elimination* reactions of b-or g-substituted a-amino acids. Some closely related pyridoxal-

5-phosphate-containing enzymes, *e.g., tryptophan synthase* and *cystathionine* b-*synthase* catalyse *replacement* reactions in which a b-or g-substituent is replaced by a second reactant without creating a double bond. Formally, these enzymes appear to be transferases rather than lyases.

However, there is evidence that in these cases the elimination of the b-or g-substituent and the formation of an unsaturated intermediate is the first step in the reaction. Thus, applying rule 14, these enzymes are correctly classified as lyases.

Class 5. Rule 27

In this class, the common names are, in general, similar to the systematic names which indicate the basis of classification.

Rule 28

Isomerase will be used as a general name for enzymes in this class. The types of isomerisation will be indicated in systematic names by prefixes, *e.g., maleate cis-trans-isomerase, phenylpyruvate keto-enol-isomerase, 3-oxosteroid* D^5-D^4-*isomerase*.

Enzymes catalysing an aldose-ketose interconversion will be known as *aldose-ketose-isomerases, e.g.*

L-*arabinose aldose-ketose-isomerase*. When the isomerisation consists of an intramolecular transfer of a group, the enzyme is named a *mutase*, *e.g.,* the *phosphomutases*; when it consists of an intramolecular lyase-type reaction, *e.g.,* it is systematically named a *lyase (decyclising)*.

Rule 29

Isomerases catalysing inversions at asymmetric centres should be termed *racemases* or *epimerases*, according to whether the substrate contains one, or more than one, centre of asymmetry: compare.

A numerical prefix to the word *epimerase* should be used to show the position of the inversion.

Class 6. Rule 30 (Common Names)

Common names for enzymes of this class were previously of the type *XY synthetase*.

However, as this use has not always been understood and synthetase has been confused with synthase, it is now recommended that as far as possible the common names should be similar in form to the systematic names.

(Systematic Names)

The class of enzymes catalysing the linking together of two molecules, coupled with the breaking of a diphosphate link in ATP, *etc*. should be known as *ligases*. These enzymes were often previously known as 'synthetases';

however, this terminology differs from all other systematic enzyme names in that it is based on the product and not on the substrate. For these reasons, a new systematic class name was necessary.

Rule 31 (Common Names)

The common names should be formed on the pattern *X-Y ligase,* where X-Y is the substance formed by linking X and Y. In certain cases, where a trivial name is commonly used for XY, a name of the type *XY synthase* may be recommended.

(Systematic Names)

The systematic names should be formed on the pattern *X:Y ligase (ADP-forming)*, where X and Y are the two molecules to be joined together. The phrase shown in parentheses indicates both that ATP is the triphosphate involved, and also that the terminal diphosphate link in broken. Thus, the reaction is X + Y + ATP = X-Y + ADP + P_i.

Rule 32 (Common Names)

In the special case where glutamine acts as an ammonia-donor, this is indicated by adding in parentheses (*glutamine-hydrolysing*) to a ligase name.

(Systematic Names)

In this case, the name *amido-ligase* should be used in the systematic nomenclature.

Enzymes Activity

Enzymes are catalysts that optimise cell activity while minimizing the amount of energy needed to achieve a specific reaction. Enzymes are also energised protein molecules found in every living cell, and are necessary for life.

There are over 2000 known enzymes, each of which is involved with one (1) specific chemical reaction.

They are any of various proteins, originating from living cells and capable of producing certain chemical changes in organic substances by catalytic action, such as digestion.

These proteins, and their function (s), are determined by their shape 1, 2. In cells and organisms, most reactions are catalysed by enzymes, which are regenerated during the course of a reaction.

Biological catalysts are physiologically important because they speed up rates of reactions that would otherwise be too slow to support life.

Our bodies naturally produce digestive and metabolic enzymes as they are needed. Specifically, the pancreas produces enzymes that break down foods into nutrients the body can use for energy and other bodily functions.The names of enzymes often include the substrate or substance on which they act, joined with an-ase ending. For example, lactase acts upon lactose and maltase acts on maltose to produce glucose.

Sometimes they are named for their reaction product, for example, sucrase is often called invertase, because invertase is the result of the reaction of sucrose. Enzymes can also bear a name the describes the reaction that is catalysed. An example of this includes the use of the name oxidase, because oxidase is involved in an oxidation reaction 8.

9

Industrial Chemistry

CONSISTS OF COMBUSTIBLE HYDROCARBONS

GAS FUELS

Natural gas consists of combustible hydrocarbons which are gaseous at ordinary temperatures and pressures, and have essentially the same origin as fluid hydrocarbons. Methane (also called marsh gas) and ethane are commonly the chief constituents. Most natural gases usually contain small and variable quantities of carbon dioxide, carbon monoxide, sulfur dioxide, hydrogen sulfide, nitrogen, hydrogen, and oxygen.

In the absence of sulfurous compounds, natural gas is colourless, nearly odourless, and, when mixed in certain proportions with air, is highly explosive. An odourant is added before gas is sold to the public to aid in detection of gas leaks. Natural gas resources can be categorized into two main types on the basis of producing rock characters: conventional natural gas and unconventional natural gas.

Conventional natural gas is produced by a well drilled into a geologic formation in which the reservoir and fluid characteristics permit the natural gas to readily flow to the wellbore. Unconventional natural gas does not exist in these conventional reservoirs—rather, this natural gas takes another form, or is present in a peculiar formation that makes its extraction quite different from conventional resources. The major unconventional gas resources in U.S. include tight gas, shale gas, and coalbed methane (CBM). The latter two types are present in Arkansas.

Natural gas was first discovered in 1887 at Fort Smith, but commercial development did not begin until 1902 when two gas wells were completed near Mansfield in Sebastian County. Gas was first discovered in southern Arkansas on April 22, 1920, when the Constantin Oil Company completed a gas well near El Dorado in Union County.

The heating value of gas varies from about 700 to 1,200 British thermal units (Btu) per cubic foot. Dry natural gas from the Arkoma basin fields has a heating value of 986 to 1,016 Btu per cubic foot, and is used principally as fuel. Major accumulations of natural gas are present in two areas in Arkansas–the Arkoma basin and the southern Arkansas oil fields in the West Gulf Coastal Plain. Natural gas is commonly discussed as either "wet" or "dry" gas. Wet gas contains some of the heavier fluid hydrocarbons as vapour, is commonly associated with petroleum, and is valuable because of the extractable hydrocarbon liquids it contains. Most gas from oil fields in southern Arkansas is of this type. Dry gas differs from wet gas in that it does not carry appreciable amounts of the heavier hydrocarbons as vapour. The gas of the Arkoma basin in west-central Arkansas is of this type.

Coalbed natural gas or coalbed methane (CBM) is the methane gas contained in coal seams. The coalification process, begins with plant material that is progressively converted to coal and this results in large quantities of methane-rich gas that is generated and stored within the coal.

The presence of this gas has been long-recognized due to explosive outbursts associated with underground coal mining. Only recently has coal been recognized as a reservoir rock and a source rock, thus representing an enormous undeveloped "unconventional" energy resource. Because of its large internal surface area, coal stores between six and seven times more gas than the equivalent rock volume of a conventional gas reservoir. The United States is expected to have 700 Tcf of CBM, compared with its conventional gas reserves of 187 TCF.

The development of Arkansas's coalbed natural gas resources began in 2001 and has yielded an approximate cumulative production of 10 Bcf. Estimated 2007 annual production of CBM is approximately 3 Bcf. CDX Gas LLC, a Texas based energy company, is currently the only producer of this resource and has drilled approximately 37 Z-pinnate horizontal wells and 15 vertical wells in Sebastian County, Arkansas. The wells are completed in the Pennsylvanian Lower Hartshorne coal and approximately 564,238 feet of horizontal pinnate lateral has been drilled in Arkansas. On average, approximately 15,000 feet of horizontal lateral is drilled for each of CDX's Z-pinnate wells in the Lower Hartshorne coal.

COAL

Coal was formed by the decomposition of vegetation that grew in prehistoric forests. At that time the climate was favorable for very rapid growth. Layer upon layer of fallen trees was covered with sediment, and after long periods of aging, the chemical and physical properties of the now ancient vegetation deposits were changed, through various intermediate processes, into coal. The process of coal formation can be observed in the various stages on the earth today. However, present-day formations are insignificant when compared with the magnitude of the great coal deposits. It is estimated that 100 years is required to deposit 1 ft of vegetation in the form known as *peat,* and 4 ft of peat is necessary for the formation of 1 ft of coal. Therefore, it requires 400 years to accumulate enough vegetable matter for a 1-ft layer of coal. The conversion from peat to coal requires ages of time. In some areas where other fuel is scarce, the peat is collected, dried, and burned. The characteristics of coal depend on the type of vegetation from which it was formed, the impurities that became intermixed with the vegetable matter at the time the peat bog was forming, and the aging, time, temperature, and pressure.

It is apparent that the characteristics of coal vary widely. For example, peat often contains partially decomposed stems, twigs, and bark. Peat is progressively transformed to lignite, which eventually can become anthracite when provided with the proper progression of geologic changes. However, this transformation takes hundreds of years to complete. Coal is a heterogeneous material that varies in chemical composition according to location. In addition to the major organic ingredients of C, H2, and 02, coal also contains impurities. The impurities that are of major concern are ash and sulfur. The ash results from mineral or inorganic material that was introduced during formation of the coal. Ash sources include inorganic substances, such as silica, which are part of the chemical structure of the plants. Dissolved mineral grains that are found in swamp water are also captured by the organic matter during the formation of coal. Mud, shale, and pyrite are deposited in pores and cracks of the coal seams. Sulfur occurs in coal in three forms:

1. Organic sulfur, which is part of the coal's molecular structure
2. Pyritic sulfur, which occurs as the mineral pyrite
3. Sulfate sulfur, which is primarily from iron sulfate

The highest sulfur source is sulfate iron, which is found in water. Fresh water has a low sulfate concentration, whereas salt water is high in sulfate. Bituminous coal is found deposited in the interior of the United States, where oceans once covered the region, and this coal has a high sulfur content.

PEAT

Peat is the first product in the formation of coal and consists of partially decomposed plant and mineral matter. Peat has a moisture content of up to 70 percent and a heating value as low as 3000 Btu/lb. Although it is not an official coal classification, it is used as a fuel in some parts of the world.

LIGNITE

Lignite is the lowest ranking of coal with a heating value of less than 8300 Btu/lb and a moisture content as high as 35 percent. The volatile content is also high, and therefore lignite ignites easily.

SUBBITUMINOUS

These coals are non-coking; i.e., they have minimal swelling on heating and have a relatively high moisture content of 15 to 30 percent. They are high in volatile matter and thus ignite easily. They also have less ash and burn cleaner than lignite. They have a low sulfur content, generally less than 1 percent, and a heating value between 8300 and 11,500 Btu/lb. Because of the low sulfur content, many power plants have changed to subbituminous coal in order to limit SO2 emissions.

BITUMINOUS

This coal is the one most commonly burned in electric utility boilers, and it has a heating value between 10,500 and 14,000 Btu/lb. As compared with lignite and subbituminous coals, the heating value is higher and the moisture content and volatile matter are lower. The high heating value and its relatively high volatile matter enable these coals to burn easily when fired as pulverized coal. Some types of bituminous coal, when heated in the absence of air, soften and release volatiles and then form coke, a porous, hard, black product. Coke is used as fuel in blast furnaces to make iron.

ANTHRACITE

This is the highest ranked coal. It has the highest content of fixed carbon, ranging from 86 to 98 percent. It has a low volatile content, which makes it a slow-burning fuel. Its moisture content is low at about 3 percent, and its heating value can be as high as 15,000 Btu/lb. Anthracite is low in sulfur and volatiles and burns with a hot, clean flame. It is used mostly for domestic heating as well as some metallurgical processes.

However, utility-type boilers are designed to burn this low-volatile coal in pulverized coal-fired boilers. Preparation plants are capable of upgrading coal quality. In this process, foreign materials, including slate and pyrites, are separated from the coal. The coal may be washed, sized, and blended to meet the most exacting power plant demands. However, this processing increases the cost of coal, and an economic evaluation is required to determine whether the cost can be justified. If the raw coal available in the area is unsatisfactory, the minimum required upgrading must be determined. Utility plants obtain the lowest steam cost by selecting their combustion equipment to use the raw coal available in the area. However, in order to meet sulfur dioxide emission limits, many utilities use low-sulfur coals that often have to be transported significant distances.

For example, many plants located in the midwest of the United States use coals shipped from Montana and Wyoming. For over 20 years, Powder River Basin (PRB) coals, where mines are located across eastern Wyoming and Montana, have led in the supply of these low-sulfur coals. The annual production of this coal has increased significantly, from less than 10 million tons per year in the mid-1970s to over 300 million tons per year today. Nearly all this coal is used in electric power plants. Most of this coal produces less than 1 lb of SO2 per million Btus, thus making it a popular choice, even though it has a relatively low heat content, ranging from 8400 to 9300 Btu/lb. The effect on the SO2 removal requirements is simplified. Since this coal can be mined efficiently and economically because its seams lie close to the surface, it also has a price advantage over coals from other locations. However, transportation costs are higher than those for coals from eastern regions. Many utilities, independent power producers (IPPs), and large industrial plants have switched from eastern bituminous coals to PRB coal for reasons of cost alone. Others are using the fuel because it reduces emissions of SO2 without the need to install costly scrubbers.

However, the use of PRB coals has some negative effects that must be handled. The fuel has a higher moisture content, which results in a lower heating value. It also has a lower ash softening temperature and a higher ash content. These characteristics often result in greater fouling and slagging of boiler heating surfaces, and dust control in and around the power plant is more diffcult.

Fugitive dust can ignite and even explode under certain conditions, and this puts plant personnel and equipment at risk, making it even more important to have good housekeeping procedures. The higher ash content places additional burden on the ash handling system and on the needed landfill to handle the waste material. However, these potential operating difficulties can be managed, and PRB coals will continue to be used because of their lower cost and lower sulfur content. All the many factors involved in obtaining the lowest-cost steam production must be taken into consideration. For a new facility, this would involve the selection of combustion equipment for the coal that is available in the region or for the coal that may be imported to the plant.

TEMPERATURE REQUIRED FOR COMBUSTION

All around us we see combustible material in intimate contact with air, and still it is not burning. Actually, a chemical reaction is taking place, but it is so slow that it is referred to not as *combustion* but as *oxidation.* The corrosion (rusting) of steel when exposed to the atmosphere is an example of this oxidation. When the combustible material reaches its ignition temperature, oxidation is accelerated and the process is called *combustion.* It is the rapid chemical combination of oxygen with the combustible elements of a fuel. Therefore, it is evident that it is important to maintain the fuel and air mixture at a temperature sufficiently high to promote combustion. When the flame comes into contact with the relatively cool boiler tubes, the carbon particles are deposited in the form of soot. When boilers are operated at a very low capacity, the temperatures are lower, which can result in incomplete combustion and excessive smoke if combustion controls are not set properly.

TIME REQUIRED FOR COMBUSTION

Air supply, mixing, and temperature determine the rate at which combustion progresses. In all cases an appreciable amount of time is required to complete the process. When the equipment is operated at an excessively high capacity, the time may be insufficient to permit complete combustion. As a result, considerable unburned fuel is discharged from the furnace. The rejected material may be in the form of solid fuel or combustible gases. The resulting loss may be appreciable and therefore must be checked and controlled.

Here the principles of combustion are applied to solid fuels burned on grates and to pulverized coal, gas, and oil burned in suspension. The coal is supplied by hand through the fire door. Air for combustion enters through both the ashpit and the fire door.

Primary air comes through the stoker grate, and secondary air enters through the fire door in this illustration. For the purpose of illustration, the fuel bed may be considered as having four zones. Coal is added to the top or distillation zone; next are the reduction and oxidation zones and, finally, the layer of ash on the grates. The primary air that enters the ashpit door flows up through the grates and ash into the oxidation zone, where the oxygen comes into contact with the hot coal and is converted into carbon monoxide (CO). As the gases continue to travel upward through the hot-coal bed, this carbon monoxide (CO) changes to carbon dioxide (CO2) as part of the combustion process.

The exposure of the coal in the upper zone to the high temperature results in distillation of hydrocarbons (chemical compounds of hydrogen and carbon), which are carried into the furnace by the upward flow of gases. Therefore, the gases entering the furnace through the fuel bed contain combustible materials in the form of carbon monoxide and hydrocarbons. The oxygen in the secondary air that enters the furnace through the fire door must combine with these combustibles to complete the combustion process before they enter the boiler tube bank and become cooled. From this discussion of the process it is evident that with hand firing a number of variables are involved in obtaining the required rate of combustion and complete utilization of the fuel with a minimum amount of excess air. The fuel must be supplied at the rate required by the steam demand.

Not only must the air be supplied in proportion to the fuel, but the amount entering through the furnace doors and the ashpit must be in correct proportions. The air that enters through the ashpit door and passes up through the fuel bed determines the rate of combustion. The secondary air that enters directly into the furnace is used to burn the combustible gases. Thorough mixing of the combustible gases and air in the furnace is necessary because of the short time required for these gases to travel from the fuel bed to the boiler tubes. Steam or high-pressure air jets (overfire air system) are used to assist in producing turbulence in the furnace and mixing of the gases and air. A failure to distribute the coal evenly on the grates, variation in the size of the coal, and the formation of clinkers result in unequal resistance of the fuel bed to the flow of gases. (A *clinker* is a hard, compact, congealed mass of fuel matter that has fused in the furnace.

It is often called *slag.*) Areas of low resistance in the fuel bed permit high velocity of gases and accelerated rates of combustion, which deplete the fuel and further reduce the resistance. These areas of low resistance have been called *holes in the fire.* As a result of these inherent shortcomings in hand firing and the physical labor involved, mechanical methods of introducing solid fuel into the furnace and automatically controlling the air supply have been developed.

This method may be utilized in the combustion of gaseous fuels without special preparations, of fuel oil by providing for atomization, and of solid fuels by pulverization. The fuel particles and air in the correct proportions are introduced into the furnace, which is at an elevated temperature. The fine particles of fuel expose a large surface to the oxygen present in the combustion air and to the high furnace temperature. The air and fuel particles are mixed either in the burner or directly after they enter the furnace. When coal is burned by this method, the volatile matter—hydrocarbons and carbon monoxide—is distilled off when the coal enters the furnace.

These combustible gases and the residual carbon particles burn during the short interval of time required for them to pass through the furnace. The period of time required to complete combustion of fuel particles in suspension depends on the particle size of the fuel, control of the flow of combustion air, mixing of air and fuel, and furnace temperatures.

Relatively large furnace volumes are required to ensure complete combustion. From these illustrations we note that the requirements for good combustion are sufficient time of contact between the fuel and air, elevated temperature during this time, and turbulence to provide thorough mixing of fuel and air. These are referred to as the three *T*'s of combustion—time, temperature, and turbulence.

CHEMICAL PROCESS OF COMBUSTION

Combustion is a chemical process that takes place in accordance with natural laws. By applying these laws, the theoretical quantity of air required to burn a given fuel can be determined when the fuel analysis is known. The air quantity used in a furnace, expressed as percentage of excess above the theoretical requirements (excess air), can be determined from the flue gas analysis. In the study of combustion we encounter matter in three forms: solid, liquid, and gas. *Melting* is the change of phase from solid to liquid. Heat must be added to cause melting. The change in the reverse direction, liquid to solid, is *freezing* or *solidifying.*

The change of phase from liquid to gas is called *vaporization,* and the liquid is said to *vaporize* or *boil.* The change from the gaseous or vapor phase to liquid is *condensation,* and during this process the vapor is said to be *condensing.* Matter in the form of a *solid* has both volume and shape. A *liquid* has a definite volume, in that it is not readily compressible, but its shape conforms to that of the container. A *gas* has neither a definite volume nor shape, since both conform to that of the container. When liquids are heated, a temperature is reached at which *vapor* will form above the surface. This vapor is only slightly above the liquid state.

When vapor is removed from the presence of the liquid and heated, a gas will be formed. There is no exact point at which a substance changes from a gas to a vapor or from a vapor to a gas.

It is simply a question of degree as to how nearly the vapor approaches a gas. Steam produced by boiling water at atmospheric pressure is vapor because it is just above the liquid state. On the other hand, air may be considered a gas because under normal conditions it is far removed from the liquid state (liquid air). Gases follow definite laws of behaviour when subjected to changes in pressure, volume, and temperature.

The more nearly a vapor approaches a gas, the more closely it will follow the laws. When considering gas laws and when making calculations in thermodynamics (i.e., the relationship between heat and other forms of energy), pressures and temperatures must be expressed in absolute units rather than in *gauge* values (read directly from gauges and thermometers). Absolute pressures greater than atmospheric are found by adding the atmospheric pressure to the gauge reading.

Both pressures must be expressed in the same units. The atmospheric pressure is accurately determined by means of a barometer, but for many calculations the approximate value of 14.7 psi is sufficiently accurate. For example, if a pressure gauge reads 150 psi, the absolute pressure is 164.7 psia (read as "pounds per square inch absolute"). Absolute pressures below zero gauge are found by subtracting the gauge reading from the atmospheric pressure.

When a gauge reads –5 psi, the absolute pressure would be 14.7 –5.0 = 9.7 psia. Pressures a few pounds above zero gauge and a few pounds below (in the vacuum range) are frequently measured by a Utube containing mercury and are expressed in inches of mercury. Many of the pressures encountered in combustion work are nearly atmospheric (zero gauge) and can be measured by a U-tube containing water.

The zero on the Fahrenheit scale is arbitrarily chosen and has no scientific basis. Experiments have proved that the true or absolute zero is 460° below zero on the Fahrenheit thermometer.

The absolute temperature on the Fahrenheit scale is found by adding 460°F to the thermometer reading. Absolute temperatures on the Fahrenheit scale are called *degrees Rankine* (°R).

On the centigrade or Celsius scale, the absolute temperature is determined by adding 273°C to the centigrade thermometer reading. Absolute temperatures on the centigrade scale are measured in *kelvins* (K). Expressed in absolute units of pressure and temperature, the three principal laws governing the behaviour of gases may be stated as follows (where $V1$ and $V2$ are, respectively, the initial and final volumes, $P1$ and $P2$ are, respectively, the initial and final absolute pressures, and $T1$ and $T2$ are, respectively, the initial and final absolute temperatures):

CONSTANT TEMPERATURE

When the temperature of a given quantity of gas is maintained constant, the volume will vary inversely as the pressure. If the pressure is doubled, the volume will be reduced by one-half:

$$\frac{V_1}{V_2} = \frac{P_2}{P_1}$$

CONSTANT VOLUME

When the volume of a gas is maintained constant, the pressure will vary directly as the temperature. When the temperature is doubled, the pressure also will be doubled:

$$\frac{P_1}{P_2} = \frac{T_1}{T_2}$$

CONSTANT PRESSURE

When a gas is maintained at constant pressure, the volume will vary directly as the temperature. If the temperature of a given quantity of gas is doubled, the volume also will be doubled:

$$\frac{V_1}{V_2} = \frac{T_1}{T_2}$$

In combustion work the gas temperature varies over a wide range. Air enters the furnace or air heater, for example, at 70°F, is heated in some instances to over 3000°F in the furnace, and is finally discharged from the stack at between 300 and 400°F.

During these temperature changes, the volume varies because the gases are maintained near atmospheric pressure. This is most important because fans, flues and ducts, boiler passes, etc., must be designed to accommodate these variations in volume.

In addition to these physical aspects of matter, we also must consider the chemical reactions that occur in the combustion process. All substances are composed of one or more of the chemical elements.

The smallest particle into which an element may be divided is termed an *atom.* Atoms combine in various combinations to form molecules, which are the smallest particles of a compound or substance. The characteristics of a substance are determined by the atoms that make up its molecules. Combustion is a chemical process involving the reaction of carbon, hydrogen, and sulfur with oxygen.

The trading about and changing of atoms from one substance to another constitute an exacting procedure. Substances always combine in the same definite proportions. The atoms of each of the elements have a weight number referred to as the *atomic weight.* These weights are relative and refer to oxygen, which has an atomic weight of 16. Thus, for example, carbon, which is three-quarters as heavy as oxygen, has an atomic weight of 12.

When oxygen and the combustible elements or compounds are mixed in definite proportions at an elevated temperature under ideal conditions, they will combine completely.

This shows that a given combustible element requires a definite amount of oxygen to complete combustion. If additional oxygen is supplied (more than necessary for complete combustion), the excess will not enter into the reaction but will pass through the furnace unchanged.

On the other hand, if there is a deficiency of oxygen, the combustible material will remain unburned. The law of combining weights states that the elements and compounds combine in definite proportions that are in simple ratio to their atomic or molecular weights.

The following is an explanation of some of the chemical reactions involved in combustion: The volume of carbon dioxide (CO2) produced is equal to the volume of oxygen (O2) used. The carbon dioxide gas is, however, heavier than the oxygen. The combining weights are 12 lb of carbon (1 x 12) and 32 lb of oxygen (2 x 16), uniting to form 44 lb of carbon dioxide, or 1 lb of carbon requires 2.67 lb of oxygen (32/12 = 2.67) and produces 3.67 lb of carbon dioxide (1 + 2.67 = 3.67).

The combustion of 1 lb of carbon produces 14,540 Btu. When carbon is burned to carbon monoxide (CO), which is incomplete combustion, the volume of oxygen used is only one-half of that required for completely burning the carbon to carbon dioxide; the volume of carbon monoxide produced is two times that of the oxygen supplied.

The heat released is only 4355 Btu/lb, but it is 14,540 Btu when 1 lb of carbon is completely burned. The net loss is, therefore, 10,185 Btu/lb of carbon and shows the importance of completely burning the combustible gases before they are allowed to escape from the furnace.

The two molecules of carbon monoxide (CO) previously produced, by the incomplete combustion of two molecules of carbon, are combined with the necessary one molecule of oxygen to produce two molecules of carbon dioxide. The 1 lb of carbon produced 2.333 lb of carbon monoxide. Finally, however, the 1 lb of carbon produces 3.67 lb of carbon dioxide regardless of whether the reaction is in one or two steps.

$$C + O_2 \rightarrow CO_2$$
$$2C + O_2 \rightarrow 2CO$$
$$2CO + O_2 \rightarrow 2CO_2$$
$$2H_2 + O_2 \rightarrow 2H_2O$$
$$S + O_2 \rightarrow SO_2 \qquad (\textit{sulfur dioxide})$$
$$2S + 3O_2 \rightarrow 2SO_3 \qquad (\textit{sulfur trioxide})$$

$$\underset{\textit{Methane}}{CH_4} + 2O_2 \rightarrow CO_2 + 2H_2O$$
$$\underset{\textit{Acetylene}}{2C_2H_2} + 5O_2 \rightarrow 4CO_2 + 2H_2O$$
$$\underset{\textit{Ethylene}}{C_2H_4} + 3O_2 \rightarrow 2CO_2 + 2H_2O$$
$$2C_2H_6 + 7O_2 \rightarrow 4CO_2 + 6H_2O$$

The total amount of oxygen required, as well as the heat liberated per pound of carbon, is the same for complete combustion in both cases. Hydrogen is a very light gas with a high heat value. The combustion of 1 lb of hydrogen gas liberates 62,000 Btu.

To develop this heat, two molecules of hydrogen combine with one molecule of oxygen to form two molecules of water. One volume of oxygen is required for two volumes of hydrogen. The weight relations are 1 lb of hydrogen and 8 lb of oxygen, producing 9 lb of water. This water appears as water vapor in the flue gases. The equations for these and some of the other reactions involved in combustion are as follows:

Sulfur is an undesirable constituent in fuels. It has a heating value of only 4050 Btu/lb, contaminates the atmosphere with sulfur dioxide unless controlled with air pollution control equipment, and causes corrosion in the flues, economizers, and air heaters. Some forms of sulfur adversely affect pulverization.

In the combustion process, 1 lb of sulfur combines with 1 lb of oxygen to form 2 lb of sulfur dioxide. (Actually, a portion of the sulfur is converted to sulfur trioxide. The summation of all the sulfur oxides in the flue gases is referred to as SO*x*..) The sulfur dioxide in the flue gases can be approximated as follows:

$$SO_2 \text{ lb/h} = K \times \text{lb/h fuel burned} \times 2 \times S/100$$

where K = the ratio of SO_2 in flue gases to a theoretical amount resulting from the combustion of the sulfur in the fuel (frequently assumed to be 0.95) and S = the percentage of sulfur in the fuel. It is customary to express sulfur oxide emission in pounds per million Btus of fuel burned.

$$SO_2\ lb\ /\ million\ Btu = \frac{SO_2\ lb\ /\ h \times 1{,}000{,}000}{lb\ fuel\ /\ h \times Btu\ /\ lb\ in\ fuel}$$

Example Coal containing 1.5 percent sulfur with a Btu content of 11,500 Btu/lb burns at the rate of 3 tons per hour. What is the sulfur emission in pounds per million Btus input to furnace?

Solution

$$SO_2\ lb\ /\ h = 0.95 \times 3 \times 2000 \times 2 \times 1.5\ /\ 100 = 171$$

$$SO_2\ lb\ /\ million\ Btu = \frac{171 \times 1{,}000{,}000}{3 \times 2000 \times 11{,}500} = 2.48$$

In practice, the oxygen supplied for combustion is obtained from the atmosphere. The atmosphere is a mixture of gases that for practical purposes may be considered as being composed of the following:

Element	Volume, %	Weight, %
Oxygen	20.91	23.15
Nitrogen	79.09	76.85

Only the oxygen enters into chemical combination with the fuel. The nitrogen combines in small amounts with the oxygen to form nitrogen oxides, commonly called NO*x*.

These are an atmospheric pollutant. The amount of NO*x* produced depends on the combustion process. The higher the temperature in the furnace, the greater the amount of nitrogen oxides. The remainder of the nitrogen passes through the combustion chamber without chemical change. It does, however, absorb heat and reduces the maximum temperature attained by the products of combustion. Since air contains 23.15 percent by weight of oxygen, in order to supply 1 lb of oxygen to a furnace it is necessary to introduce

$$\frac{1}{0.2315} = 4.32\ lb\ of\ air$$

Since 1 lb of carbon requires 2.67 lb of oxygen, we must supply

- 4.32 x 2.67 = 11.53 lb of air per pound of carbon

The 11.53 lb of air is composed of 2.67 lb of oxygen and 8.86 lb of nitrogen.

$$\underset{lb\,air}{11.53} - \underset{lb\,O_2}{2.67} = \underset{lb\ N_2}{8.86}$$

By referring to the equation for the chemical reaction of carbon and oxygen, we find that 1 lb of carbon produces 3.67 lb of carbon dioxide. Therefore, the total products of combustion formed by burning 1 lb of carbon with the theoretical amount of air are 8.86 lb of nitrogen and 3.67 lb of carbon dioxide.

In a similar manner it can be shown that 1 lb of hydrogen requires 34.56 lb of air for complete combustion. The resulting products of combustion are 9 lb of water and 26.56 lb of nitrogen. Also, 1 lb of sulfur requires 4.32 lb of air, and therefore, the products of combustion are 3.32 lb of nitrogen and 2 lb of sulfur dioxide.

Constituent	Weight per lb
Carbon	0.75
Hydrogen	0.05
Nitrogen	0.02
Oxygen	0.09
Sulfur	0.01
Ash	0.08
Total	1.00

The condition under which, or degree to which, combustion takes place is expressed as *perfect, complete,* or *incomplete. Perfect combustion,* which we have been discussing, consists of burning all the fuel and using only the calculated or theoretical amount of air.

Complete combustion also denotes the complete burning of the fuel but by supplying more than the theoretical amount of air. The additional air does not enter into the chemical reaction. *Incomplete combustion* occurs when a portion of the fuel remains unburned because of insufficient air, improper mixing, or other reasons.

GEOLOGICAL EXPLORATION METHODS IN PETROLEUM

A petroleum geologist's main job is to select promising site for the drilling of exploratory wells based on his prediction of an area's subsurface stratigraphy and structure.

Subsurface maps include the following basic forms:

- Structural contour maps: maps composed of lines connecting points of equal elevation above or below datum (normally sea level)
- Isopachous maps:–maps composed of lines connecting points of equal bed thickness.
- Cross sections: a form of subsurface presentation which depicts the position and thickness of various strata.

Subsurface maps are a necessary part of any reservoir engineering study; and petroleum engineers, as well as geologists, must be completely familiar with their construction and interpretation.

The data for subsurface maps are obtained from a number of sources, such as:

- *Well logs*: Representations of some rock property of properties versus depth.
- *Core drilling*: Shallow, small hole drilling for information purposes only. The formations encountered are core, *i.e.*, obtained as small cylindrical samples which are readily and accurately identified.

- *Strat tests*: Deep exploratory holes drilled primarily for information.

The construction of subsurface maps requires great interpretive skill

GEOPHYSICAL EXPLORATION

The methods are gravitational method, magnetic method and seismic.The gravitational method is based on Newton's hypothesis that every particle in the universe attracts every other particles in the manner defined by the equation,

where F = attractive force

$m_1 . m_2$ = masses of particles in question

$$F = \gamma \frac{m_1 m_2}{r^2}$$

r = distance between particles

= gravitational constant (6.67×10^{-8}) in cgs units.

SEISMIC EXPLORATION

Seismic exploration uses vibrations such as sound waves and shock waves in order to map the different layers of the ground, thus enabling the operator to predict the earth's density at varying depths. It is able to map the subsurface and to show in a 2-D, 3-D or even 4-D maps the explored region thus suggesting the locations of the oil or gas "traps" for drilling purposes.Seismic surveying uses tools such explosives or vibroseis trucks in order to explore on land, or a tool called the airgun in order to explore offshore (ocean floor). Here is the explanation of seismic surveying by Utah BLM Stone Cabin:"Seismic survey methodologies are tools for analysis of geologic formations and features in the subsurface.

The process consists of using a source of energy that is directed into the subsurface and then recorded back at the surface (with geophones) as the energy waves travel through the subsurface and reflect back to the surface. Various types of rock reflect the energy waves differently, and these differences are measured. Data helps show the tops and bottoms of formations, thickness, and structural configurations. It cannot identify pools of oil and gas, but rather, conditions favourable for the possible accumulation of oil and gas.

Explosives

Seismic geophones are able to collect their data from many sources that generate shock waves. Explosives method is one of those sources. By drilling small holes into the ground, approx. 12 meters deep, and packing them with 10 pounds of capped explosives (directed towards the center of earth), followed by detonation of those explosives, the geophones will be able to get sufficient data to map the are underneath the receivers. The grid lines of the explosives vary because of the different ground composure.

Computer based software is often used to calculate the distance needed between the location of such explosive holes. Based on the study assembled by USGS, the grid of the explosives used in the seismic surveying that was done in ANWR 1002, was about 300 feet between the charges.

Geophones were positioned in groups of 24 geophones per group while the interval between the groups was about 100-160 feet. In overall there were 120 groups in use. Due to the breaking of the ice immediately after detonation of the charges, the geologists encountered many problems, such as picking up wrong vibration data (vibration caused by the breaking and not by the blast) such secondary data affected the precision of the survey."

Thumper" trucks, Vibroseis

30,000 pounds trucks generate vibrations underneath the ground by elevating themselves above the ground on a short pole, thus concentrating their entire weight on a platter and "shaking" for several second per location, thus sending vibrations through the ground.

The rest of the process is very similar to the explosive process because all that is left is the data gathering phase. This process is the most precise process as it uses controlled vibrations that are spread over period of time, as oppose to the explosion vibration that is just a giant burst of energy. Those trucks are able to operate even inside major cities because the vibration they are causing is negligible due to the spread of vibration over a period of time.

Airgun

Airgun is an excellent example of an offshore method. This exploration technique is used with assistance from a ship that actually carries all the equipment necessary to both send signal and to analyse. Such ship will carry an airgun and many receivers at greater distance that actually read the sound data as it is reflected from different rock formations and layers beneath the ocean floor. Then, just as the land surveys, the data is being processed by a computer which is later on able to generate a detailed map of several layers underneath the sea bottom.

Data Processing

The data is being received by geophones, which are relatively small devices placed on the ground. Those devices are synchronized with computer and placed on the ground using DGPS equipment (more precise than GPS, 2 meters to 30 cm precision). This way the computer that analyses the data has all the variables, the time of the vibrations, the relative distances at which the reflections are measured, and the strength (wavelength) of the vibrations. Different rocks and layers give different reflections of vibration, different changes in frequencies in the wavelength of the vibration.

Using this data, gathered from the sensors, the main computer unit is able to constructed a detailed map, thus enabling it's user to analyse the map for possible oil location. It is worth mentioning that the computer is in most cases a dedicated super computer and not just a regular PC/MACThe computer is able to generate a 3-D map of the subsurface, enabling future analysis of the region by experts and recommending a possible drill site.Another capability is a generation of 4-D map, which is a relatively new concept. Basically 4-D is a 3-D map repeated over time period.

This way you can also see any changes in the ground versus time. For example shift of layers or flow of material inreservoirs. The truck on the left is the vibroseis truck sending vibrations as waves into the ground which are reflected of the different layer, recorder by geophones and transmitted to the data recording truck

Non-Seismic Exploration Techniques Electrical Resistivity

Overview: This process involves placing probes in the ground and passing a current between them. By measuring the resistance of this current you are able to tell within a degree of certainty whether there is oil or not. "Electrical and electromagnetic data are analysed primarily to yield the electrical resistivity of the rock formation where currents have been injected or induced to flow. The resistivity is in turn a strong function of the porosity and pore fluid saturation. Potential Problems: The steel casing of the pipes underground can act as a barrier to the electrical signals. "Steel casing severely attenuates electromagnetic signals transmitted or received from within the pipe. The casing typically acts as a low-pass filter, attenuating signal above 10 Hz and virtually eliminating signals above a few hundred Hz.

This means that field measurements are more difficult in steel casing, and field data predominantly reflects casing effects. Solutions: By taking into account the effects that the pipe has on the signal, you can separate it when you interpret the data and just get the information you want. "Current research suggests, however, that the shielding effect of the steel pipe is a fairly simple function of the thickness, electrical conductivity, and magnetic permeability of the steel pipe segment surrounding the sensor. Although EM fields are severely attenuated by the steel pipe, the response may be calculated with fairly simple numerical models and separated from the formation effect using straightforward techniques. Thus, if the properties are obtained, the response due to the casing may be easily separated from the total field, leaving the formation response as a residual.

Experimental Methods Earth's Field NMR

By using the earth's magnetic field, you can disrupt water molecules. By measuring this disruption you can have a good idea of whether there is water in an area. This technique is used for finding groundwater but it might be able to be used to find oil as well."This technique involves locally perturbing the direction and amplitude of the earth's ambient magnetic field to affect the dipole moment of hydrogen-based molecules (*i.e.* water or oil) within the pore structures.

After the perturbing field is shut off, a decay signal is generated in regions containing mobile hydrogen atoms. These data can be used to estimate the porosity, saturation, and possibly even permeability of these volumes. ...Recently, a project has been initiated through the DeepLook consortium to study the feasibility of extending this technology to oil and gas exploration.

Electrical Vs. Seismic

Electric is better at telling if there is oil and Seismic is better at telling if there is gas. Seismic is also better at characterizing the structure of the reservoir."Although electrical data are sensitive to variations in storage and saturation of reservoir liquids, seismic techniques are more sensitive to the presence of the reservoir gases. In addition, the higher resolution offered by seismic techniques is superior in mapping reservoir structure.

Magnetic

Overview: By measuring the magnetic field, you can tell where there is likely to be oil because the rocks that may contain oil have very low magnetic readings. The magnetic field can be measured with an instrument called a magnetometer which can be flown over an area or used on the ground.

"Magnetic surveys are usually made with magnetometers borne by aircraft flying in parallel lines spaced two to four kilometers apart at an elevation of about 500 meters when exploring for petroleum deposits. ...Ground surveys are conducted to follow up magnetic anomaly discoveries made from the air. Such surveys may involve stations spaced only 50 meters apart. ...Magnetic effects result primarily from the magnetization induced in susceptible rocks by the Earth's magnetic field. Most sedimentary rocks have very low susceptibility and thus are nearly transparent to magnetism. Accordingly, in petroleum exploration magnetics are used negatively: magnetic anomalies indicate the absence of explorable sedimentary rocks.

Methods

Proton-precession magnetometer: One type of magnetometer which utilizes the disruption of protons in oil to measure magnetism."One such method involves the proton-precession magnetometer, which makes use of the magnetic and gyroscopic properties of protons in a fluid such as gasoline. In this method, the magnetic moments of protons are first aligned by a strong magnetic field produced by an external coil.

The magnetic field is then turned off abruptly, and the protons try to align themselves with the Earth's field. However, since the protons are spinning as well as magnetized, they precess around the Earth's field with a frequency dependent on the magnitude of the latter. The external coil senses a weak voltage induced by this gyration. The period of gyration is determined electronically with sufficient accuracy to yield a sensitivity between 0.1 and 1.0 nanotesla. Schmidt vertical-field balance: Another type of magnetometer that measures the relative magnetic field by observing the torque produced by the earth's magnetic field on the instrument.

Using this instrument involves setting up observation stations along the region of interest approximately a half mile apart. "The Schmidt vertical-field balance, a relative magnetometer used in geophysical exploration, uses a horizontally balanced bar magnet equipped with mirror and knife edges.Field procedure consisted of observation stations located at 0.5 mile intervals for reconnaissance surveys and 0.025 mile intervals for detail surveys.

Evaluation

The field balance is a more accurate way of exploring for oil when the region of interest is shallow. "Field experiments showed that aerial magnetometers, proton precession magnetometers, and gravity surveys were not sufficiently accurate to map the small topographic lows of the Precambrian granite. Our conclusion was that only the field balance, ...could acquire the necessary data.

Gravitational

Overview: This procedure involves taking reading about a kilometer apart throughout the region with a device called a gravimeter. The gravimeter measures the gravitational field and this reading correlates with the density of the region. By studying the differences in the density, you can predict which areas of the region might contain oil "Gravity differences occur because of local density differences. Anomalies of exploration interest are often about 0.2 mgal. ...Gravity surveys on land often involve meter readings every kilometer along traverse loops a few kilometers across. ...In most cases, the density of sedimentary rocks increases with depth because the increased pressure results in a loss of porosity.

Uplifts usually bring denser rocks nearer the surface and thereby create positive gravity anomalies. Faults that displace rocks of different densities also can cause gravity anomalies. Salt domes generally produce negative anomalies because salt is less dense than the surrounding rocks. Such folds, faults, and salt domes trap oil, and so the detection of gravity anomalies associated with them is crucial in petroleum exploration. Conventional Gravimeter Vs. Gravity Gradiometry: A gravity gradiometer is another type of gravimeter that can give more information about the gravitational field of a region.

This information can give more accuracy in oil exploration but interpretation of it is underdeveloped and is usually done using techniques for interpreting conventional gravimeter data. "The conventional gravimeter measures a single component (the vertical component) of the gravity field vector. In contrast, a gravity gradiometer can measure up to five of the nine terms in the gravity field's gradient tensor which completely describes the anomalous gravity field gradient.

...Note how the two gradiometer measurements better emphasize the structural highs and lows as well as the bounding fault zones. ... interpretation of gravity gradiometry data is presently immature in practice and application. ...However, many existing gravity and magnetic interpretation algorithms are easily and naturally adapted to the interpretation of gravity gradiometer data. Conclusion: Each of these methods is unique and would provide valuable information in the exploration stage. Based on my research, it is my recommendation that we utilize each of these techniques, in addition to seismic technology, in order to get a comprehensive view of the region. I believe that it is crucial to have as much information as we can from exploration so that we can reduce the impact of drilling needlessly.

NATURALLSY OCCURRING POLYMERS

There are a number of naturally occurring polymers which find technical application, including cellulose and its derivatives, starch, and rubber. In addition, a number of important biological materials, most notably the proteins, are made up of macromolecules. These will be considered briefly in the sections which follow.

Cellulose

This is a very widely available polymer, since it is the main component of the cell walls of all plants. It is a carbohydrate of molecular formula $(C_6H_{10}0_5)_n$, where n runs to thousands. The cellulose 'monomer' is D-glucose, and the cellulose molecules are built up from this substance, effectively by condensation and removal of the elements of water.D-Glucose itself is highly soluble in water, but cellulose is not.

This is essentially a kinetic phenomenon; the hydroxy-groups in cellulose would, in principle, readily form hydrogen bonds with water molecules, and hence the cellulose macromolecule would be carried off into aqueous solution. But these hydroxy-groups interact with neighbouring cellulose molecules, making it impossible for water molecules to penetrate, or solvate, the individual molecules.

Cellulose will, however, dissolve in the somewhat strange medium aqueous ammoniacal cupric hydroxide, $Cu(NH_3)_4(OH)_2$.Cellulose is a linear polymer. Despite this, it is not thermoplastic, essentially because of its extensive intermolecular hydrogen bonding which never allows the molecules to move sufficiently for the polymer to melt.

Cellulose may be solubilised by treatment with sodium hydroxide and carbon disulfide. It can be regenerated by acidification of the solution. This is the basis of the production of regenerated cellulose fibre, so-called 'viscose rayon', which is a major textile fibre.

The technique is also used for the production of continuous cellulose-derived film, so-called 'cellophane' (from 'cellulose' and 'diaphane', the latter being French for transparent).Cellulose is also commercially modified by acetylation to produce a material suitable for X-ray and cine film. Commercially cellulose ethers are also prepared, such as methylcellulose.

This material is water-soluble and gives a highly viscous solution at very low concentrations. Hence it is widely used as a thickener in latex paints and adhesives, in cosmetics and for coating pharmaceutical tablets.

Starch

Starch is a widely distributed material which occurs in roots, seeds, and fruits of plants. For commercial use, corn is the principal source, though wheat and potatoes are also used. Starch is extracted by grinding with water, filtering, centrifuging, and drying, a process which yields starch in a granular form.

Starch consists of two components, amylose and amylopectin. The ratio of these varies with the source of the starch, with amylopectin usually predominating and representing some 70-85per cent of the total mass of the starch. Amylopectin is the polymeric component of starch and consists mainly of glucose units joined at the 1,4-positions. Relative molar mass tends to be very high, e.g. between 7 and 70 million.

A variety of modified starches are used commercially which are produced by derivatisation to give materials such as ethanoates (acetates), phosphates, and hydroxyalkyl ethers. Modified and unmodified starches are used in approximately equal tonnages, mainly in paper-making, paper coatings, paper adhesives, textile sizes, and food thickeners.

Natural Rubber

Rubber is obtained from the juice of various tropical trees, mainly the tree Hevea brasiliensis. The juice is a latex consisting of a dispersion of polymer phase at a concentration of about 35per cent by mass, together with traces of pro-teins, sterols, fats, and salts. The rubber is obtained either by coagulation of the latex with acid, either ethanoic or methanoic, or by evaporation in air or over a flame.

The material that results from this process is a crumbly, cheese-like substance, sometimes called raw rubber or caoutchouc. In order to develop the mechanical properties that are considered characteristic of rubber, i.e. so-called rubberlike elasticity, this raw rubber needs further processing, and in particular lightly crosslinking. This is achieved in the process known as vulcanisation, as will be discussed later.

The polymer in natural rubber consists almost entirely of cis-poly(isoprene). The molecules are linear, with relative molar mass typically lying between 300 000 and 500 000. The macromolecular nature of rubber was established mainly by Staudinger in 1922, when he hydrogenated the material and obtained a product that retained its colloidal character, rather than yielding fragments of low relative molar mass.

Vulcanisation is the term used for the process in which the rubber molecules are lightly crosslinked in order to reduce plasticity and develop elasticity.

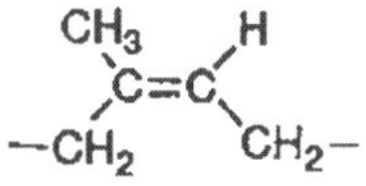

$$\begin{array}{ccc} CH_3 & & H \\ & \diagdown C{=}C \diagup & \\ -CH_2 \diagup & & \diagdown CH_2- \end{array}$$

It was originally applied to the use of sulfur for this purpose, but is now used for any similar process of cross-linking. Sulfur, though, remains the substance most widely used for this purpose.Sulfur reacts very slowly with rubber, and so is compounded with rubber in the presence of accelerators and activators. Typical accelerators are thia-zoles and a typical activator is a mixture of zinc oxide and a fatty acid. The chemistry of the vulcanisation reactions is complicated, but generates a three-dimensional network in which rubber molecules are connected by short chains of sulfur atoms, with an average of about five atoms in each chain.A much more heavily crosslinked material can be obtained by increasing the amount of sulfur in the mixture, so that it represents about a third of the mass of the product. Heating such a mixture of raw rubber and sulfur at 150 °C until reaction is complete gives a hard, thermoset material that is not at all elastic. This material is called ebonite and is used to make car battery cases.

Proteins

The proteins are a group of macromolecular substances of great importance in biochemistry. Their very name provides testimony to this - it was coined by Mulder in 1838 from the Greek word 'proteios', meaning 'of first impor-tance'. They appear in all cells, both animal and plant, and are involved in all cell functions. Proteins are linear polyamides formed from ?-amino acids.

An ?-amino acid is one in which carboxylic acid and the amino group reside on the same carbon atom.In nature, there are 20 amino acids available for incorporation into the protein chain. They are arranged in a specific and characteristic sequence along the molecule. This sequence is generally referred to as the 'primary structure' of the protein. Also part of the primary structure is the relative molar mass of the macromolecule.

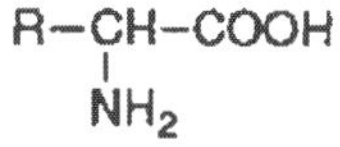

$$\begin{array}{c} R-CH-COOH \\ | \\ NH_2 \end{array}$$

As a result of the particular amino acid sequence, the protein molecule adopts a characteristic arrangement, such as symmetrical coils or orderly foldings. This is known as the 'secondary structure'. Such individual coils or folded structures bring about a longer range stable arrangement called the 'tertiary structure'. Lastly, several protein molecules may join together to form a complex, this final grouping being known as the 'quaternary structure'.

Proteins themselves consist of large molecules with several hundred amino acids in the primary structure. Smaller units, known as polypeptides, may be formed during physiological processes.

There is no clear distinction between a protein and a polypeptide; about 200 amino acid units in the primary structure is usually taken to be the minimum for a substance to be considered a protein.The diversity in primary, secondary, tertiary, and quaternary structures of proteins means that few generalisations can be made concerning their chemical properties.

Some fulfil structural roles, such as the collagens (found in bone) and keratin (found in claws and beaks), and are insoluble in all solvents. Others, such as albumins or globulins of plasma, are very soluble in water. Still others, which form part of membranes of cells, are partly hydrophilic ('water-loving', hence water-soluble) and partly lipophilic ('lipid-loving', hence fat-soluble).

Because proteins are not composed of identical repeating units, but of different amino acids, they are nevertheless macromolecular and techniques developed for the study of true polymers have been applied to them with success.

Poly(3-hydroxybutyrate)

Poly(3-hydroxybutyrate) is a bacterial polyester that behaves as an acceptable thermoplastic, yet can be produced from renewable agricultural feedstocks and is biodegradable. It is typically produced not in the pure state,but formed alongside minor amounts of poly(3-hydroxyvalerate). The ratio of these two polymers in a given sample is determined by the ratio of glucose and propionic acid in the medium in which the bacteria live and carry out their metabolic processes.

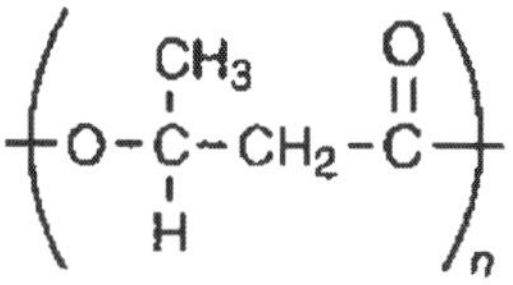

$$\left(-O-\underset{H}{\overset{CH_3}{C}}-CH_2-\overset{O}{\overset{\|}{C}}- \right)_n$$

The carbon atom which carries the methyl group is chiral, but biosynthesis is stereoselective, and gives rise to a natural polymer with the R configuration. The polymer is a partially crystalline thermoplastic which melts at about 80 °C. Poly(3-hydroxybutyrate) has attracted interest as an environmentally degradable thermoplastic that can be used in packaging, agriculture and medicine. It undergoes enzymic degradation quite readily, becoming less crystalline and rapidly decreasing in molar mass. The rate of degradation depends on a number of factors, such as moisture level, nutrient supply, temperature, and pH.

Polymer Solution

The importance assigned to polymer solutions, a topic whose discussion has evolved from a mere informative mention in textbooks to whole books exclusively devoted to that subject, has become increasingly notorious. The reasons are based on key factors. In the first place, the understanding of the behaviour and both physical and chemical properties of macromolecules has been mainly sustained in studies carried out in solution, like for example the determination of the relative molecular mass, made by viscometry or gel permeation chromatography (GPC). On the other hand, since polymer solutions are highly viscous even at low concentrations, their commercial application includes a wide range of products, from paintings to processed foods.Therefore, we can then consider polymer solutions as *liquid mixtures made of long macromolecular chains, and small, light molecules of solvent* (Grosberg and Khokhlov, 1997). This, by the way, is not a usual situation. The large size of this chains implies the employ of certain theoretical models, which should take into account, among other things, the numerous and diverse conformations that these flexible structures may assume.

This particularity is not consistent with a behaviour that could be regarded as an "ideal" behaviour. In addition to these features, it is easily understandable that the studies performed in the evaluation of the physical-chemical properties of macromolecules are focused on dilute solutions, where the chains are separated by long distances, and therefore the interaction between them is reduced to a minimum.

This is not taken for the sake of simplicity, but also because the properties of dilute solutions are governed by the properties of the individual macromolecules. In the case of concentrated solutions, the chains are entangled each other, their interaction increases, and in such conditions, the system is no longer suitable to evaluate the contribution of each macromolecule in particular.

CATALYTIC HYDROTREATING

DESCRIPTION

Catalytic hydrotreating is a hydrogenation process used to remove about 90per cent of contaminants such as nitrogen, sulfur, oxygen, and metals from liquid petroleum fractions. These contaminants, if not removed from the petroleum fractions as they travel through the refinery processing units, can have detrimental effects on the equipment, the catalysts, and the quality of the finished product. Typically, hydrotreating is done prior to processes such as catalytic reforming so that the catalyst is not contaminated by untreated feedstock.

Hydrotreating is also used prior to catalytic cracking to reduce sulfur and improve product yields, and to upgrade middle-distillate petroleum fractions into finished kerosene, diesel fuel, and heating fuel oils. In addition, hydrotreating converts olefins and aromatics to saturated compounds.

Catalytic Hydrodesulfurization Process

Hydrotreating for sulfur removal is called hydrodes-ulfurization. In a typical catalytic hydrodesulfurization unit, the feedstock is deaerated and mixed with hydrogen, preheated in a fired heater (600°-800°F) and then charged under pressure (up to 1,000 psi) through a fixed-bed catalytic reactor. In the reactor, the sulfur and nitrogen compounds in the feedstock are converted into H_2S and NH_3. The reaction products leave the reactor and after cooling to a low temperature enter a liquid/gas separator. The hydrogen-rich gas from the high-pressure separation is recycled to combine with the feedstock, and the low-pressure gas stream rich in H_2S is sent to a gas treating unit where H_2S is removed. The clean gas is then suitable as fuel for the refinery furnaces. The liquid stream is the product from hydrotreating and is normally sent to a stripping column for removal of H_2S and other undesirable components. In cases where steam is used for stripping, the product is sent to a vacuum drier for removal of water. Hydrodesulfurized products are blended or used as catalytic reforming feedstock.

Other Hydrotreating Processes

- Hydrotreating processes differ depending upon the feedstock available and catalysts used. Hydrotreating can be used to improve the burning characteristics of distillates such as kerosene. Hydrotreatment of a kerosene fraction can convert aromatics into naphthenes, which are cleaner-burning compounds.
- Lube-oil hydrotreating uses catalytic treatment of the oil with hydrogen to improve product quality. The objectives in mild lube hydrotreating include saturation of olefins and improvements in colour, odour, and acid nature of the oil. Mild lube hydrotreating also may be used following solvent processing. Operating temperatures are usually below 600°F and operating pressures below 800 psi. Severe lube hydrotreating, at temperatures in the 600°-750° F range and hydrogen pressures up to 3,000 psi, is capable of saturating aromatic rings, along with sulfur and nitrogen removal, to impart specific properties not achieved at mild conditions.
- Hydrotreating also can be employed to improve the quality of pyrolysis gasoline (pygas), a by-product from the manufacture of ethylene. Traditionally, the outlet for pygas has been motor gasoline blending, a suitable route in view of its high octane number. However, only small portions can be blended

untreated owing to the unacceptable odour, colour, and gum-forming tendencies of this material. The quality of pygas, which is high in diolefin content, can be satisfactorily improved by hydrotreating, whereby conversion of diolefins into mono-olefins provides an acceptable product for motor gas blending.

Table: Hydrodesulfurization Process

Feedstock	From	Process	Typical product	To
Naphthas, distillates sourgas oil, residuals	Atmospheric & vacuum tower, catalytic & thermal cracker	Treating, hydrogenation	Naphtha	Blending
			Hydrogen	Recycle
			Distillates	Blending
			H_2S, ammonia	Sulfure plant, treater
			Gas	Gas plant

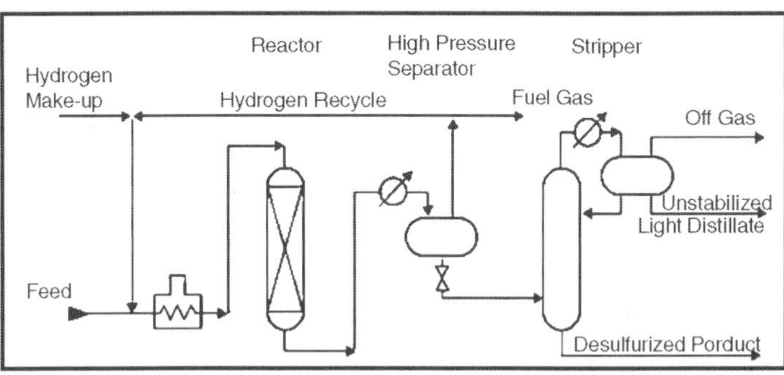

Fig. Distillate Hydrodesulfurization

Health and Safety Considerations

- *Fire Prevention and Protection:* The potential exists for fire in the event of a leak or release of product or hydrogen gas.
- *Safety:* Many processes require hydrogen generation to provide for a continuous supply. Because of the operating temperatures and presence of hydrogen, the hydrogen sulfide content of the feedstock must be strictly controlled to a minimum to reduce corrosion. Hydrogen chloride may form and condense as hydrochloric acid in the lower-temperature parts of the unit. Ammonium hydrosulfide may form in high-temperature, high-pressure units. Excessive contact time and/or temperature will create coking. Precautions need to be taken when unloading coked catalyst from the unit to prevent iron sulfide fires. The coked catalyst should be cooled to below 120°F before removal, or dumped into nitrogen-inerted bins where it can be cooled before

further handling. Special antifoam additives may be used to prevent catalyst poisoning from silicone carryover in the coker feedstock.

- *Health:* Because this is a closed process, exposures are expected to be minimal under normal operating conditions. There is a potential for exposure to hydrogen sulfide or hydrogen gas in the event of a release, or to ammonia should a sour-water leak or spill occur. Phenol also may be present if high boiling-point feedstocks are processed. Safe work practices and/or appropriate personal protective equipment may be needed for exposures to chemicals and other hazards such as noise and heat; during process sampling, inspection, maintenance, and turnaround activities; and when handling amine or exposed to catalyst.

ISOMERIZATION

- Isomerization converts n-butane, n-pentane and n-hexane into their respective isoparaffins of substantially higher octane number. The straight-chain paraffins are converted to their branched-chain counterparts whose component atoms are the same but are arranged in a different geometric structure. Isomerization is important for the conversion of n-butane into isobutane, to provide additional feedstock for alkylation units, and the conversion of normal pentanes and hexanes into higher branched isomers for gasoline blending. Isomerization is similar to catalytic reforming in that the hydrocarbon molecules are rearranged, but unlike catalytic reforming, isomerization just converts normal paraffins to isoparaffins.
- There are two distinct isomerization processes, butane (C_4) and pentane/hexane (C_5/C_6). Butane isomerization produces feedstock for alkylation. Aluminum chloride catalyst plus hydrogen chloride are universally used for the low-temperature processes. Platinum or another metal catalyst is used for the higher-temperature processes. In a typical low-temperature process, the feed to the isomerization plant is n-butane or mixed butanes mixed with hydrogen (to inhibit olefin formation) and passed to the reactor at 230°-340°F and 200-300 psi. Hydrogen is flashed off in a high-pressure separator and the hydrogen chloride removed in a stripper column. The resultant butane mixture is sent to a fractionator (deisobutanizer) to separate n-butane from the isobutane product.
- Pentane/hexane isomerization increases the octane number of the light gasoline components n-pentane and n-hexane, which are found in abundance in straight-run gasoline. In a typical C_5/C_6 isomerization process, dried and desulfurized feedstock is mixed with a small amount of organic chloride and recycled hydrogen, and

then heated to reactor temperature. It is then passed over supported-metal catalyst in the first reactor where benzene and olefins are hydrogenated. The feed next goes to the isomerization reactor where the paraffins are catalytically isomerized to isoparaffins. The reactor effluent is then cooled and subsequently separated in the product separator into two streams: a liquid product (isomerate) and a recycle hydrogen-gas stream. The isomerate is washed (caustic and water), acid stripped, and stabilized before going to storage.

Table. Isomerization Processes

Feedstock	From	Process	Typical	To products
n-Butane	Various	Rearrangement Processes	Isobutane	Alkylation
n-Pentane			Isopentane	Blending
n-Hexane			Isohexane	Blending
			Gas	Gas Plant

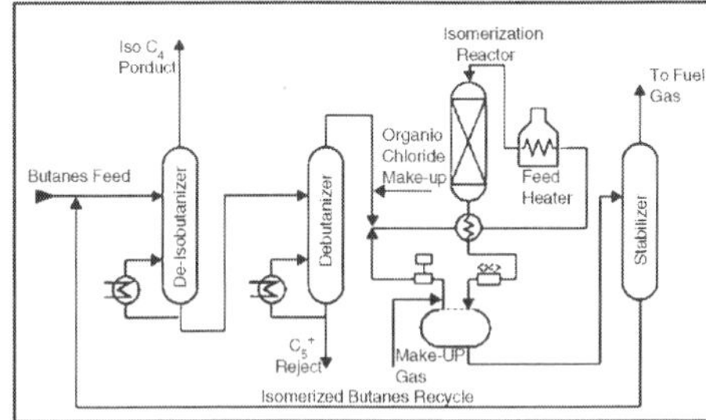

Fig. Isomerization.

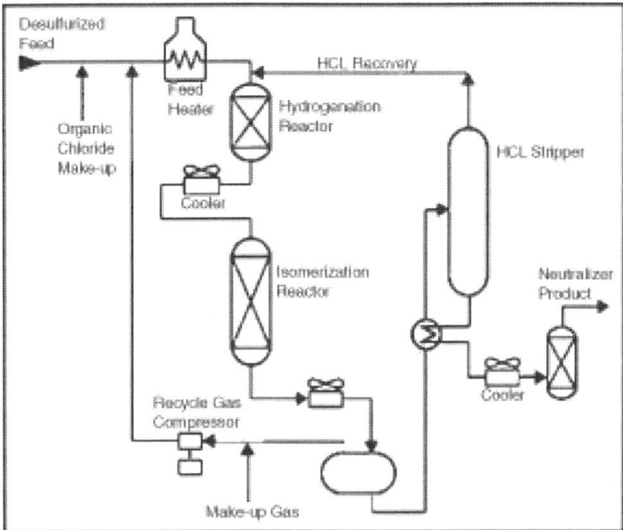

Fig. C_5 and C_6 Isomerization.

Health and Safety Considerations

- *Fire Protection and Prevention:* Although this is a closed process, the potential for a fire exists should a release or leak contact a source of ignition such as the heater.
- *Safety:* If the feedstock is not completely dried and desulfurized, the potential exists for acid formation leading to catalyst poisoning and metal corrosion. Water or steam must not be allowed to enter areas where hydrogen chloride is present. Precautions are needed to prevent HCl from entering sewers and drains.
- *Health:* Because this is a closed process, exposures are expected to be minimal during normal operating conditions. There is a potential for exposure to hydrogen gas, hydrochloric acid, and hydrogen chloride and to dust when solid catalyst is used. Safe work practices and/or appropriate personal protective equipment may be needed for exposures to chemicals and other hazards such as heat and noise, and during process sampling, inspection, maintenance, and turnaround activities.

THERMAL CRACKING PROCESSES

With the advent of mass production and World War I, the number of gasoline-powered vehicles increased dramatically and the demand for gasoline grew accordingly. However, distillation processes produced only a certain amount of gasoline from crude oil. In 1913, the thermal cracking process was developed, which subjected heavy fuels to both pressure and intense heat, physically breaking the large molecules into smaller ones to produce additional gasoline and distillate fuels. Visbreaking, another form of thermal cracking, was developed in the late 1930s to produce more desirable and valuable products.

CATALYTIC PROCESSES

Higher-compression gasoline engines required higher-octane gasoline with better antiknock characteristics. The introduction of catalytic cracking and polymerization processes in the mid- to late 1930s met the demand by providing improved gasoline yields and higher octane numbers.

Alkylation, another catalytic process developed in the early 1940s, produced more high-octane aviation gasoline and petrochemical feedstocks for explosives and synthetic rubber. Subsequently, catalytic isomerization was developed to convert hydrocarbons to produce increased quantities of alkylation feedstocks. Improved catalysts and process methods such as hydrocracking and reforming were developed throughout the 1960s to increase gasoline yields and improve antiknock characteristics. These catalytic processes also produced hydrocarbon molecules with a double bond (alkenes) and formed the basis of the modern petrochemical industry.

TREATMENT PROCESSES

Throughout the history of refining, various treatment methods have been used to remove non-hydrocarbons, impurities, and other constituents that adversely affect the properties of finished products or reduce the efficiency of the conversion processes. Treating can involve chemical reaction and/or physical separation. Typical examples of treating are chemical sweetening, acid treating, clay contacting, caustic washing, hydrotreating, drying, solvent extraction, and solvent dewaxing. Sweetening compounds and acids desulfurize crude oil before processing and treat products during and after processing.

Following the Second World War, various reforming processes improved gasoline quality and yield and produced higher-quality products. Some of these involved the use of catalysts and/or hydrogen to change molecules and remove sulfur. A number of the more commonly used treating and reforming processes are described.

HISTORY OF REFINING

Year Process name Purpose By-products, etc. 1862 Atmospheric distillation Produce kerosene Naphtha, tar, etc. 1870 Vacuum distillation Lubricants (original) Asphalt, residual Cracking feedstocks coker feedstocks. Thermal cracking Increase gasoline Residual, bunker fuel 1916 Sweetening Reduce sulfur andamp; odour Sulfur 1930 Thermal reforming Improve octane number Residual 1932 Hydrogenation Remove sulfur Sulfur 1932 Coking Produce gasoline Coke basestocks 1933 Solvent extraction Improve lubricant Aromatics viscosity index 1935 Solvent dewaxing Improve pour point Waxes Cat.

Polymerization Improve gasoline Petrochemical yield; octane feedstocks number 1937 Catalytic cracking Higher octane Petrochemical gasoline feedstocks 1939 Visbreaking Reduce viscosity Increased distillate, tar 1940 Alkylation Increase gasoline High-octane aviation octan; yield gasoline 1940 Isomerization Produce alkylation Naphtha feedstock 1942 Fluid catalytic Increase gasoline Petrochemical cracking yield andamp; octane feedstocks 1950 Deasphalting Increase cracking Asphalt feedstock 1952 Catalytic reforming Convert low-quality Aromatics naphtha 1954.

Hydrodesulfurization Remove sulfur Sulfur 1956 Inhibitor sweetening Remove mercaptan Disulfides 1957 Catalytic Convert to molecules Alkylation isomerization with high octane feedstocks number 1960 Hydrocracking Improve quality and Alkylation reduce sulfur feedstocks 1974 Catalytic dewaxing Improve pour point Wax 1975 Residual Increase gasoline Heavy residuals hydrocracking yield from residual.

BASICS OF CRUDE OIL

Crude oils are complex mixtures containing many different hydrocarbon compounds that vary in appearance and composition from one oil field to another.

Crude oils range in consistency from water to tar-like solids, and in colour from clear to black. An average crude oil contains about 84per cent carbon, 14per cent hydrogen, 1-3per cent sulfur, and less than 1per cent each of nitrogen, oxygen, metals, and salts. Crude oils are generally classified as paraffinic, naphthenic, or aromatic, based on the predominant proportion of similar hydrocarbon molecules. Mixed-base crudes have varying amounts of each type of hydrocarbon. Refinery crude base stocks usually consist of mixtures of two or more different crude oils.

Relatively simple crude-oil assays are used to classify crude oils as paraffinic, naphthenic, aromatic, or mixed. One assay method is based on distillation, and another method (UOP K factor) is based on gravity and boiling points. More comprehensive crude assays determine the value of the crude (*i.e.*, its yield and quality of useful products) and processing parameters. Crude oils are usually grouped according to yield structure.

Crude oils are also defined in terms of API. The higher the API gravity, the lighter the crude. For example, light crude oils have high API gravities and low specific gravities. Crude oils with low carbon, high hydrogen, and high API gravity are usually rich in paraffins and tend to yield greater proportions of gasoline and light petroleum products; those with high carbon, low hydrogen, and low API gravities are usually rich in aromatics.

Crude oils that contain appreciable quantities of hydrogen sulfide or other reactive sulfur compounds are called sour. Those with less sulfur are called sweet. Some exceptions to this rule are West Texas crudes, which are always considered sour regardless of their H_2S content, and Arabian high-sulfur crudes, which are not considered sour because their sulfur compounds are not highly reactive.

BASICS OF HYDROCARBON CHEMISTRY

Crude oil is a mixture of hydrocarbon molecules, which are organic compounds of carbon and hydrogen atoms that may include from one to 60 carbon atoms. The properties of hydrocarbons depend on the number and arrangement of the carbon and hydrogen atoms in the molecules. The simplest hydrocarbon molecule is one carbon atom linked with four hydrogen atoms: methane. All other variations of petroleum hydrocarbons evolve from this molecule.

Hydrocarbons containing up to four carbon atoms are usually gases; those with five to 19 carbon atoms are usually liquids; and those with 20 or more are solids. The refining process uses chemicals, catalysts, heat, and pressure to separate and combine the basic types of hydrocarbon molecules naturally found in crude oil into groups of similar molecules. The refining process also rearranges their structures and bonding patterns into different hydrocarbon molecules and compounds. Therefore it is the type of hydrocarbon, (paraffinic, naphthenic, or aromatic) rather than its specific chemical compounds that is significant in the refining process.

POLYMERIZATION

- Polymerization in the petroleum industry is the process of converting light olefin gases including ethylene, propylene, and butylene into hydrocarbons of higher molecular weight and higher octane number that can be used as gasoline blending stocks. Polymerization combines two or more identical olefin molecules to form a single molecule with the same elements in the same proportions as the original molecules. Polymerization may be accomplished thermally or in the presence of a catalyst at lower temperatures.
- The olefin feedstock is pretreated to remove sulfur and other undesirable compounds. In the catalytic process the feedstock is either passed over a solid phosphoric acid catalyst or comes in contact with liquid phosphoric acid, where an exothermic polymeric reaction occurs. This reaction requires cooling water and the injection of cold feedstock into the reactor to control temperatures between 300° and 450°F at pressures from 200 psi to 1,200 psi. The reaction products leaving the reactor are sent to stabilization and/or fractionator systems to separate saturated and unreacted gases from the polymer gasoline product.

NOTE: In the petroleum industry, polymerization is used to indicate the production of gasoline components, hence the term "polymer" gasoline. Furthermore, it is not essential that only one type of monomer be involved. If unlike olefin molecules are combined, the process is referred to as "copolymerization."

Polymerization in the true sense of the word is normally prevented, and all attempts are made to terminate the reaction at the dimer or trimer (three monomers joined together) stage. However, in the petrochemical section of a refinery, polymerization, which results in the production of, for instance, polyethylene, is allowed to proceed until materials of the required high molecular weight have been produced.

HEALTH AND SAFETY CONSIDERATIONS

- *Fire Prevention and Protection:* Polymerization is a closed process where the potential for a fire exists due to leaks or releases reaching a source of ignition.
- *Safety:* The potential for an uncontrolled exothermic reaction exists should loss of cooling water occur. Severe corrosion leading to equipment failure will occur should water make contact with the phosphoric acid, such as during water washing at shutdowns.

Corrosion may also occur in piping manifolds, reboilers, exchangers, and other locations where acid may settle out.

- *Health:* Because this is a closed system, exposures are expected to be minimal under normal operating conditions. There is a potential for exposure to caustic wash (sodium hydroxide), to phosphoric acid used in the process or washed out during turnarounds, and to catalyst dust. Safe work practices and/or appropriate personal protective equipment may be needed for exposures to chemicals and other hazards such as noise and heat, and during process sampling, inspection, maintenance, and turnaround activities.

Table: Polymerization Process

Feedstock	From	Process	Typical products	To
Olefins	Cracking processe	Unification	High octane naphtha	Gasoline blending
			Petrochem. feedstock	Petrochemical
			Liquefied petro. gas	Storage

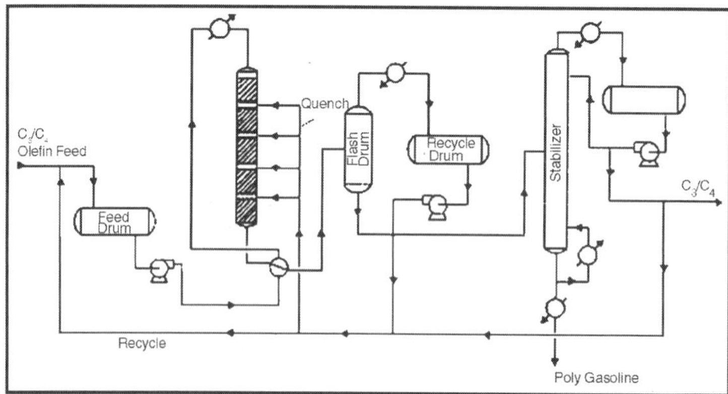

Fig. Polymerization Process.

ALKYLATION

Alkylation combines low-molecular-weight olefins (primarily a mixture of propylene and butylene) with isobutene in the presence of a catalyst, either sulfuric acid or hydrofluoric acid.

The product is called alkylate and is composed of a mixture of high-octane, branched-chain paraffinic hydrocarbons. Alkylate is a premium blending stock because it has exceptional antiknock properties and is clean burning. The octane number of the alkylate depends mainly upon the kind of olefins used and upon operating conditions.

Sulfuric Acid Alkylation Process

- In cascade type sulfuric acid (H_2SO_4) alkylation units, the feedstock (propylene, butylene, amylene, and fresh isobutane) enters the reactor and contacts the concentrated sulfuric acid catalyst (in concentrations of 85per cent to 95per cent for good operation and to minimize corrosion). The reactor is divided into zones, with olefins fed through distributors to each zone, and the sulfuric acid and isobutanes flowing over baffles from zone to zone.
- The reactor effluent is separated into hydrocarbon and acid phases in a settler, and the acid is returned to the reactor. The hydrocarbon phase is hot-water washed with caustic for pH control before being successively depropanized, deisobutanized, and debutanized. The alkylate obtained from the deisobutanizer can then go directly to motor-fuel blending or be rerun to produce aviation-grade blending stock. The isobutane is recycled to the feed.

Hydrofluoric Acid Alylation Process

Phillips and UOP are the two common types of hydrofluoric acid alkylation processes in use. In the Phillips process, olefin and isobutane feedstock are dried and fed to a combination reactor/settler system. Upon leaving the reaction zone, the reactor effluent flows to a settler (separating vessel) where the acid separates from the hydrocarbons. The acid layer at the bottom of the separating vessel is recycled.

The top layer of hydrocarbons (hydrocarbon phase), consisting of propane, normal butane, alkylate, and excess (recycle) isobutane, is charged to the main fractionator, the bottom product of which is motor alkylate. The main fractionator overhead, consisting mainly of propane, isobutane, and HF, goes to a depropanizer.

Propane with trace amount of HF goes to an HF stripper for HF removal and is then catalytically defluorinated, treated, and sent to storage.

Isobutane is withdrawn from the main fractionator and recycled to the reactor/settler, and alkylate from the bottom of the main fractionator is sent to product blending. The UOP process uses two reactors with separate settlers. Half of the dried feedstock is charged to the first reactor, along with recycle and makeup isobutane.

The reactor effluent then goes to its settler, where the acid is recycled and the hydrocarbon charged to the second reactor. The other half of the feedstock also goes to the second reactor, with the settler acid being recycled and the hydrocarbons charged to the main fractionator. Subsequent processing is similar to the Phillips process. Overhead from the main fractionator goes to a depropanizer. Isobutane is recycled to the reaction zone and alkylate is sent to product blending.

Table. Alkylation Process

Feedstock	From	Process	Typical products	To
Petroleum gas	Distillation or cracking	Unification	High octane gasoline	Blending
Olefins	Cat. or hydro cracking		n-Butane & propane	Stripper or blender
Isobutane	Isomerization			

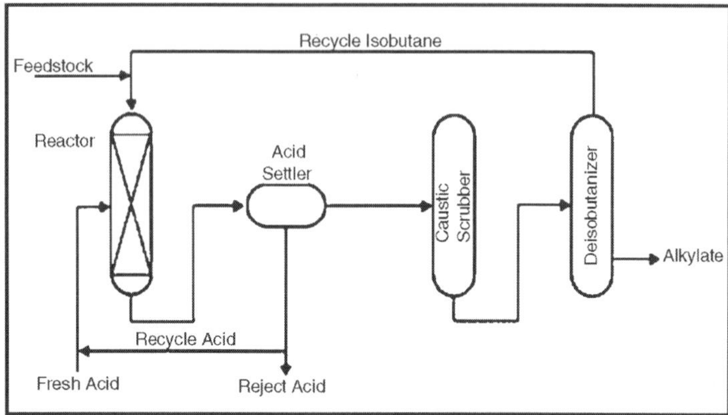

Fig. Sulfuric Acid Alkylation.

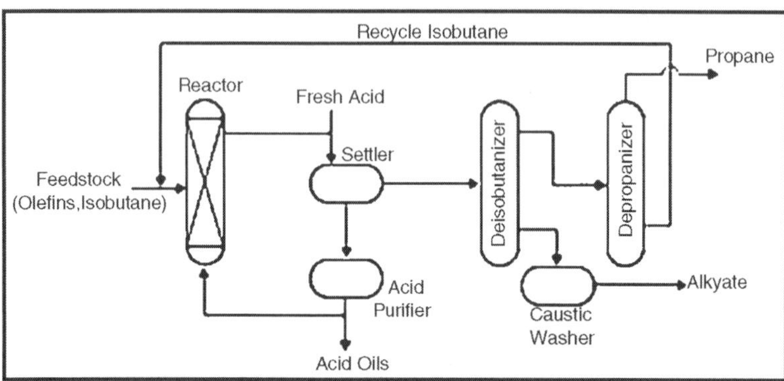

Fig. Hydrogen Fluoride Alkylation.

Health and Safety Considerations

- *Fire Protection and Prevention:* Alkylation units are closed processes; however, the potential exists for fire should a leak or release occur that allows product or vapour to reach a source of ignition.
- *Safety:* Sulfuric acid and hydrofluoric acid are potentially hazardous chemicals. Loss of coolant water, which is needed to maintain process temperatures, could result in an upset. Precautions are necessary to ensure that equipment and materials that have been in contact with

acid are handled carefully and are thoroughly cleaned before they leave the process area or refinery. Immersion wash vats are often provided for neutralization of equipment that has come into contact with hydrofluoric acid.

- Hydrofluoric acid units should be thoroughly drained and chemically cleaned prior to turnarounds and entry to remove all traces of iron fluoride and hydrofluoric acid. Following shutdown, where water has been used the unit should be thoroughly dried before hydrofluoric acid is introduced.
- Leaks, spills, or releases involving hydrofluoric acid or hydrocarbons containing hydrofluoric acid can be extremely hazardous. Care during delivery and unloading of acid is essential. Process unit containment by curbs, drainage, and isolation so that effluent can be neutralized before release to the sewer system is considered. Vents can be routed to soda-ash scrubbers to neutralize hydrogen fluoride gas or hydrofluoric acid vapors before release. Pressure on the cooling water and steam side of exchangers should be kept below the minimum pressure on the acid service side to prevent water contamination.
- Some corrosion and fouling in sulfuric acid units may occur from the breakdown of sulfuric acid esters or where caustic is added for neutralization. These esters can be removed by fresh acid treating and hot-water washing. To prevent corrosion from hydrofluoric acid, the acid concentration inside the process unit should be maintained above 65per cent and moisture below 4per cent.
- *Health:* Because this is a closed process, exposures are expected to be minimal during normal operations. There is a potential for exposure should leaks, spills, or releases occur. Sulfuric acid and (particularly) hydrofluoric acid are potentially hazardous chemicals. Special precautionary emergency preparedness measures and protection appropriate to the potential hazard and areas possibly affected need to be provided.
- Safe work practices and appropriate skin and respiratory personal protective equipment are needed for potential exposures to hydrofluoric and sulfuric acids during normal operations such as reading gauges, inspecting, and process sampling, as well as during emergency response, maintenance, and turnaround activities. Procedures should be in place to ensure that protective equipment and clothing worn in hydrofluoric acid activities are decontaminated and inspected before reissue.
- Appropriate personal protection for exposure to heat and noise also may be required.

THERMOCHEMICAL LIQUEFACTION OF MICROALGAE

LIQUID FUELS FROM MICROALGAL BIOMASS

It is well known that microalgae can assimilate CO_2 gas as a carbon source for growth. However, if the resulting cell mass is not suitably treated, CO_2 will be evolved and diluted into the environment by decomposition, thus preventing CO_2 fixation from contributing to a reduction in atmospheric CO_2.

Petroleum is widely believed to have its origins in kerogen, which is easily converted to an oily substance under conditions of high pressure and temperature.

Kerogen is formed from algae, biodegraded organic compounds, plankton, bacteria, plant material, etc., by biochemical and/or chemical reactions such as diagenesis and catagenesis. Several studies have been conducted to simulate petroleum formation by pyrolysis, some of which used the marine alga *Fucus* sp.as the base material.

Recently, activated sludge and fungi were converted to oily substances at relatively low temperatures as compared with those used in previous experimental simulations. On the basis of these findings, it is assumed that algae grown in CO_2-enriched air can be converted to oily substances, and that such an approach can contribute to solving two major problems: air pollution resulting from CO_2 evolution, and future crises due to a shortage of energy sources.

Use of thermochemical liquefaction of organisms in the production of alternative fuels, would reduce CO_2 evolution into the atmosphere since such fuels would indeed be produced from CO_2.

Apart from the experimental simulation discussed above, other work has also been conducted with the objective of producing fuel from microalgae. Feinberg reported that diesel fuel and gasoline were produced through the transesterification and catalytic cracking of lipids accumulated in algal cells. However, the raw material utilized in their work was restricted to microalgae of high lipid content. A process for the production of fuel oil from microalgae by pyrolysis has been proposed. The pyrolysis usually requires a drying procedure in which large amounts of energy are required to vaporize water. An alternative technique involving the direct thermochemical liquefaction of biomass of high moisture content, such as wood and sewage sludge, has been proposed and applied to the production of fuel oils from microalgae.

This liquefaction is carried out in an aqueous solution of either alkali or NaCl at a temperature of about 300°C and pressure of 10 MPa in the absence of reducing gases such as hydrogen and/or carbon monoxide. Since drying is not required, energy consumption for water vaporization is avoided. Microalgal cell precipitates derived from centrifugation, which are of a high moisture content, are thus good raw materials for liquefaction.

Cultivation Of Microalgae

The cultivation and liquefaction of microalgae has been recently investigated. The green microalga *Dunaliella tertiolecta* ATCC 30929 was cultured in a medium of the following composition: NaCl 58.5 g; $MgCl_2 \bullet 6H_2O$ 1.5 g; KNO_3 1.0 g; $MgSO_4 \bullet 7H_2O$ 0.5 g; KCl 0.2 g; $CaCl_2 \bullet 2H_2O$ 0.2 g; $NaHCO_3$ 43 mg; KH_2PO_4, 40.8 mg; K_2HPO_4 0.495 g; $FeCl_3$ solution 1.0 ml; metal solution 1.0 ml.

The $FeCl_3$ solution consisted of the following (per liter): $FeCl_3$ 0.03 g; EDTA •2Na 5.84 g. The metal solution consisted of the following (per liter): H_3BO_4 0.61 g; $MnCl_2 \bullet 4H_2O$ 23 mg; $ZnSO_4 \bullet 7H_2O$ 87 mg; $CuSO_4 \bullet 7H_2O$ 0.06 g; $(NH_4)_6Mo_7O_{24} \bullet 4H_2O$ 21 mg; $CoCl_2 \bullet 5H_2O$ 15 mg; EDTA •2Na 1.89 g. The main culture was conducted in 3 L of medium at 27°C, using a 5-L fermenter. The aeration rate and speed of agitation were maintained constant at 1 vvm and 200 rpm, respectively. Sparging gas was produced by mixing CO_2 with air at a fixed ratio of between 0 and 0.1. Four fluorescent lamps were used for external irradiation of the fermenter; the degree of irradiation at the inner surface of the fermenter vessel was 10,000 1x.

Algae were also cultured in 20-L box-type water tanks (366 × 216 × 250 mm) containing 12 L of medium as illustrated in Fig. 6-4. The medium was not sterilized prior to cultivation. The aeration rate was maintained constant at 0.25 vvm and the CO_2 content of the aeration gas was fixed at approximately 3per cent, throughout the culture experiment.

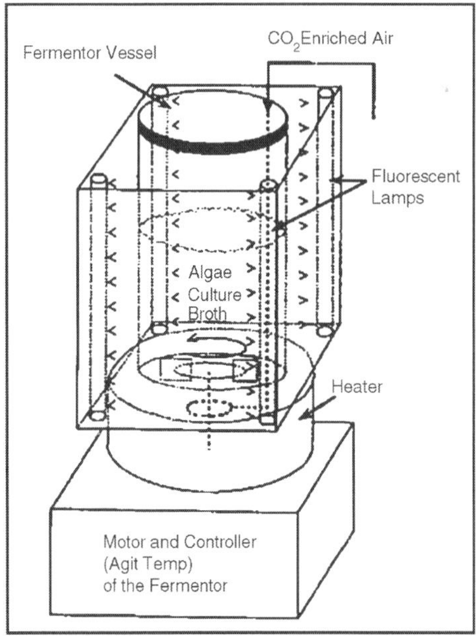

Fig. Fermentation Apparatus (5 L Fermentor) used in the Culture of Mmicroalgae.

The water tanks were placed either in a bio-photo chamber (max 20,000 Ix, type LX-2100, Taitec Co. Ltd., Tokyo), or on shelves with fluorescent lamps

(Iwasaki Co. Ltd., Tokyo), enabling the irradiation intensity to be altered, as shown in Fig. The initial pH of the culture broth was adjusted to 8.0 in the main culture. The innoculum volume of 400 ml, was equivalent to 3.3per cent of the working volume of the main culture.

The effect of NaCl concentration on growth was examined using 5-L jar fermentors. Optimal growth was obtained at an NaCl concentration of 1.0 M. In order to ascertain the effect of contamination under saline conditions, algae were also cultured at a 1M NaCl concentration in unsterilized box-type water tanks containing unsterilized medium. The vessels were not airtight, thus exposing the culture broth to contaminating microorganisms. *D. tertiolecta* grew to levels of 1.2 g/L and was not affected by the non-sterility of growth conditions. Figure shows cell growth under 5,000 and 10,000 Ix illumination in the fermenter. Growth under 10,000 Ix was almost twice that under 5,000 Ix, indicating that growth is limited by insufficient intensities of light irradiation. Light intensity is thus an important factor in scale up of the cultivation process. Cell growth within the fermenter occurred normally within a CO_2 concentration range of 3 to 10per cent. However, limited cell growth occurred at very low CO_2 concentrations (0.03per cent CO_2).

Liquefaction of Microalgae

Liquefaction was performed using a conventional stainless steel autoclave of 100-ml capacity with mechanical mixing. The autoclave was charged with algal cells (about 20 g), following which nitrogen was introduced to purge the residual air. The nitrogen pressure was then elevated to 3 MPa, in order to prevent water present from vaporizing.

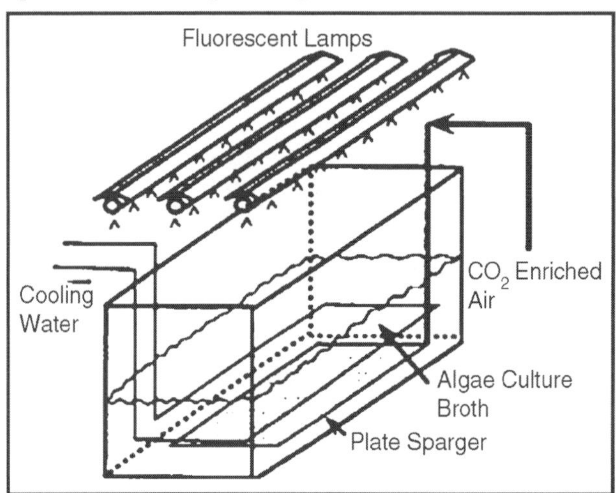

Fig. Box-type Vessel (20-L capacity) for the Culture of Microalga

The reaction was initiated by heating the autoclave to a fixed temperature, using an electric heater. The temperature of the heated autoclave was maintained constant for a 5 to 60 minute period, following which it was cooled with the use of an electric fan. The separation procedure is schematically presented in Figure. The autoclave was opened, and the reaction mixture which was removed for separation and analysis consisted of a tar-like material, floating on the surface of a water phase. This reaction mixture was extracted with dichloromethane in order to separate out the oil fraction. The dichloromethane extract was filtered from the reaction mixture, following which residual dichloromethane was evaporated at 35°C under reduced pressure, yielding a dark-brown viscous material (hereafter referred to as the oil). oduct yield in thermo-chemical liquefaction was thus calculated using the following equation:

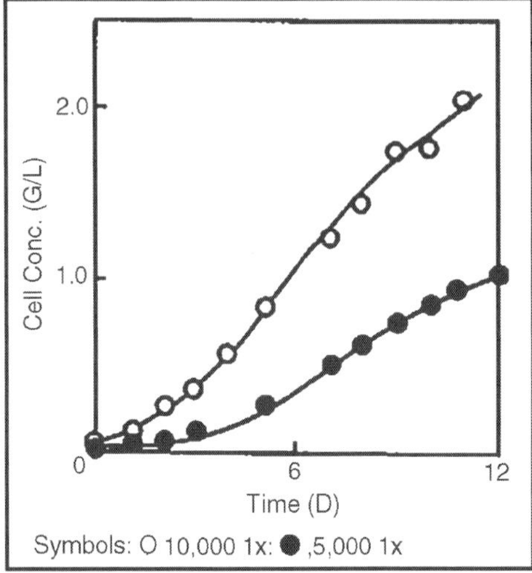

Fig. Effect of Light Intensity on Cell Growth Within a Fermentor

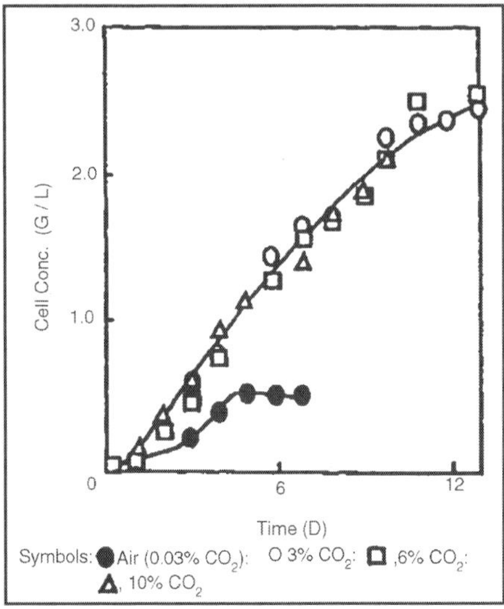

Fig. Effect of CO_2 Concentration (as a spa rged gas) on Cell Growth Within a Fermentor

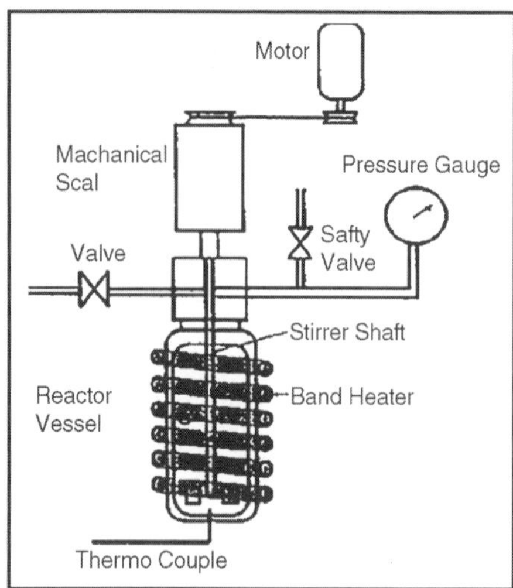

Fig. Schematic of Autoclave Equipment Used for Thermo-Chemical Liquefaction Reactions

A heavy oil yield of 35.6per cent was obtained. This heavy oil consisted of carbon (73per cent), hydrogen (9per cent), nitrogen (5per cent), and oxygen (13per cent). The heating volume of the heavy oil was 34.7 kJ/g, which is almost the same as that of C heavy oil. This heavy oil had a viscosity of 860 cps, which was similar to that of castor oil. This heavy oil was of a higher nitrogen content than ordinary petroleum, thus necessitating requirements for flue gas treatment in order to prevent the formation of NOx.

These experimental results indicate that oil can be produced from CO_2 gas using microalgae. However, two problems remain to be solved before such a system can be put to practical use. The first is the existence of a severe growth limiting factor, namely, light intensity. The effect of light intensity on growth is clearly shown in figure; light could not penetrate deeply into the culture broth, owing to its obstruction by the microalgal body. In order to scale up algal cultivation while maintaining good growth levels, an extremely powerful light source and/or large surface area for irradiation will be required. The second problem is the high nitrogen content of the heavy oil which could be a potential source of Nox synthesis if the oil were used as a fuel. These two problems need to be addressed in the scale-up of CO_2 fixation and/or oil production using microalgae.

Bibliography

Hema Gautam: *Agricultural and Industrial Applications of Biotechnology* , Rajat, 2006.

Padma S Vankar: *Handbook on Natural Dyes for Industrial Applications*, National Institute of Industrial Research, 2007.

V.R. Moorthi: *Power Electronics : Devices, Circuits, and Industrial Applications*, Oxford University Press, 2005.

Alkali and Alkaline: *A Handbook of Inorganic Chemistry : Nuclear, Atomic, Aqueous*, ISPA, 2005, .

R.K. Yadav: *A Textbook of Objective Inorganic Chemistry*, Wisdom Press, 2011.

P.N. Mukherjee: *A Textbook of Objective Organic Chemistry*, Wisdom Press, 2011.

K S Tewari and N K Vishnoi: *A Textbook Of Organic Chemistry*, Vikas Publishing House, 2006.

K Ishita; *Fritz Helmet and Morris Sylvin: Advance Inorganic Chemistry, Vol. I*, Ivy Pub, 2007.

Amit Arora: *Advance Practical Organic Chemistry*, Discovery Publishing House, 2009.

Ranpal Singh: *Advanced Inorganic Chemistry*, RBSA Pub, 2006.

D.P. Sharma: *Advanced Inorganic Chemistry*, Pearl Books, 2010.

Arun Rastogi: *Advanced Inorganic Chemistry*, Anmol, 2010.

R K Acharya: Advanced Inorganic Chemistry, Campus Books International, 2009.

Wilkinson, Murillo and Bochmann: *Advanced Inorganic Chemistry*, Anmol, 2012.

Sujata Pattanaik: Advanced Inorganic Chemistry, Swastik Publications, 2012.

R. Ramanujam and Giridhar Sharma: *Advanced Inorganic Chemistry*, Pacific Books International, 2012.

Arun Bahl and B. S. Bahl: *Advanced Organic Chemistry*, S. Chand Publisher.

Akhilesh K. Verma: *Advanced Organic Chemistry* , Shree Pub, 2005.

R. Ramanujam and Giridhar Sharma: Advanced Organic Chemistry, Pacific Books International, 2012.

P. Sharma: Advanced Organic Chemistry, Pearl Books, 2010.

Reinhard Bruckner: Advanced Organic Chemistry : Reaction Mechanism, Elsevier pbk, 2008.

Arun Sarkar: *Advanced Organic Chemistry : Reactions and Mechanisms*, Swastik Publications, 2011.

Maya Shankar Singh: *Advanced Organic Chemistry : Reactions And Mechanisms*, Dorling Kindersley Publication, 2004.

Arun Rastogi: *Advanced Organic Chemistry : Reactions, Mechanisms and Structure*, Anmol, 2011.

N K Vishnoi: *Advanced Practical Organic Chemistry*, Vikas Publishing House, 2010.

Wai-Kee Li, Gong-Du Zhou and Thomas Chung Wai Mak: *Advanced Structural Inorganic Chemistry*, Oxford University Press, 2008.

K.R. Desai: *Advances in Inorganic Chemistry*, Oxford Book Company, 2008 .

Raghavendra Reddy and Narayan Goswami: *Applied Inorganic Chemistry*, Pacific Books International, 2012.

Raghavendra Reddy and Narayan Goswami: *Applied Organic Chemistry*, Pacific Books International, 2012.

A K Goel: *Basic Concept of Organic Chemistry*, Pearl Books, 2008.

Rajbir Singh: *Basic Concepts in Organic Chemistry*, Book Enclave Publication,2010.

D N Singh. *Basic Concepts Of Inorganic Chemistry*, Pearson Education, 2009.

f. Albert cotton, geoffrey Wilkinson: *Basic Inorganic Chemistry, 3rd ed*, Pearson Education, 2009.

Amar Tyagi: Basic Study of Organic Chemistry, Anmol, 2006.

Prabir Pujhari: *Basics of Advanced Inorganic Chemistry*, Anmol, 2008.

Prabir Pujhari: *Basics of Inorganic Chemistry*, Anmol, 2008.

Durgesh Kumar Maurya: *Basics of Organic Chemistry*, Anmol, 2007.

Jagbir Sharma: *Bio Inorganic Chemistry*, RBSA, 2006.

R.K. Soni and Preeti Sharma: *Bio-Organic Chemistry*, Shree Pub, 2007.

P K Shivaraj Kumar: *Bio-organic Chemistry*, RBSA Pub, 2007.

V.K. Puri: Bio-organic Chemistry, ALP Books, 2010.

Ashok Kumar Sharma: *Bioinorganic Chemistry*, Random Pub, 2011.

Irena Kostova and R.K. Soni: *Bioinorganic Chemistry*, Shree Pub, 2012.

Rosette M. Roat Malone: *Bioinorganic Chemistry : A Short Course*, Wiley, 2013.

Index

I

K

L

M

N

O

P

R

S

T

U

V

W

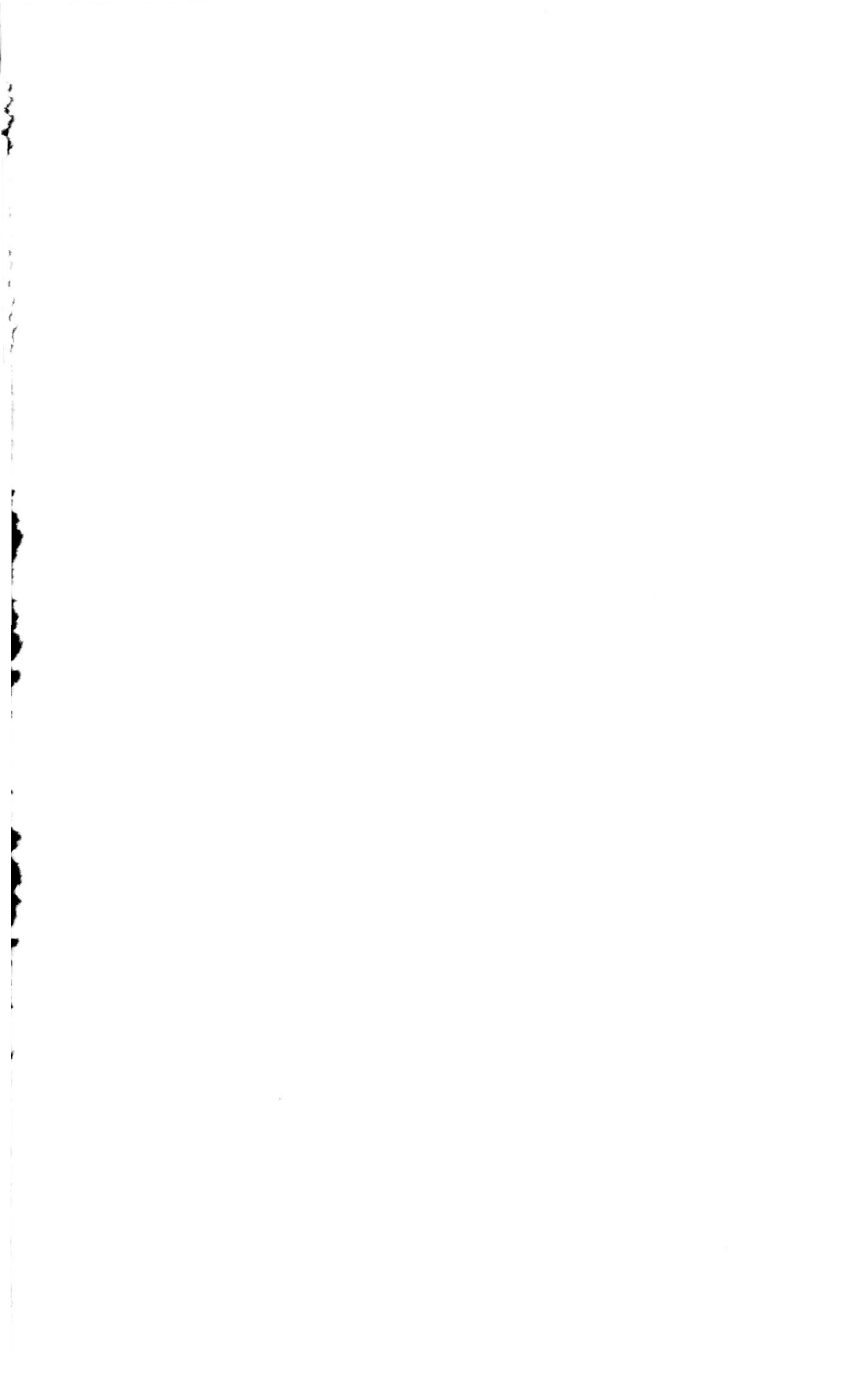